Chemical Reactor
Design in Practice

CHEMICAL ENGINEERING MONOGRAPHS

Edited by Professor S.W. CHURCHILL, Department of Chemical Engineering,
University of Pennsylvania, Philadelphia, Pa. 19104, U.S.A.

Chemical Reactor Design in Practice

L.M. ROSE

Privatdozent, Technisch-Chemisches Laboratorium,
Eidgenössische Technische Hochschule, Zürich, Switzerland

ELSEVIER SCIENTIFIC PUBLISHING COMPANY
Amsterdam — Oxford — New York   1981

CHEMISTRY
6525-8046

ELSEVIER SCIENTIFIC PUBLISHING COMPANY
1, Molenwerf
P.O. Box 211, 1000 AE Amsterdam, The Netherlands

Distributors for the United States and Canada:

ELSEVIER/NORTH-HOLLAND INC.
52, Vanderbilt Avenue
New York, N.Y. 10017

ISBN 0-444-42018-5 (Vol. 13)
ISBN 0-444-41295-6 (Series)

Printed in The Netherlands

To Jane

Preface

This text was compiled as a result of an invitation from Helsinki University of Technology to give a lecture course in "Practical Reactor Design", to follow on from their existing undergraduate introductory course as a senior course, also open to industrial participation.

There are a number of texts on reactor design now available, all of which cover the basic principles of the subject and discuss the present state of the art in various aspects of reactor engineering research. However, since reactor engineering research has moved from solving problems of practical interest to solving problems demanding high intellectual challenge, the available texts do not contain enough practical information to teach students how to design and specify reactors, or even to know the techniques by which this is done in technically advanced companies.

It is this void that this text is trying to fill.

In practice, reactor design usually starts by obtaining laboratory data from a new reaction system. It then sorts out the important factors, predicts a full-scale performance, and if this is attractive the pilot reactor is designed and operated. Data from the pilot reactor are then used in addition to laboratory data to design and predict the performance of a full-scale reactor that will work as optimally as possible, based on an economic criterion, and be easily operable and safe.

The engineer is best confronted with reactor problems at an early stage, when the chemist is going to look for good operating conditions, or even before, when he is thinking about the "reaction mechanism". The engineer can contribute to the search for optimum conditions, since he can bring economic factors into the picture and often has a numerate training which helps in the experimental planning and analysis of results.

The engineer produces a model from the laboratory results and uses this model for the design of his pilot and full-scale plant, for the choice of optimum conditions and for the stability analysis.

Reactor design in practice is concerned with the laboratory determination of data, their analysis, the development of models of the chemical and physical processes occurring, their optimization, the hardware dimensioning, and the definition of the necessary control and safety systems.

A text describing these activities touches on a multitude of disciplines - statistics, economics, optimization, control, and safety, as well as those subjects traditionally thought of as reactor design. This is an advantage for a senior course in that it is bringing together separate earlier courses, to show how, in practice, they are all needed to solve real problems.

The text was prepared for a 36 hours course plus nine two-hour exercise sessions. Since much of the course emphasized computer methods, the majority of the exercises were computer-based and run interactively with three or four students at each terminal. The nine exercises are included as Appendix 3, together with the programs used for the computer-based exercises.

The questions included at the end of each chapter are simply to emphasize the main point of that chapter and can be used by the reader to check that he has grasped the points that were being made.

It was the intention to compile a text of about 450 pages surveying the whole area of reactor design. When this is divided into topics and then further subdivided to produce a balanced text, it is surprising how few pages remain for the subjects that occupy volumes in the research literature because of the few recommendations that result from this activity. Whenever possible, only industrially tested techniques are described. When untried though promising methods are presented, a warning is given in the text. Since this is not a research text, no effort is made to substantiate the points made by reference to original literature. "Further Reading" lists are given where most points can be found discussed in more detail, and references are given only when the work would otherwise be difficult to find.

This "textbook" appears as a monograph because of the general feeling that not many teachers will be able to integrate this type of material into their present "reactor engineering" courses. If those teachers who do achieve this would contact the publishers, this would give some indication of the need for a textbook edition.

I would like to record my thanks to the staff of TKK Helsinki who made my stay in Finland so interesting and enjoyable - particularly to Professors B-son Bredenberg and Järveläinen, who arranged for the reactor design course to take place, and J. Aitamaa, who put much of his time into preparing the programs for the computer exercises.

Particular thanks are due to Elfriede Kilian for taking the major typing load and to Sirpa Pauni, who transferred some of my hand-written notes to typed text by a very novel, though somewhat laborious process.

Zürich, May 1981   L.M. Rose

CONTENTS

Notation

a	empirical constant	-
a	interfacial area/unit volume	m^{-1}
A	cross-section area	m^2
A_i	pre-exponential factor (of reaction i)	$K mol^{(1-n)} m^{3(n-1)} s^{-1}$
A'	heat-transfer area	m^2
b	empirical constant	-
c	empirical constant	-
c_i	cash flow for year i	$
c_p	mass specific heat at constant pressure	$kJ kg^{-1} K^{-1}$
C_A	molar concentration (of component A)	$kmol m^{-3}$
$d (d_p, d_B, d_t, d_I, d_e)$	diameter of particle, bubble, tube, impeller or equivalent diameter	m
D	reactor diameter	m
D_A	molecular diffusivity (of component A)	$m^2 s^{-1}$
E_i	activation energy (of reaction i)	$kJ kmol^{-1}$
E'	enhancement factor	-
f	fanning friction factor	-
f	fraction capital costs incurred annually	-
F	molar flowrate	$kmol s^{-1}$
F	objective function	-
F'	tax factor	-
g	gravitational constant	$9.81 m s^{-2}$
g_g	heat generation/unit reactor volume	$kJ m^{-3}$
g_r	heat removed/unit reactor volume	$kJ m^{-3}$
G	mass flowrate (gas)	$kg s^{-1}$
G	free energy	$kJ kmol^{-1}$
h	single film heat-transfer coefficient	$kW m^{-2} K^{-1}$
H	enthalpy	$kJ kmol^{-1}$
H'	Henry's law solubility constant	$bar m^3 kmol^{-1}$
ΔH_i	heat of reaction (of ith reaction)	$kJ kmol^{-1}$
I	investment cost	$
k_i	reaction rate constant (of ith reaction)	$kmol^{(1-n)} m^{3(n-1)} s^{-1}$
k_g	gas film mass-transfer coefficient	$kmol m^{-2} bar^{-1} s^{-1}$
k_L	liquid film mass-transfer coefficient	$m s^{-1}$
$K (K_a, K_c)$	equilibrium constant (based on activity, concentration)	-
K_g	overall gas mass-transfer coefficient	$kmol m^{-2} bar^{-1} s^{-1}$

Symbol	Description	Units
K_A	absorption coefficient	bar^{-1}
K', K'', K'''	miscellaneous constants	-
L	length	m
$L_{\frac{1}{2}}$	catalyst life (number of half lives)	-
L	project life	years
L'	mass flowrate (liquid)	$kg\ s^{-1}$
m	lifetime for taxation purposes	years
M_A	molecular weight (of A)	$kg\ kmol^{-1}$
n_A	number of kmoles (of A)	kmol
$\dot{n}_A$	molar flowrate (of A)	$kmol\ s^{-1}$
N	stirrer speed (RPS)	s^{-1}
N	number of tubes, number of beds, number of tanks	-
N_{C_a}	dimensionless group in Calderbank's surface area equations	-
N_P	agitator power number	-
N_{Re}	Reynolds number	-
p_A	partial pressure (of A)	bar
P	total pressure	bar
P', P'_g	agitator power (ungassed and gassed)	kW
P''	power supplied by gas	kW
Q	volumetric flowrate	$m^3\ s^{-1}$
r	fractional rate of interest	-
r	recycle rate	$kmol\ s^{-1}$
r_i	molar reaction rate/unit reactor volume	$kmol\ m^{-3}\ s^{-1}$
r_t	fractional taxation rate	-
R	gas constant	8.314 kJ $kmol^{-1}$ K^{-1} or 0.08314 bar m^3 $kmol^{-1}$ K
s	tube pitch	m
s'	stirrer tip speed	$m\ s^{-1}$
S	entropy	$kJ\ K^{-1}\ kmol^{-1}$
S	selectivity	-
t	time	s
$t_{\frac{1}{2}}$	half-life time	s
t_F	residence time in the film	s
t_R	residence time of the reaction	s
$\bar{t}$	mean residence time of a distribution	s
T	temperature	K
T'	production	$kmol\ s^{-1}$ or tons $year^{-1}$
u	velocity	$m\ s^{-1}$
u_s, u_t, u_m	superficial, terminal, minimal velocity	$m\ s^{-1}$

U	overall heat-transfer coefficient	$kW\ m^{-2}\ K^{-1}$
V	volume	m^3
V_R	reactor volume	m^3
w	width	m
W	weight	tons
x	independent variable	-
x_A	mole fraction in liquid (of A)	-
X_A	fractional conversion (of A)	-
y	dependent variable	-
y_A	mole fraction in gas (of A)	-
Z	film thickness	m

GREEK LETTERS

α	empirical constant	-
ε_g	gas hold up in gas/liquid dispersions	-
ε_p	void fraction in packed bed	-
λ	thermal conductivity	$kW\ m^{-1}\ K^{-1}$
θ	reduced time	-
μ	dynamic viscosity	$kg\ m^{-1}\ s^{-1}$
ν	stoichiometric coefficient (-ve for reactants)	-
σ	surface tension	$kg\ s^{-2}$
π	3.1416	-
ρ	mass density	$kg\ m^{-3}$
τ	space time	s
η	effectiveness factor (gas/solid reactions)	-
β	pellet temperature rise factor	-
ϕ	Thiele modulus	-

SUBSCRIPTS

A, B ...	for components A, B ...
amb	ambient conditions
b	in bulk (fluid)
g	in gas phase (or in presence of gas)
i	for i^{th} reaction
in	of inlet
j	for j^{th} species
J	of jacket
L	in liquid phase
0	at initial (or reference) condition

out	of outlet
R	of reactor
s	at solid surface, of solid
S	for standard case
U	for utilities
w	at wall

GENERAL NOTE ON UNITS

SI units have been used throughout, but in order to be able to deal with conveniently sized numbers the following have been consistently used: m, kg, kmol, kJ, kW, s, K, bar.

Chapter 1

CHEMICAL KINETICS AND REACTOR DESIGN PRINCIPLES

1.1   THE SINGLE CHEMICAL REACTION

Substances are converted into other substances as a result of chemical reactions. These other substances - the products of the reactions - are a result of individual molecules converting themselves to the product, and this can occur only to molecules which:

(a) have sufficient energy to overcome the energy barrier associated with the formation of the product molecules,

(b) "collide" with a molecule of the second reactant necessary to form the product. In a mixture of reactants there is a distribution of energies amongst the molecules, and a probability that in a certain time a number with sufficient energy will have collided and so have met all the necessary conditions to react. Reaction is therefore not a spontaneous occurrence, but takes time, depending on the energy distribution of the molecules and the chance of the required molecules colliding.

There are, at present, two theories of reaction rates which differ in detail, but for engineering purposes lead to the same conclusion.

The COLLISION THEORY assumes that reaction occurs when two appropriate molecules collide together with sufficient energy to penetrate the molecular van der Waals repulsive forces and therefore are able to combine together. When two molecules are forced together the total energy in the system depends on the distance between them. This energy reaches a maximum at approximately the sum of the molecular radii. Further compression involves formation of the product and a lower energy, as shown by Figure 1.1. Note that by definition the heat of reaction is +ve for an endothermic reaction (as shown by Fig. 1.1) and -ve for an exothermic reaction.

From the kinetic theory of gases it is possible to calculate the fraction of molecules containing energy in excess of E, their absolute number and the number of collisions of such molecules per second. However, the observed reaction rates are always many factors lower than such calculated rates. To explain this discrepancy the concept of steric hindrance is introduced which claims that only the fraction of molecules colliding together with a specific orientation to each other will react. Either the reacting groups must be in close proximity, or the potential energy barrier is lower at certain points around the molecule. There is no method of predicting the steric hindrance factor.

The TRANSITION STATE THEORY again assumes that the reacting molecules must collide and have sufficient energy to reach the top of this potential energy barrier, but in fact the state at the top of the energy barrier is a defined state, a transition state of a definite form that the molecules are existing in (see Fig. 1.2).

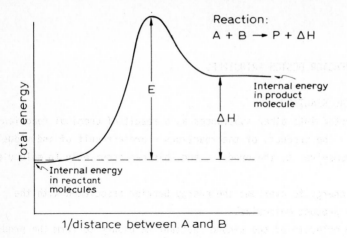

Fig. 1.1. Energy levels between approaching molecules.

This transition state then breaks down into more stable states, which could either
be a reversion to the reactants or the formation of the reaction products. With
this theory, reaction rates can be predicted once the form of the transition state
is defined from potential energy surface studies. Although the predictions are
valuable contributions to the understanding of reaction rate theory, their accuracy
is inadequate for use in any engineering study.

Hence, at the present time we are unable to predict reaction rates from first
principles, and must always resort to measurements in the laboratory for our data.

Both theories of reaction rates predict that the rate is proportional to the
number of collisions between appropriate molecules. The probability of molecules
A and B colliding (i.e. being within a certain volume at the same time) is directly
proportional to the product of the numbers of molecules of A and B per unit volume.
Since the reaction rate is proportional to the number of collisions, we can expect

$$\frac{1}{V_R} \frac{dn_P}{dt} = k_1 \, C_A C_B = r_1 \quad \text{kmol m}^{-3} \text{ s}^{-1} \qquad (1.1)$$

to hold for reaction A + B → P,
where C represents molar concentrations (kmol m^{-3}), n_P is the number of moles of
component P in the system (kmol), and V_R is the volume of the reacting system (m^3).
The relationship described by equation 1.1 has been observed for simple reactions
and is known as the Law of Mass Action. k_1 is known as the reaction velocity
constant and r_1 is defined as the reaction rate.

Both theories of reaction rate also claim that only those molecules possessing
more than a certain energy level can react together. The Boltzmann distribution

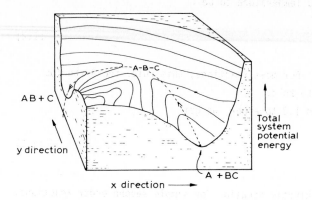

Fig. 1.2. Potential energy contours for A + BC $\rightleftharpoons$ AB + C.

of energy levels in a sample of molecules resulting from energy exchanges due to random collisions has an exponential form (see Fig. 1.3). The formation of molecules possessing more than a definite energy, E, is given by the area in the hatched tail of the distribution. The displacement of this distribution by increasing the temperature produces a different fraction of molecules possessing more than energy E. Figure 1.3 shows that because of the shape of the distribution there is a disproportionately large increase in this fraction compared to the increase in the mean energy. It can be shown that this fraction is proportional to $e^{-E/RT}$, and so from both reaction rate theories one would predict the relation between

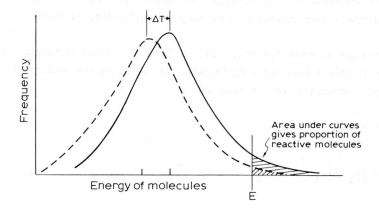

Fig. 1.3. Energy distribution of gas molecules at two temperatures.

reaction velocity constant and temperature to be

$$k_1 = A_1 \, e^{-E_1/RT} \tag{1.2}$$

where A_1 and E_1 are constants. For most reactions, this rule, the Arrhenius Relationship, has been shown to hold.

Combining equations 1.1 and 1.2 together, we get:

$$\frac{1}{V_R} \frac{dn_P}{dt} = C_A C_B \, A_1 \, e^{-E_1/RT} = r_1 \tag{1.3}$$

which is the normal chemical kinetic equation for simple second-order reactions. E_1 is known as the activation energy, and A_1 is the pre-exponential or frequency factor for reaction 1.

The prediction of E from theory is as inaccurate as the prediction of the reaction rate. The value of E depends upon the mapping of potential energy surfaces, a subject which is presently being tackled only for the very simplest molecules. As with the reaction rates, the engineer must obtain his data from laboratory experiments on the reaction in question.

1.1.a   Prediction of heats of reaction

As Figure 1.1 shows, chemical reactions are associated with changes in internal energies of the molecules. This means that energy is required or liberated as a result of the reaction taking place. This energy, the heat of reaction (ΔH), is very important information for the engineer because of the scale on which he works. Even small heats of reaction, often considered "negligible" in laboratory work, lead to unexpected temperature changes or even dangerous situations in plant-scale reactors.

Heats of reaction can be predicted from heats of formation of the reactants and products according to Hess's Law. For example, when the reactants and products are all at the standard conditions, for the reaction:

$$A + B \xrightarrow{25°C} P + R + \Delta H,$$
$$g \quad g \qquad g \quad g$$

$$\Delta H = - (H_f^o)_A - (H_f^o)_B + (H_f^o)_P + (H_f^o)_R \tag{1.4}$$

where $(H_f^o)_P$ is the heat of formation of gaseous P at standard conditions of 25°C. Note that from this definition of ΔH an exothermic reaction has a negative heat of reaction.

If, however, the reaction does not take place at 25°C, but at T°C, corrections

must be made by involving thermal capacities of reactants and products, again
using Hess's Law.

$$\Delta H_T = (H_f^\circ + \int_{25}^{T} Mc_p dT)_P + (H_f^\circ + \int_{25}^{T} Mc_p dT)_R - (H_f^\circ + \int_{25}^{T} Mc_p dT)_A - (H_f^\circ + \int_{25}^{T} Mc_p dT)_B \qquad (1.5)$$

where Mc_p is the molar specific heat of the appropriate component. Equation 1.5
in general reduces to

$$\Delta H = \sum_{i=1}^{n} \nu_i (H_f^\circ)_i + \sum_{i=1}^{n} \int_{25}^{T} (\nu Mc_p)_i \, dT \qquad (1.5a)$$

where ν_i is the stoichiometric coefficient for component i.

Collections of values for heats of formation and thermal capacities are available
in the literature, or from one of a number of thermodynamic data bank organizations
that collect such data.

When the phases of the components in the reactor do not agree with those of the
standard state, further terms must be included in equation 1.5 to represent the
heats of vaporization or fusion. These can be straightforwardly introduced using
Hess's Law.

However, heats of reaction are, in practice, not often as easy to predict as
equation 1.5 would suggest.

o If the reaction is occurring in solution, heats of solution must be known, and
these data are rarely available.

o The heats of formation of all components in the reaction are known only when
simple molecules are involved. In complex organic chemistry, for instance,
the data are often not available.

o All reaction products must be definable. The identification of all by-products
is often not possible. "Unknown impurities", "polymeric material", or even
"coke formation" make heat of reaction predictions uncertain.

For these reasons the engineer is often faced with having to measure heats of
reaction in the laboratory.

It may be possible to obtain orders of magnitude estimates for heats of reaction
from comparisons with similar reactions whose heats of reaction are known. It may
also be possible to make an estimate from the bond energies of the chemical bonds
that are broken and formed during the reaction. However, this method is very
unreliable when the heat of reaction is small compared with the bond energies.
In particular, modification of bond energies by the presence of other groups in the

molecule make this calculation very inaccurate.

1.2   SIMULTANEOUS REACTION SYSTEMS

Having discussed the simple single chemical reaction, and shown the basic equations by which it can be described and the kinetic and thermodynamic data that are necessary to describe it, we will now look at the effect of multiple reactions occurring together.

When looked at in sufficient detail, for instance in the detail considered by the chemist when he is trying to explain why the reaction proceeds at all, all reactions are the result of a chain of more elementary reactions involving electron transfers and intermolecular rearrangements. However, a simple reaction between simple components can often be adequately treated as a single reaction behaving according to equation 1.3, as long as one of the steps is predominantly slower than the others (i.e. it contains a "rate determining step"). At the other extreme are reactions involving more than one product which must have arisen from different reaction routes. In these cases the real system contains a number of different reactions which are occurring together and it is not possible to simplify its treatment to a single reaction.

The single reaction system is comparatively simple to handle in reactor design. Optimum conditions all lie on upper constraints - maximum concentrations, temperature and time. Laboratory data are particularly easy to analyse to obtain k_1 and E_1 and scale-up problems reduce themselves to mixing and heat removal.

However, most reaction systems cannot be considered to be single reactions and this brings with it many of the complications associated with reactor design.

We shall now consider the various types of simultaneous reaction systems and much of the rest of the book will be concerned with the handling of mixtures of these types of basic reaction systems.

1.2.a   Reversible reactions

$$A + B \xrightarrow{k_1} P + R$$

$$P + R \xrightarrow{k_2} A + B$$

This is the most common form of simultaneous reaction system, and is easily understandable when one considers either of the theories of reaction rates. There remains for all reactions the possibility of products of reaction colliding and having sufficient energy to achieve the transition state and to revert to raw materials.

In a gas mixture, the forward rate r_1 and the backward rate r_2 are given by:

$$r_1 = k_1 \, C_A C_B \quad \text{and} \quad r_2 = k_2 \, C_p C_R$$

At equilibrium

$$r_1 = r_2$$

therefore

$$\frac{k_1}{k_2} = K = \frac{C_p C_R}{C_A C_B} \tag{1.6}$$

where K is known as the equilibrium constant for the reaction. Equation 1.6 describes the concentration of one component in terms of the other components present in the equilibrium mixture and shows how the equilibrium can be "displaced" by adding excess of one component. K can be experimentally determined from compositions of reaction mixtures at equilibrium.

Since it can be shown that

$$\Delta G^\circ = -RT \ln K_a$$

where ΔG° is the difference in Gibb's free energies of the reactants and products and K_a is the equilibrium constant based on activities, it is possible to predict the value of K from thermodynamic data. ΔG° can be evaluated from heats of formation and entropy data which are to be found in thermodynamic tables, since

$$\Delta G^\circ = \Delta H_f^\circ + T\Delta S^\circ \tag{1.8}$$

K_a values at other than the standard temperature of 25°C can be calculated using the van't Hoff Equation:

$$\frac{d(\ln K_a)}{dT} = \frac{\Delta H_f}{RT^2} \tag{1.9}$$

Hence, if ΔH_f is independent of temperature,

$$(\ln K_a)_{T_2} = -\frac{\Delta H_f}{R}\left(\frac{1}{T_2} - \frac{1}{T_1}\right) + \ln (K_a)_{T_1} \tag{1.10}$$

If there is a considerable difference between reactant and product thermal capacities, ΔH_f is temperature dependent, and this relationship must be defined before the van't Hoff Equation is integrated:

$$(H_f)_T = (H_f)_{T_1} + \int_{T_1}^{T} Mc_p \, dT \qquad (1.11)$$

if

$$c_p = a + bT \qquad (1.12)$$

then

$$(H_f)_T = (H_f)_{T_1} + \int_{T_1}^{T} M(a + bT) \, dT$$

$$= (H_f)_{T_1} + (T - T_1) \, aM + \frac{1}{2}(T^2 - T_1^2) \, bM \qquad (1.13)$$

Summing for all components gives

$$(\Delta H_f)_T = (\Sigma H_f)_{T_1} + (T - T_1) \sum_{i=1}^{n}(v_i \, a_i \, M_i) + \frac{1}{2}(T^2 - T_1^2) \sum_{i=1}^{n}(v_i \, b_i \, M_i) \qquad (1.14)$$

where v_i are the stoichiometric coefficients for the reaction.

Setting this in the van't Hoff Equation gives:

$$\ln (K_a)_{T_2} = \ln (K_a)_{T_1} + \frac{1}{R}\int_{T_1}^{T_2} \frac{(\Delta H_f)_T}{T^2} \qquad (1.15)$$

which can then be integrated.

The relation between K_a and K is given by the activity coefficients (f_i) of the n components in the mixture.

Since, in general,

$$K = \prod_{i=1}^{n} x_i^{v_i} \qquad (1.16)$$

and

$$K_a = \prod_{i=1}^{n} (f_i \, x_i)^{v_i} \qquad (1.17)$$

then

$$K = K_a \prod_{i=1}^{n} (f_i)^{-v_i} \qquad (1.18)$$

Hence, given H_f^o or ΔH^o, S^o, a and b of the c_p-relationship and f_i for all the components in the mixture, the equilibrium conditions of that mixture and the effect of temperature on it can be calculated. The method can be applied to gaseous systems because the data are usually available in thermodynamic tables. Liquid systems are not usually treated in this way, because data are rarely available.

This procedure is very useful in initial screening of possible reaction schemes to show which routes to a product are thermodynamically feasible. Proposed reactions with very low equilibrium constants can be immediately discarded from further study. Reactions with high equilibrium constants can then be studied further. However, it is not possible to conclude that they will be good synthetic routes, because this study determines only the equilibrium conditions. It does not give information on how quickly the reaction proceeds. The reaction could be infinitely slow, i.e. the "reactants" to all intents and purposes do not react.

1.2.b   Consecutive reactions

When the product of one reaction can react further we have a consecutive reaction system, e.g.

$$A + B \xrightarrow{k_1} C$$
$$C + B \xrightarrow{k_2} D$$

When they are both simple reactions, the rate of change of number of moles of each component is given by

$$\frac{1}{V_R} \frac{dn_A}{dt} = k_1 \, c_A c_B \tag{1.19}$$

$$\frac{1}{V_R} \frac{dn_C}{dt} = k_1 \, c_A c_B - k_2 \, c_C c_B \tag{1.20}$$

$$\frac{1}{V_R} \frac{dn_D}{dt} = k_2 \, c_C c_B \tag{1.21}$$

The rate of formation of the individual components, particularly the intermediate components, is now quite a complicated function, and this leads to some concentration/time profiles giving a higher selectivity for a particular component than for others.

This type of reaction is very often met in organic reactions, where a mono-substituted compound (be it halide, sulphide, nitro, etc.) is required as product, but this can undergo further substitution to unwanted di- and tri-substituted compounds.

It is often possible to simplify the kinetics of consecutive reactions if one of the steps has a considerably lower rate constant than the other steps. In this case the behaviour of the system resembles the behaviour of a system containing only that single step. In our example, if k_1 were much less than k_2, we could represent the system by equation 1.21a.

$$\frac{1}{V_R} \frac{dn_D}{dt} = k_1' \, C_A C_B \qquad\qquad\qquad (1.21a)$$

The system would have simple second-order characteristics, because there is a
"rate-controlling step" present. k_1', however, is not a measure of the real k_1,
particularly if the rate-controlling step is not the first step in the chain.

1.2.c Parallel Reactions

Parallel reactions are those in which a reactant can, in addition to reacting
to give the required product, also undergo a reaction to give a second compound,
e.g:

$$A + B \xrightarrow{\ k_1\ } P$$
product

$$A + B \xrightarrow{\ k_2\ } R$$
by-product

When both are simple reactions, the rate of change of concentration is given by:

$$\frac{1}{V_R} \frac{dn_A}{dt} = -k_1 \, C_A C_B - k_2 \, C_A C_B \qquad\qquad (1.22)$$

$$\frac{1}{V_R} \frac{dn_P}{dt} = k_1 \, C_A C_B \qquad\qquad\qquad (1.23)$$

$$\frac{1}{V_R} \frac{dn_B}{dt} = -k_1 \, C_A C_B - k_2 \, C_A C_B \qquad\qquad (1.24)$$

$$\frac{1}{V_R} \frac{dn_R}{dt} = k_2 \, C_A C_B \qquad\qquad\qquad (1.25)$$

Parallel reactions result in a reduction in yield of product from the reaction.
To improve the yield, methods have to be found which favour the required reaction
over the unwanted reaction. They may have different temperature dependences which
can be exploited. The unwanted reaction may have only one common reactant, not
two as in this example, and then there is the opportunity to modify concentrations
to favour the desired reaction. When a catalyst is involved it may be possible to
modify the catalyst to change the ratio of k_1 to k_2 and so improve the selectivity.

1.2.d Catalytic reactions

Catalytic reactions are defined as those reactions where a compound must be
present for the reaction to proceed, although it is not consumed by the reaction.
This non-consumed compound is called the catalyst, and often it need only be
present in very small quantities in order to make a reaction proceed at a reasonable

rate, whereas without the catalyst the reaction may not take place at all.

Catalysts are involved in the reaction by reacting with one of the components to convert it into an active form. The active form then reacts more easily with the second reactant, and by this process the catalyst is liberated and is available for a further cycle. The activated form may be a loosely bound complex, or it may be a strained adsorbed species, as is the case for adsorbed catalyst gas phase reactions.

Catalytic processes are therefore always simultaneous reaction systems involving consecutive reactions. Homogeneous liquid phase catalysed reactions are usually the simplest form of catalysed reaction and heterogeneous (particularly solid-catalysed, gas phase reactions) the most complicated because, in addition to the consecutive reactions being necessary to represent the chemical system, there are mass-transfer processes involved to get the reactants to the catalyst, and to transfer the products from the catalyst.

Solid-catalysed gas reactions are by far the most significant industrial reaction type, and there is continued research effort in trying to fully understand the mechanisms involved. The solid surface adsorbs one or more of the reactants on "sites" in such a way that the molecule is under strain and more able to react. Sometimes this reaction is thought to take place with the second reactant in the gas phase, sometimes the second reactant must also adsorb before reaction can occur, and sometimes the catalyst is involved in providing oxidation/reduction cycles with the metal oxides in the catalyst which enable the reaction to proceed.

Though the mechanism of catalysis is far from well understood, it appears that the first step is a dynamic equilibrium between adsorbing and desorbing molecules on the catalyst surface. A number of different models have been suggested to explain the shape of the experimental adsorption isotherms of gas components on the catalyst in the absence of reaction.

Of the models proposed, the first, the Langmuir model form appears reasonable for modelling purposes (Carberry, 1976). This model assumes that there are a certain number of equally attractive sites available on the surface, and once these are full, no more can adsorb. This would be the case if a monomolecular layer formed with all sites being far enough apart to have no influence on each other.

For the catalysed reaction

$$A + B \rightarrow P$$

with both A and B being adsorbed on different sites, the concentrations of A and B on the solid are given by the equilibrium set up by molecules desorbing from the surface and those being adsorbed. Let fraction θ_A of the A sites be occupied, then $1-\theta_A$ are vacant. Assuming adsorption and desorption rates to be proportional to the fraction of the surface unoccupied and occupied respectively, with adsorption also being proportional to partial pressure p_A, at equilibrium:

$$(1 - \Theta_A) K' \, p_A = K'' \, \Theta_A$$

hence:

$$\Theta_A = \frac{K' \, p_A}{K'' + K' \, p_A} = \frac{K_A \, p_A}{1 + K_A \, p_A} \tag{1.26}$$

(defining K_A as $\dfrac{K'}{K''}$).

A similar expression can be derived for Θ_B, and so when the reaction rate is proportional to <u>surface concentrations</u>, Θ_A and Θ_B:

$$r_1 = k_1 \, \Theta_A \, \Theta_B = \frac{K''' \, p_A \, p_B}{(1 + (K_A)_A \, p_A)(1 + (K_A)_B \, p_B)} \tag{1.27}$$

When there is competition for sites in that both species adsorb on the same site, then:

$$(1 - \Theta_A - \Theta_B) K'_A \, p_A = K''_A \, \Theta_A$$

and

$$(1 - \Theta_A - \Theta_B) K'_B \, p_B = K''_B \, \Theta_B$$

which can be solved simultaneously to give

$$\Theta_A = \frac{(K_A)_A \, p_A}{(1 + (K_A)_A \, p_A + (K_A)_B \, p_B)} \tag{1.28}$$

and

$$\Theta_B = \frac{(K_A)_B \, p_B}{(1 + (K_A)_A \, p_A + (K_A)_B \, p_B)} \tag{1.29}$$

resulting in:

$$r_1 = \frac{K''' \, p_A \, p_B}{(1 + (K_A)_A \, p_A + (K_A)_B \, p_B)^2} \tag{1.30}$$

Other expressions can be derived assuming that there is competition between reactant and inerts or reactants and products.

Since catalytic reaction occurs on the catalyst surface it is reasonable to define reaction velocity constants in terms of kmoles sec^{-1} (unit of catalyst surface area)$^{-1}$. Since this cannot be defined without having made specific surface area measurements, a more useful definition is kmoles sec^{-1} (unit weight of catalyst)$^{-1}$, and reaction rates for catalytic reactions are often quoted in these units. However, in this text the convention will be maintained of using the rate as kmol sec^{-1} (unit reactor volume)$^{-1}$, where the volume is the bulk volume of the catalyst bed (particles plus voidage between the particles), to enable the same equations to represent homogeneous and heterogeneous reactions.

1.3   REACTION MECHANISMS

Only the simplest of molecules reacting together can be treated as simple single reactions and even these are generally two reversible consecutive reactions according to the transition theory. Most reactions occur as a result of a number of reaction stages, particularly commercially interesting reactions. When no single stage is predominantly slower than the rest, i.e. when there is not a single rate-determining step, the chemical kinetics observed for such reactions do not correspond to simple first- or second-order kinetics, but fractional orders or orders greater than unity can result.

Because of this effect on the form of the kinetic equation it is useful to postulate and experimentally confirm a reaction mechanism so that a satisfactory kinetic representation, and hence a reliable reactor design, can be made. What exactly is meant by a reaction mechanism must be discussed in some detail, because it has a different meaning to the chemist and to the chemical engineer.

The chemist's reaction mechanism is an explanation of why and how the reactants react; how the electron densities around the molecule move so as to provide charged areas onto which the second reactant can attach because of some induced opposite charge. This activated complex then has a modified electron structure which results in a part of the complex being weakly attached which can then detach. A final rearrangement of the charges around the molecule results in the reaction product being formed.

Such a description for a typical organic reaction of commercial interest may involve 20 or so individual steps. Clearly, such a complex set of reactions cannot be handled quantitatively, and this leads to a second definition of reaction mechanism which is understood by the chemical engineer: to him a reaction mechanism involves defining the reaction steps, where each step is the change from one distinct chemical species (which has a distinct observable life time) to another. These distinct chemical species could be normal chemical compounds which appear as intermediates and which could be involved in competitive side reactions. The species could be less stable complexes which defy isolation but can nonetheless be shown to exist in solution. The species could also be particular definable reactive states such as free radicals. When restricted in this way, the number of individual stages in any "mechanism" is considerably reduced, and comes within the bounds of quantitative treatment.

The engineer includes in his mechanism the mass-transfer steps involved, whereas the chemist usually excludes them from his reaction mechanism.

It is important to recognize that there are two meanings of the "mechanism" of the reaction: the micro- and the macro-kinetic mechanisms, because this is often a source of misunderstanding between the chemist and the chemical engineer.

Example 1

What is the mechanism for the chlorination of benzene to monochlorbenzene ?

(1)   Mechanism as defined by the Chemist

$$Cl_2 + FeCl_3 \rightleftharpoons Cl^+FeCl_4^-$$

The monochlorbenzene can react in a similar way to form ortho- and para-dichlor-benzenes, and from these, higher chlorinated derivatives.

(2)   Mechanism as defined by the Chemical Engineer

$$Cl_{2_g} \rightleftharpoons Cl_{2_L}$$

$$C_6H_6 + Cl_2 \longrightarrow C_6H_5Cl + HCl \longrightarrow HCl$$

L $\quad\quad\quad\quad\quad\quad$ L $\quad\quad\quad\quad$ g

$$\Big\downarrow \; \begin{array}{c} + Cl_2 \\ L \end{array}$$

$$C_6H_4Cl_2 + HCl \longrightarrow HCl$$

$\quad\quad\quad\quad\quad$ L $\quad\quad\quad\quad$ g

$$\Big\downarrow \; \begin{array}{c} + Cl_2 \\ L \end{array}$$

$$C_6H_3Cl_3 + HCl \longrightarrow HCl$$

$\quad\quad\quad\quad\quad$ L $\quad\quad\quad\quad$ g

where L and g denote in liquid or gas phase.

Example 2

Sulphonation of anthraquinone in the σ position by means of a mercury catalyst and oleum.

(1)   Mechanism as defined by the Chemist

$$H_2S_2O_7 \rightleftharpoons H_2SO_4 + SO_3$$

$$S_2O_6 \rightleftharpoons 2SO_3$$

$$2H_2SO_4 \rightleftharpoons HSO_4^- + H_2O + HSO_3^+$$

$$2H_2SO_4 \rightleftharpoons SO_3 + H_3O^+ + HSO_4^-$$

(2) Mechanism as defined by the Chemical Engineer

$$R + H_2SO_4 \xrightarrow[k_1]{\text{Hg}} RHSO_3 \quad \text{(1-sulphonic acid)}$$

$$\xrightarrow{k_2} RHSO_3 \quad \text{(2-sulphonic acid)}$$

$$r_1 = k_1\, c_R^\alpha\, c_{H_2SO_4}^\beta\, c_{Hg^{++}}^\gamma$$

$$\text{Selectivity} = \frac{k_1}{k_2}\, c_{Hg^{++}}^\delta$$

where α to δ are empirical constants to be determined by experiment.

Example 3

The substitution of a side chain onto an aromatic ring with aluminium chloride as catalyst (Friedel-Crafts reaction).

A step in the manufacture of anthraquinone is the reaction between phthalic anhydride and benzene:

(1) Mechanism as defined by the Chemist

+ HCl + Al(OH)₃   further complexing

(2)   Mechanism as defined by the Chemical Engineer

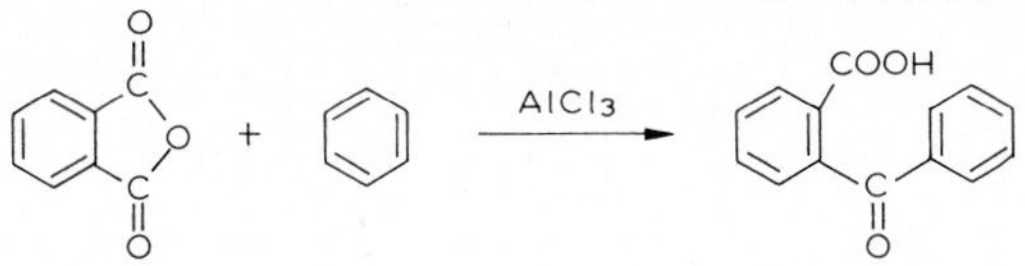

A   B

$r_1 = k_1 C_A C_B$ (AlCl$_3$ must be present in 2 moles/mole of organic.)

This mechanism is adequate when the reaction goes virtually to completion; there
is no need to introduce simultaneous reaction, and so the simple second-order
formulation suffices. For stoichiometric reasons, 2 moles of aluminium chloride
"catalyst" are required, and there are no technical grounds for considering
changes to this ratio.

Example 4
    The chlorination of methyl chloride in the gas phase is a free radical reaction
initiated at high temperatures (350°C). The reaction proceeds as a chain reaction,
with most frequent reactive collisions being between radical and molecular species,
according to the following reaction scheme.

(1)   Mechanism according to the Chemist

$Cl_2 \longrightarrow Cl\cdot + Cl\cdot$

$CH_3Cl + Cl\cdot \longrightarrow CH_2Cl\cdot + HCl$

$CH_2Cl\cdot + Cl_2 \longrightarrow CH_2Cl_2 + Cl\cdot$

$CH_2Cl_2 + Cl\cdot \longrightarrow CHCl_2\cdot + HCl$

$CHCl_2\cdot + Cl_2 \longrightarrow CHCl_3 + Cl\cdot$

$CHCl_3 + Cl\cdot \longrightarrow CCl_3\cdot + HCl$

$CCl_3\cdot + Cl_2 \longrightarrow CCl_4 + Cl\cdot$

$CH_2Cl\cdot + CH_2Cl\cdot \longrightarrow C_2H_4Cl_2$

$CH_2Cl\cdot + CHCl_2\cdot \longrightarrow CH_2ClCHCl_2$ etc.

(2) Mechanism according to the Chemical Engineer

$$CH_3Cl + Cl_2 \longrightarrow CH_2Cl_2 + HCl$$

$$CH_2Cl_2 + Cl_2 \longrightarrow CHCl_3 + HCl$$

$$CHCl_3 + Cl_2 \longrightarrow CCl_4 + HCl$$

Such a simple second-order reaction scheme has produced a perfectly adequately fitting model for the product distribution of di-, tri- and tetrachlormethanes.

The model was not tested in the prediction of chlorine concentration and chlorinated ethane by-products. It may well be that such simple models fail to predict these components. However, this example shows how gross assumptions are tolerable when the model is only required to predict in a limited way.

1.3.a Partial-order kinetics as a result of simultaneous reactions

Although the Law of Mass Action suggests that all reactions should be first- or second-order, in practice reaction orders are often non-integral, sometimes less than 1 and sometimes between 1 and 2.

The explanation for this is that although individual reaction steps themselves have first- or second-order kinetics, when a complex set of reaction steps is occurring the overall equation defining the reaction order loses its simple form. Two such examples will be given; one the overall equation for a chain reaction, and a second the result of a catalytic adsorption, reaction, desorption system.

Example 1

The classical example of the reaction between H_2 and Br_2 to form HBr is an excellent demonstration of the strange apparent kinetics that occur for a complex system of reactions.

For the overall reaction

$$H_2 + Br_2 \longrightarrow 2HBr$$

experimental work in 1906 by Bodenstein showed that it could not be described by simple kinetics, but by the empirical relationship

$$\frac{dC_{HBr}}{dt} = \frac{K' \, C_{H_2} \, C_{Br_2}^{0.5}}{\left(K'' + \dfrac{C_{HBr}}{C_{Br_2}}\right)} \tag{1.31}$$

where K' and K'' are constants. The explanation for this was found in 1920.

The proposed mechanism is that the Br_2 molecules must first dissociate into atoms, and then the reaction proceeds as a chain reaction:

$$Br_2 \xrightarrow{k_1} Br + Br$$

$$Br + H_2 \xrightarrow{k_2} HBr + H$$

$$H + Br_2 \xrightarrow{k_3} HBr + Br$$

$$H + HBr \xrightarrow{k_4} H_2 + Br$$

$$Br + Br \xrightarrow{k_5} Br_2$$

Hence:

$$\frac{dC_{HBr}}{dt} = k_2 \, C_{Br}C_{H_2} + k_3 \, C_H C_{Br_2} - k_4 \, C_H C_{HBr} \tag{1.32}$$

$$\frac{dC_{Br}}{dt} = 2k_1 \, C_{Br_2} + k_3 \, C_H C_{Br_2} + k_4 \, C_H C_{HBr} - k_2 \, C_{Br}C_{H_2} - 2k_5 \, C_{Br}^2 \tag{1.33}$$

and

$$\frac{dC_H}{dt} = k_2 \, C_{Br}C_{H_2} - k_3 \, C_H C_{Br_2} - k_4 \, C_H C_{HBr} \tag{1.34}$$

Since the concentrations of hydrogen and bromine atoms are very low, we can assume the rate of change of their concentrations to be zero (Bodenstein's stationary concentration).

Hence, from equations 1.33 and 1.34,

$$2k_1 \, C_{Br_2} + k_3 \, C_H C_{Br_2} + k_4 \, C_H C_{HBr} = k_2 \, C_{Br}C_{H_2} + 2k_5 \, C_{Br}^2 \tag{1.35}$$

and

$$k_2 \, C_{Br}C_{H_2} = k_3 \, C_H C_{Br_2} + k_4 \, C_H C_{HBr} \tag{1.36}$$

Adding these two equations gives:

$$k_1 \, C_{Br_2} = k_5 \, C_{Br}^2$$

Hence

$$C_{Br} = \sqrt{k_1 \, C_{Br_2}/k_5} \tag{1.37}$$

and from equation 1.36,

$$C_H = \frac{k_2\, C_{H_2}\, C_{Br}}{k_3\, C_{Br_2} + k_4\, C_{HBr}} \qquad (1.38)$$

Substituting for C_{Br} and C_H in equation 1.32 and simplifying gives:

$$\frac{dC_{HBr}}{dt} = \frac{2k_2 k_3\, \sqrt{k_1/k_5}\, C_{H_2}\, C_{Br_2}^{1.5}}{k_3\, C_{Br_2} + k_4\, C_{HBr}}$$

which can be re-written as

$$\frac{dC_{HBr}}{dt} = \frac{2k_2\, \sqrt{k_1/k_5}\, C_{H_2}\, (C_{Br_2})^{0.5}}{\left(1 + \dfrac{k_4\, C_{HBr}}{k_3\, C_{Br_2}}\right)} \qquad (1.39)$$

which is identical to equation 1.31, which had been determined 14 years previously as an empirical fit to the experimental results.

Kinetics of chain reactions are always complex, and often it can be shown that under specific conditions reaction orders of 1.5 are obtained. It is interesting to note that for chain reactions the simple Arrhenius equation also does not hold, presumably because the species are so active they are not at the tail of the energy distribution (see Fig. 1.3).

Example 2

A second example of the effect of simultaneous reactions causing fractional orders in the chemical kinetic equation is gas-phase reactions on a solid catalyst. Consider, for example, the partial oxidation of propylene to acrolein over a solid catalyst.

$$C_3H_6 + O_2 \longrightarrow H_2C{=}CH{-}CHO + H_2O$$

In fact, this reaction occurs between adsorbed species of C_3H_6 and O_2, and so the full mechanism is:

$$C_3H_6 \text{ (g)} \longrightarrow C_3H_6 \text{ (s)}$$
$$O_2 \text{ (g)} \longrightarrow O_2 \text{ (s)}$$
$$C_3H_6 \text{ (s)} + O_2 \text{ (s)} \longrightarrow C_3H_4O \text{ (s)} + H_2O$$

$$C_3H_4O \longrightarrow C_3H_4O$$
$$ s \qquad\qquad g$$

Assuming the adsorption occurs according to the Langmuir isotherm, the relation between the surface concentration and the gas concentration is given by the equilibrium set up by the molecules leaving the surface and those attaching, which, as shown in section 1.2.d , leads to expression 1.39 for adsorption on separate sites.

$$r_1 = \frac{K''' \, P_{O_2} P_{Pr}}{(1 + K_{A_{O_2}} P_O)(1 + K_{A_{Pr}} P_{Pr})} \qquad\qquad (1.39)$$

This equation is typical of solid catalysed gas reactions. The order of reaction is dependent on the size of the adsorption coefficients K_A. When these are low, the reaction is first-order in both components. When one is very high, the reaction is zero-order in that component. Intermediate values of K_A give fractional orders.

If both species adsorb on the same type of site, the same analysis can result in a negative order appearing; if the product adsorbs and blocks sites for the reactants, the product appears in the rate equation with a negative order.

1.4   TYPES OF REACTOR

There are four basic forms of chemical reactor:

o the batch reactor

o the semi-batch reactor

o the continuous-stirred tank reactor (CSTR)

o the plug-flow continuous reactor (PFR)

These reactors differ in their mixing patterns, that is the extent to which the reaction products mix with the feed. This results in different concentrations of the various components in the different reactors, which results in different reaction rates. In the case of complex reactions these different reaction rates result in different selectivities. Depending on the reaction mechanism, one reactor type will be more suitable than another. The different reaction mechanisms introduce degrees of freedom that the designer can use to advantage in reactor selection. If a reaction is fast, leading to heat removal problems, these can be alleviated by a careful choice of reactor. If the reaction is slow, then another type of reactor will be the most suitable.

1.4.a   The batch reactor (Fig. 1.4)

In the batch reactor the reactants are mixed together in a vessel at an instant in time, and the reaction proceeds, quickly at first because of high concentrations of all reactants, and reducing as the reaction proceeds further.

For a second-order reaction:

$$A + B \xrightarrow{k_1} P$$

the rates are given by the equations:

$$\frac{1}{V_R}\frac{dn_A}{dt} = -k_1\, C_A C_B \qquad (1.40)$$

$$\frac{1}{V_R}\frac{dn_B}{dt} = -k_1\, C_A C_B \qquad (1.41)$$

and

$$\frac{1}{V_R}\frac{dn_P}{dt} = k_1\, C_A C_B \qquad (1.42)$$

where

$$\frac{1}{V_R}\frac{dn_i}{dt} = \frac{dC_i}{dt}$$

assuming there is no volume change. Hence, the time t to reduce the concentration of A to C_A from C_{A_0}, assuming the initial solution had equimolar concentrations of A and B, is given by:

$$t = -\frac{1}{k_1}\int_{C_{A_0}}^{C_A}\frac{dC_A}{C_A^2} = \frac{1}{k_1}\left(\frac{1}{C_A} - \frac{1}{C_{A_0}}\right) \qquad (1.43)$$

Notice that the mathematical description of the process occurring results in a set of differential equations, which sometimes can be solved analytically, but often have to be solved by computer.

1.4.b   The semi-batch reactor (Fig. 1.5)

Semi-batch reactors have some reactants charged into the reactor at time zero, while other reactants are added during the reaction. The reactor has no outlet. This type of reaction is inevitable when a batch reactor is required but one of the components is a sparingly soluble gas. This reactor is also to be preferred to the batch reactor when initial reaction rates are very high, which would result in uncontrollable temperature rises or gas evolution in a batch reactor, where all reaction concentrations are high initially.

Given the reaction $A + B \xrightarrow{k_1} P$, with B being added continuously at a rate F_B kmol s^{-1} for a V_R m³ reactor, the change in concentration of B is given by the molar balance of rates of feeding and reaction:

$$V_R \frac{dC_B}{dt} = F_B - V_R r_1 \quad \text{kmol s}^{-1} \tag{1.44}$$

where the rate of reaction is given by:

$$r_1 = k_1 C_A C_B \tag{1.45}$$

and the change in A is given by:

$$\frac{dC_A}{dt} = -r_1 \tag{1.46}$$

Given concentrations defining the initial conditions in the reactor, such a set of differential equations can easily be solved by numerical integration with a computer.

Notice that to preserve the simplicity of the example it has been assumed that the feed flow does not increase the volume of the reactor contents. When this is not so, differential equations defining the change in reactor volume with time must be included and all concentrations recalculated with the new volume. This presents no equation-solving problems, but it does make the whole set of equations look unwieldy.

1.4.c   The continuous stirred tank reactor (CSTR) (Fig. 1.6)

In order to build a stirred tank reactor into a continuous process it must have a continuous outflow Q equal to its total feed stream.

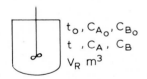

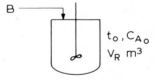

Fig. 1.4. The batch reactor.   Fig. 1.5. The semi-batch reactor.

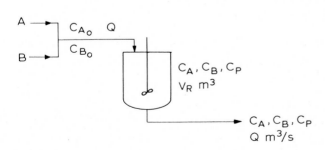

Fig. 1.6. The CSTR reactor.

24

Because the vessel is well mixed, the outlet product concentration is equal to that of the reactor contents, C_p, and after a time, steady state is attained where the rate of reaction is equal to the rate of P leaving the reactor in the outlet stream.

$Q C_p$ kmol s^{-1} of P are being continuously produced, hence:

$$V_R k_1 C_A C_B = Q C_p \tag{1.47}$$

and by mass balance on A:

$$C_{A_0} Q = Q C_p + Q C_A \tag{1.48}$$

and on B:

$$C_{B_0} Q = Q C_p + Q C_B \tag{1.49}$$

Hence, from equations 1.47, 1.48, and 1.49:

$$(C_{A_0} - C_p)(C_{B_0} - C_p) V_R k_1 = Q C_p \tag{1.50}$$

From this equation, the reactor volume to achieve conversion X_A can be determined. Since

$$X_A = (C_{A_0} - C_A) / C_{A_0} = C_p / C_{A_0}$$

$$V_R = \frac{Q X_A}{k_1 (1 - X_A)(C_{B_0} - C_{A_0} X_A)} \tag{1.51}$$

Notice that this type of reaction is defined by a set of algebraic equations, built up from mass balance and chemical kinetic equations. When the system is fairly simple, these equations can be solved analytically, but if it is complicated it is necessary to resort to a numerical routine for the solution of sets of simultaneous non-linear equations.

Notice that the reactants in the reactor are all at the exit concentration level i.e. hopefully very low concentrations. Hence reaction rates are low and so reactor volumes are large. To produce higher conversions or smaller reactors, a number of CSTRs should be installed in series. The first reactor could be run to give a 50% conversion, giving a high rate of reaction and so reducing the total reactor volume. The next might run from 50 to 75% conversion, and the third from 75% to 85%, and so on. This produces a continuous reaction system that has a much lower volume but has more equipment items because of the "cascade" of reactor vessels required. The mathematical analysis takes the form of repeated sets of algebraic equations, exactly as for the single CSTR. Analytic solutions of these equations can be difficult but they are ideally suited to computer solution.

The limiting case of a cascade of reactors is when there is an "infinite" number of tanks, each of negligible volume. At this point the "cascade" becomes a tube with no back mixing, which is in fact the ideal form of the fourth type of reactor.

1.4.d   The plug-flow reactor (PFR) (Fig. 1.7)

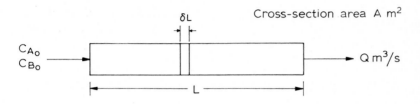

Fig. 1.7. The plug-flow reactor (PFR).

Here the reactants flow through the tube without back-mixing, with concentrations changing down the tube as a result of the reaction.

Taking a small element δL, cross-section area A, gives a volume $A\delta L$ in which reaction is occurring in a steady state producing P, so that when $\dot{n}$ kmol s^{-1} of P are flowing into the element, $\dot{n} + \delta\dot{n}$ kmol s^{-1} flow out. Hence:

$$\delta\dot{n} = k_1 C_A C_B A\delta L \qquad (1.52)$$

This causes a concentration change in the fluid flowing through this section of:

$$\delta C_P = \frac{\delta\dot{n}}{Q}$$

Also we can replace tube length by space time τ since

$$\delta\tau = \frac{A\delta L}{Q}$$

Hence equation 1.52 can be re-written as:

$$Q\delta C_P = k_1 C_A C_B Q \delta\tau$$

or, in the limit as $\delta\tau \longrightarrow 0$

$$\frac{dC_P}{d\tau} = k_1 C_A C_B \qquad (1.53)$$

Similarly

$$\frac{dC_A}{d\tau} = -k_1 \, C_A C_B \tag{1.54}$$

and

$$\frac{dC_B}{d\tau} = -k_1 \, C_A C_B \tag{1.55}$$

Notice that this limiting case of a cascade of CSTRs has resulted in a set of different equations similar to those of the batch reactor, except that reaction time t is replaced by reactor space time τ. It can be stated in general that mathematical treatment of the kinetics of a plug-flow reactor are the same as those of a batch reactor. This can be seen to be so if one pictures a packet of the reacting fluid moving down the tube. Since there is no mixing with other packets, this packet is experiencing the same concentration change with time as the contents of the batch reactor; therefore the mathematical descriptions must correspond. As with batch reators, therefore, the reaction rates are highest at the beginning of the tube, where concentrations are highest.

1.5   THE SELECTIVITIES SHOWN BY THE FOUR REACTOR TYPES IN SIMULTANEOUS REACTION
        SYSTEMS

1.5.a   Consecutive reactions

Consider the reaction scheme

$$A \; + \; B \; \xrightarrow{\;k_1\;} P$$

$$P \; + \; B \; \xrightarrow{\;k_2\;} D$$

with P being the required product. The selectivity to P of the reaction, S, is define as:

$$S = \frac{C_P - C_{P_0}}{(C_P - C_{P_0}) + (C_D - C_{D_0})} \tag{1.56}$$

the conversion of reactant A, X_A, is given by:

$$X_A = \frac{C_{A_0} - C_A}{C_{A_0}} = \frac{(C_P - C_{P_0}) + (C_D - C_{D_0})}{C_{A_0}} \tag{1.57}$$

and the yield of P based on A is

$$\frac{C_P - C_{P_0}}{C_{A_0}} \tag{1.58}$$

These definitions assume no change in reactor volume. More strictly, these definitions should use moles and not concentrations.

To maximize the selectivity to P, the reactor conditions must lead to the first reaction being preferred. This will be so if C_A is high when C_B is high, and when C_P becomes high, C_B is low. This concentration pattern holds for batch reactors and plug-flow reactors.

For a continuous stirred tank reactor, owing to the mixing, C_P is present at the same concentration during the whole reaction, and so the first reaction is not favoured. Hence the selectivity will be less than that of the batch reactor.

In the semi-batch reactor, the concentration of B (assuming it to be the component added continuously) is not high at the beginning when A is high, and so again a high selectivity is not to be expected.

Hence, for consecutive reactions it is easy to argue that differences in selectivity can be expected from different reactor types.

1.5.b   Parallel reactions

$$A + B \xrightarrow{\ k_1\ } P$$

$$A + B \xrightarrow{\ k_2\ } D$$

with P being the required product.

For this reaction scheme the concentrations of A and B have the same effect on reaction rates of both reactions. Therefore, whatever the concentration profiles, the relative rates of P and D will be the same. Therefore, for all reactor types the selectivities will be the same, and will in fact be:

$$S = \frac{P}{P + D} = \frac{k_1}{k_1 + k_2} \qquad\qquad (1.59)$$

If, however, the kinetic rate expressions of the competing reaction were not the same as those for the main reaction, the conclusion would be different.

Assume B tended to polymerize:

$$A + B \xrightarrow{\ k_1\ } P$$

$$B + B \xrightarrow{\ k_2\ } D$$

Now a reactor which held concentrations of B low whilst A was high would favour the formation of P and so increase selectivity. A semi-batch reactor would, in fact, do this effectively. The other reactors would be expected to give lower selectivities because their concentration profiles do not hinder the second reaction to the same extent.

These examples of simultaneous reactions show that the best type of reactor for a particular reaction depends on the reaction mechanism, and that for simple cases it is possible to argue which types of reactor would be expected to be most suitable. In other cases, particularly as the mechanisms become more complex, it is not possible to decide by inspection which reactor type would be the best, and this must come from a quantitative study of the different types of reactor with the specific reaction kinetics in question.

1.6   MODELLING OF SIMULTANEOUS REACTION SYSTEMS

Take the mixed simultaneous reaction scheme:

$$A + B \xrightarrow{k_1} C$$

$$C + B \xrightarrow{k_2} D$$

$$B + B \xrightarrow{k_3} E$$

$$A + B \xrightarrow{k_4} F$$

Let us see how this can be modelled with the various types of reactor.

1.6.a   Batch and plug-flow reactors

If this is carried out in a batch reactor (or a plug-flow tube reactor), it is possible to write down the differential equations defining the concentration changes directly, assuming there is no volume change:

$$\frac{dC_A}{dt} = -k_1 \, C_A C_B - k_4 \, C_A C_B \tag{1.60}$$

$$\frac{dC_B}{dt} = -k_1 \, C_A C_B - k_2 \, C_C C_B - 2k_3 \, C_B^2 - k_4 \, C_A C_B \tag{1.61}$$

$$\frac{dC_C}{dt} = k_1 \, C_A C_B - k_2 \, C_C C_B \tag{1.62}$$

$$\frac{dC_D}{dt} = k_2 \, C_C C_B \tag{1.63}$$

$$\frac{dC_E}{dt} = k_3 \, C_B^2 \tag{1.64}$$

$$\frac{dC_F}{dt} = k_4 \, C_A C_B \tag{1.65}$$

All concentrations can be defined at time zero (or at the tube entrance in the case of the plug-flow reactor), and so there is usually no problem in integrating such a set of equations numerically.

Numerical integration problems, however, do occur when the various components have <u>very</u> different concentrations. For example, if C was a free radical of very low concentration, but nevertheless important to keep in as a reaction step, this could cause integration problems with most numerical integration techniques because of the wide range of time constants for the differential equations in the system.

Such difficulties can be avoided by making the assumption that since C_C is very small, $\frac{dC}{dt}$ must also be very small and we can assume that C reaches a steady state concentration in the reacting mixture. This is known as Bodenstein's theory of standing concentrations.

$$\frac{dC_C}{dt} = k_1\ C_A C_B - k_2\ C_C C_B \approx 0 \qquad (1.66)$$

Therefore

$$k_1\ C_A C_B = k_2\ C_C C_B \qquad (1.67)$$

and

$$C_C = \frac{k_1\ C_A}{k_2} \qquad (1.68)$$

Hence the value of the standing concentration of C can be calculated without a differential equation, reducing both the dimension and degree of difficulty of the numerical integration.

1.6.b   Continuous stirred tank reactors

The reaction scheme described at the beginning of this paragraph would be represented by a set of algebraic equations if the reactor were perfectly mixed. Defining variables C_A-C_F as the outlet (and therefore reactor content) concentrations (kmol m^{-3}) and C_{A_0} and C_{B_0} as the feed concentrations, with a feed flow rate of Q m^3 s^{-1} and a reactor volume V_R m^3, then at steady state the following mass balances hold for each component:

kmoles in outlet s^{-1} = kmoles in inlet s^{-1} + kmoles produced by reaction s^{-1}

Hence, the set of equations becomes:

$$Q\ C_A = Q\ C_{A_0} - V_R\ k_1\ C_A C_B - V_R\ k_4\ C_A C_B \tag{1.69}$$

$$Q\ C_B = Q\ C_{B_0} - V_R\ C_B\ (k_1\ C_A + k_2\ C_C + 2k_3\ C_B + k_4\ C_A) \tag{1.70}$$

$$Q\ C_C = Q\ C_{C_0} + V_R\ (k_1\ C_A C_B - k_2\ C_C C_B) \tag{1.71}$$

$$Q\ C_D = Q\ C_{D_0} + V_R\ k_2\ C_C C_B \tag{1.72}$$

$$Q\ C_E = Q\ C_{E_0} + V_R\ k_3\ C_B^2 \tag{1.73}$$

$$Q\ C_F = Q\ C_{F_0} + V_R\ k_4\ C_A C_B \tag{1.74}$$

This is a set of six simultaneous non-linear equations to be solved to determine the six unknowns for the system C_A-C_F, the outlet concentrations for each component from a reactor of volume V_R and total feed flow Q.

For a cascade of CSTR reactors, the outlet concentrations from the first reactor are taken as the inlet concentrations (C_{A_0} - C_{F_0}) for the next, and so on to the end of the cascade. Such a set of equations can be solved by standard equation-solving packages, when given reasonable starting estimates.

Sets of non-linear simultaneous equations can be difficult to solve; the main problem is the location of the wrong root of the equation set, because non-linear systems can have more than one root, and often only one of these will exist in practice. A second difficulty is the possibility of posing a problem that has no real root.

These problems are minimized by solving reactor problems in the simulation mode, that is, by defining the feeds and reactor dimensions and calculating the outlet concentrations. Such a system must have a real solution, and sensible initial estimates can be given. It would be very convenient, and possible in a purely mathematical sense, to carry out a "design" calculation, in which for a required output, quantity, and conversion, the reactor feeds and dimensions are calculated. However, such calculations can only be made for the simplest systems because of the difficulty in defining outlet concentrations which give a real solution. In practice, therefore, reactor calculations are generally done in a simulation mode.

1.6.c Semi-batch reactors

The reaction system under discussion could be carried out in a semi-batch reactor, feeding B continuously to a vessel containing A. For a feed rate of F_B kmol s^{-1} of B, in a reactor volume of V_R, assuming no reaction volume increase (say D is a gas), the change in concentration of all components is given by the same differential equations as those for the batch reactor with the exception of the equation defining B, which has a term showing the continuous feed of B:

$$V_R \frac{dC_B}{dt} = - (k_1\, C_A C_B + k_2\, C_C C_B + 2k_3\, C_B^2 + k_4\, C_A C_B)\, V_R + F_B \qquad (1.75)$$

The method of solution of this set of differential equations is exactly the same as for the batch reactor. Again, if necessary, assumptions of steady state concentrations of low concentration intermediates can be used to reduce the severity of the integration problem.

1.7 SOME ANALYTICAL SOLUTIONS

It is always convenient to have analytical solutions to the various reactor models, expressing conversions in terms of reactor velocity constants, reaction times, and initial concentrations. Such expressions can be derived for simple reactions, as in fact was done in section 1.4 for a second-order reaction.

Solutions are available for various first- and second-order simple reactions for batch and plug-flow reactors, and these are given in Table 1.1.

Table 1.2 shows some analytical solutions to two-reaction schemes and one three-reaction consecutive first-order reaction. Notice that there are few analytical solutions available for simultaneous reaction systems, and those available are, in the main, for first-order systems.

Because analytical solutions are derivable for only the simplest systems, in practice it is usually not worth spending more than a few minutes attempting to develop an analytical solution for a new system, because it is likely that one does not exist. Numerical solution, on the other hand, is a general method that has a much greater chance of succeeding, and is capable of being expanded and modified without requiring a complete revision of the solution.

Tables 1.1 and 1.2 are useful checks to see whether a convenient analytical solution exists. It should always be remembered that a second-order reaction has first-order characteristics if one of the components is in excess. For example, if one of the components is also the solvent, the concentration throughout the reaction is fairly constant, and

$$\frac{dC_C}{dt} = k_1\, C_A C_B \qquad (1.76)$$

can be approximated by

$$\frac{dC_C}{dt} = k'\, C_A \qquad (1.77)$$

where k' is $k_1 \bar{C}_B$ and $\bar{C}_B$ is the average concentration of B. In this way, first-order analytical solutions may sometimes be used to represent second-order reactions.

TABLE 1.1

Analytical solutions for some single reactions for batch and plug-flow reactors

Reaction	Rate equation	To determine time required t for a defined fraction conversion X_A
$A \rightarrow P$	$r_1 = k_1 C_A$	$t = \dfrac{1}{k_1} \ln\left[\dfrac{1}{1-X_A}\right] = \dfrac{1}{k_1} \ln \dfrac{C_{A_0}}{C_A}$
$2A \rightarrow P$	$r_1 = k_1 C_A^2$	$t = \dfrac{1}{k_1 C_{A_0}}\left[\dfrac{X_A}{1-X_A}\right]$
$A \rightarrow P$	$r_1 = k_1 C_A^{\,n}$	$t = \dfrac{1}{k_1 C_{A_0}^{n-1}(n-1)}[(1-X_A)^{1-n}-1], \; n\neq1$
$A + B \rightarrow P$ $M_{BA} \neq 1$	$r_1 = k_1 C_A C_B$	$t = \dfrac{1}{k_1 C_{A_0}(M_{BA}-1)}\ln\left[\dfrac{M_{BA}-X_A}{M_{BA}(1-X_A)}\right]$
$A + B \rightarrow P$ $M_{BA} = 1$	$r_1 = k_1 C_A C_B$	$t = \dfrac{1}{k_1 C_{A_0}}\left[\dfrac{X_A}{1-X_A}\right] = \left(\dfrac{1}{C_A} - \dfrac{1}{C_{A_0}}\right)\dfrac{1}{k_1}$
$aA + bB \rightarrow P$ $M_{BA} = b/a$	$r_1 = k_1 C_A C_B$	$t = \dfrac{1}{k_1 M_{BA} C_{A_0}}\left[\dfrac{X_A}{1-X_A}\right]$
$A + 2B \rightarrow P$ $M_{BA} = 2$	$r_1 = k_1 C_A C_B^2$	$t = \dfrac{1}{8k_1 C_{A_0}^2}\left[\dfrac{1}{(1-X_A)^2} - 1\right]$
$A + 2B \rightarrow P$ $M_{BA} \neq 2$	$r_1 = k_1 C_A C_B^2$	$t = \dfrac{1}{k_1 C_{A_0}^2 (2-M_{BA})^2}\left[\ln\dfrac{M_{BA}-2X_A}{M_{BA}(1-X_A)} + \dfrac{2X_A(2-M_{BA})}{M_{BA}(M_{BA}-2X_A)}\right]$
$A + B \rightarrow P$ $M_{BA} = 1$	$r_1 = k_1 C_A C_B^2$	$t = \dfrac{1}{2k_1 C_{A_0}^2}\left[\dfrac{1}{(1-X_A)^2} - 1\right]$
$A + B \rightarrow P$ $M_{BA} \neq 1$	$r_1 = k_1 C_A C_B^2$	$t = \dfrac{1}{k_1 C_{A_0}^2 (1-M_{BA})^2}\left[\ln\dfrac{M_{BA}-X_A}{M_{BA}(1-X_A)} + \dfrac{X_A(1-M_{BA})}{M_{BA}(M_{BA}-X_A)}\right]$
$aA + bB \rightarrow P$ $M_{BA} \neq b/a$	$r_1 = k_1 C_A C_B$	$t = \dfrac{1}{k_1 C_{A_0}[M_{BA}-(b/a)]}\ln\left[\dfrac{M_{BA}-(b/a)X_A}{M_{BA}(1-X_A)}\right]$
$aA + bB \rightarrow P$ $M_{BA} = b/a$	$r_1 = k_1 C_A^{\alpha} C_B^{\beta}$	$t = \dfrac{1}{k_1 M_{BA}(\alpha+\beta-1)C_{A_0}^{\alpha+\beta-1}}\left[\dfrac{1}{(1-X_A)^{\alpha+\beta-1}} - 1\right]$

where $M_{BA} = \dfrac{C_{B_0}}{C_{A_0}}$, and $X_A = \dfrac{C_A}{C_{A_0}}$

TABLE 1.2

Analytical solutions to some simultaneous reaction systems

Reaction	Rate equation	Integration equation time/concentration relationship	Selectivities
Equilibrium: $A \underset{k_2}{\overset{k_1}{\rightleftharpoons}} B$	$\dfrac{dC_B}{dt} = k_1 C_A - k_2 C_B$ where $\dfrac{k_1}{k_2} = K$ and $M_{BA} = C_{B_0}/C_{A_0}$	$t = \dfrac{K}{k_1(K+1)} \ln\left[\dfrac{K - M_{BA}}{K - M_{BA} - (K+1)X_A} \right]$	
First-order competing: $A \overset{k_1}{\nearrow} P$ $A \underset{k_2}{\searrow} R$	$r_1 = k_1 C_A$ $r_2 = k_2 C_A$	$\dfrac{C_A}{C_{A_0}} = e^{-k_c t}$ where $k_c = k_1 + k_2$	$\left. \begin{array}{l} \dfrac{C_P - C_{P_0}}{C_{A_0} - C_A} = \dfrac{k_1}{k_c} \\[2ex] \dfrac{C_R - C_{R_0}}{C_{A_0} - C_A} = \dfrac{k_2}{k_c} \\[2ex] \dfrac{C_P - C_{P_0}}{C_R - C_{R_0}} = \dfrac{k_1}{k_2} \end{array} \right\}$ for batch, plug-flow and CSTR

TABLE 1.2

Analytical solutions to some simultaneous reaction systems (cont'd.)

Reaction	Rate equation	Integration equation time/concentration relationship	Selectivities
Second-order competing: $A + B \xrightarrow{k_1} P$ $A + C \xrightarrow{k_2} R$	$r_1 = k_1 C_A C_B$ $r_2 = k_2 C_A C_C$		$\dfrac{(1-X_B)}{(1-X_C)} = e^{k_1/k_2}$ for batch and plug-flow $\dfrac{X_B(1-X_B)}{X_C(1-X_C)} = \dfrac{k_1}{k_2}$ CSTR where: $X_B = C_B/C_{B_0}$ $ X_C = C_C/C_{C_0}$
First-order 2 consecutive reactions: $A \xrightarrow{k_1} B \xrightarrow{k_2} C$	$r_1 = k_1 C_A$ $r_2 = k_2 C_B$	$\dfrac{C_A}{C_{A_0}} = e^{-k_1 t}$ $C_B = C_{A_0} \dfrac{k_1}{k_2 - k_1} [e^{-k_1 t} - e^{-k_2 t}]$	

TABLE 1.2
Analytical solutions to some simultaneous reaction systems (cont'd.)

Reaction	Rate equation	Integration equation time/concentration relationship	Selectivities
First-order **3 consecutive reactions:** $A \xrightarrow{k_1} B \xrightarrow{k_2} C \xrightarrow{k_3} D$	$r_1 = k_1 C_A$ $r_2 = k_2 C_B$ $r_3 = k_3 C_C$	$\dfrac{C_A}{C_{A_0}} = e^{-k_1 t}$ $C_C = C_{A_0} k_1 k_2 \left[\dfrac{e^{-k_1 t}}{(k_1-k_2)(k_1-k_3)} + \dfrac{e^{-k_2 t}}{(k_2-k_1)(k_2-k_3)} + \dfrac{e^{-k_3 t}}{(k_3-k_1)(k_3-k_2)} \right]$	

1.8 INTRODUCTION OF HEAT EFFECTS INTO REACTOR MODELS

So far all discussion has centred around the isothermal operation of various types of reactor. In practice, most reactions are associated with the liberation or absorption of heat, and this leads to temperature changes in the reactor. Hence the reaction velocity constants can no longer be represented by a constant, but must now be a function of temperature.

Thus the reaction velocity constants must be represented by the Arrhenius expression:

$$k_i = A_i \, e^{-E_i/RT} \tag{1.78}$$

1.8.a Batch, semi-batch and plug-flow reactors

In batch and plug-flow reactors the rate of heat evolution is the sum of the heats evolved from all the reactions occurring:

$$\frac{dH}{dt} = - \sum_{i=1}^{n} \Delta H_i \, r_i \, V_R \quad (\text{kJ s}^{-1} \text{ or kW}) \tag{1.79}$$

where the heat of reaction is ΔH_i and the rate of reaction r_i for reaction i.

With the reactor volume of V_R litres (or flow of V_R m^3 s^{-1} in the case of a plug-flow reactor), density ρ and specific heat c_p, the resulting temperature change is given by

$$\frac{dT}{dt} = \frac{1}{V_R \, \rho \, c_p} \frac{dH}{dt} \tag{1.80}$$

The inclusion of these two differential equations together with the relationship between k_i and T (equation 1.78) and the previously discussed equations 1.60 – 1.65 describes the reaction occurring in an adiabatically operated reactor.

If the reactor is not adiabatically operated and heat H_J kJ s^{-1} is removed by heat exchange with the walls or with heat exchange surfaces in the reactor, this heat flow is a function of the heat exchange medium temperature, T_J, the reactor temperature, T, the heat exchange area, A', and the overall heat exchange coefficient, U:

$$\frac{dH_J}{dt} = U \, A' \, (T - T_J) \quad \text{kW} \tag{1.81}$$

Hence the change in heat content of the reactor is given by inclusion of this equation in equation 1.80, resulting in the following expression for temperature change:

$$\frac{dT}{dt} = \frac{1}{V_R \, \rho \, c_p} \left\{ -V_R \sum_{i=1}^{n} (\Delta H_i \, r_i) - U \, A' \, (T - T_J) \right\} \tag{1.82}$$

In the semi-batch reactor this equation should also include a term for the heat required to heat the continuous feed up to reactor temperature if this is likely to be of significance.

1.8.b   Continuous stirred tank reactors

In this type of reactor a steady-state temperature is achieved, whereby the heat from the reactor is balanced by the sensible heat removed by the outlet stream, plus the heat removed by the heat exchange device.

The rate of heat gain, q_g, is given by

$$q_g = -V_R \sum_{i=1}^{n} \Delta H_i \, r_i \qquad (1.83)$$

The sensible heat removed, q_s, by the outlet stream is

$$q_s = Q \, \rho \, c_p \, (T - T_{in}) \qquad (1.84)$$

where Q is the inlet reactor flow rate ($m^3 \, s^{-1}$), T_{in} is the temperature of that stream, and T is the reactor (and also outlet stream) temperature.

The rate of heat removal by heat exchange, q_e, for a given exchange area A' and coolant temperature T_J is given by

$$q_e = U \, A' \, (T - T_J) \qquad (1.85)$$

Hence, since

$$q_g = q_s + q_e \qquad (1.86)$$

$$-V_R \sum_{i=1}^{n} \Delta H_i \, r_i = Q \, \rho \, c_p \, (T - T_{in}) + U \, A' \, (T - T_J) \qquad (1.87)$$

Hence, the general treatment for a CSTR reactor with multiple reactions and heat effects results in a set of non-linear simultaneous equations: in the example begun in paragraph 1.6 this would involve equations 1.69 - 1.74, describing the chemical kinetics, equation 1.78 giving the temperature relationship of the reaction velocity constants, and equation 1.87 defining the temperature.

Such a set of equations can only be solved by a numerical method, and so computer solution is inevitable. The equations are highly non-linear and such complex systems can lead to multiple solutions, a point that will be discussed in more detail in Chapter 10 on reactor control.

1.9   THE ROLE OF MASS TRANSFER

In single-phase homogeneous reactions no mass-transfer problems arise in the reactor design after the initial mixing problems have been solved. However, as soon as two or more phases are present, mass-transfer operations are introduced and this has an effect on the course of the reaction.

When a liquid reactor is fed with one gaseous feed, the gas must firstly dissolve, and then this dissolved gas can react with the second component. If the reaction is fast compared with the mass transfer, the reactor performance is limited by its performance as a mass-transfer device.

As the mass transfer rate increases, the steady state gas concentration in the liquid increases, until, when the mass-transfer rate is fast compared with the reaction rate, it can be assumed that the gas "saturates" the liquid.

Hence different mass-transfer rates cause different gas concentrations to be obtained in the reactor. Changing mass-transfer rates, e.g. changing the agitation or pressure, changes concentrations.

If the reaction is complex, and side reactions can occur with the gas when in solution, then we can see that a change in stirrer rate can affect reactor selectivity as well as conversion.

When the product of a liquid reaction is a gas, then again a mass-transfer step is involved. If the reaction is complex,   different mass-transfer rates, and hence liquid concentration of the product, could affect yield or selectivity.

In gas/solid catalysed reactions, there are again serious mass-transfer steps involved. The reactants must diffuse from the bulk gas to the catalyst surface to be adsorbed and then react. The product of the reaction must then diffuse away to clear the catalyst for further reactant.

Such mass-transfer effects are of utmost importance to the chemical engineering of the reactor, and so are considered part of the "mechanism of the reaction" by the chemical engineer. The reaction mechanism described by the chemist is unlikely to contain such steps. In fact, the chemist would hope to so construct his apparatus that mass-transfer effects were made negligible so that his results measured only the chemical kinetics occurring. On industrial equipment the economic operating conditions usually include a degree of mass-transfer resistance. In the laboratory it is possible to crush a solid catalyst or mix gas/liquid reactors intimately to remove diffusional resistances. On the plant one must use a size of catalyst granule that gives a reasonable pressure drop and determine mixing levels bearing in mind the power that will be consumed.

1.10   THE NEED FOR COMPLEX MODELS

For all basic reactor types we have now developed complex models which require computers for their solution. This is in contradiction to the modelling principle that models should be kept as simple as possible. However, in reactor modelling, any practical application usually requires the use of a complex model to obtain a worthwhile analysis.

Reaction systems cannot be reduced to a single overall reaction, ignoring mechanisms, because simple reactions have optima at extremes and real reactions have optima within possible operating ranges. A model for optimising purposes must represent the competing forces that give rise to such optima.

Temperature cannot be ignored on full-scale reactions, because heat removal is a major objective of the reactor design. When the heat is evolved, its effect on the reaction rates and resulting effect on selectivities must be included to produce anything other than an elementary design.

Reactor design should include stability and control analysis, as reactor instability is the result of the strong interplay between heats of reaction, reactor temperature and reaction rate. Under some circumstances this can represent an unstable system. For such an analysis, clearly the model must accurately predict the reactor temperature throughout the whole reaction time.

For these reasons reactor engineering in practice requires the modelling of complex reaction systems, the obtaining of experimental data and analysis of this data in terms of a proposed model, the testing of this model in a scaled-up system, and the use of this model for the design of an efficient, operable, safe plant. These topics therefore form the basis of the remainder of this book.

FURTHER READING

This chapter has been a revision of those parts of fundamental chemical reactor theory that are most important in practical reactor design. These points are treated more thoroughly in most reactor design texts. The following books are recommended for more details of the principles of chemical reactor theory.

Carberry, J.J., 1976. Chemical and Catalytic Reaction Engineering. McGraw-Hill, New York. - Strong on heterogeneous catalysis, delightful prose, the mathematics can be ignored.
Hill, C.G., 1977. An introduction to Chemical Engineering Kinetics and Reactor Design. Wiley, New York. - Good description of kinetics, mechanisms and fundamentals.
Levenspiel, O., 1972. Chemical Reaction Engineering, 2nd ed. Wiley, New York. - No comment!

QUESTIONS TO CHAPTER 1

(1) Why is the chemical engineer interested in reaction mechanisms ?
(2) What are the four types of reactor and what are their fundamental differences ?
(3) What is the difference between the chemist's and chemical engineer's concept of reaction mechanism ?
(4) What forms of equation are obtained with the four types of reactor and what methods are available for their solution ?
(5) What reactor types are preferred for the following cases:
    (a) an unwanted consecutive reaction occurs;
    (b) initial rate of heat evolution is high;
    (c) a consecutive reaction occurs and one of the feeds polymerizes ?
(6) Why are complex reactor models necessary for reactor design purposes ?
(7) Where can the engineer obtain his kinetic and thermodynamic data for his design

Chapter 2

MODELLING OF REACTORS

    Modelling (mathematical modelling) implies the representation of a physical system by a set of equations which in a limited way can represent the system under study. Relevant in reactor design is a mathematical description that can predict reactor outlet concentrations and temperature from inlet concentrations, flows, and reactor dimensions. These equations are usually, though not always, solved by computer.

    Of particular interest is the form of equation that represents the physical system. A standard form of equation could be chosen, for example an equation linear both in parameters (θ) and independent variables (x).

$$y = \theta_0 + \theta_1 x_1 + \theta_2 x_2 \tag{2.1}$$

    If this cannot properly represent the reactor performance because it is too linear, a quadratic or cubic polynomial, still linear in parameters, is possible:

$$y = \theta_0 + \theta_1 x_1^3 + \theta_2 x_1^2 + \theta_3 x_1 + \theta_4 x_2^3 + \theta_5 x_2^2 + \theta_6 x_2 \tag{2.2}$$

However, generally the linear nature of the parameters cannot properly represent the reactor performance. An equation of the type:

$$y = \theta_0 x_1^{\theta_1} x_2^{\theta_2} \tag{2.3}$$

or, in a linearized form:

$$\ln(y) = \ln(\theta_0) + \theta_1 \ln(x_1) + \theta_2 \ln(x_2) \tag{2.4}$$

would be more appropriate.

    We are discussing here purely _empirical_ models. That is, models whose form has been chosen for convenience - convenience in effort required to create the model, and effort required to carry out analysis with the model. The opposite approach is to create a model by describing the physical and chemical processes occurring within the reactor and build these together as a complete model of the reactor. Such a model is described as a mechanistic model and would consist of the sets of equations developed for the various reactors in Chapter 1. Such a model is difficult to create because the physical and chemical steps must first be identified before they can be modelled, and there is a considerable research effort needed to define the elementary steps occurring in reaction processes. Secondly, the models are difficult to analyze because they can no longer be linearized, and may contain too many parameters for convenient handling with non-linear methods.

In practice, therefore, a compromise is used in which the equations are to some extent a representation of the process occurring in reality. We have already met this philosophy in Chapter 1 where we admit that the chemist's explanation of a reaction is accurate, but far too detailed for a quantitative analysis, and we are then satisfied with a simplified reaction mechanism for modelling use.

The resulting models contain some degree of truth, but are generally gross simplifications of the reality. To avoid endless discussion with theorists, they should be called "semi-empirical" models, admitting that they are not intended to be full descriptions of real processes, and the temperature exponent likewise be called "apparent activation energy", to clearly show that we are not trying to define energy levels in molecular interactions.

Naturally one always prefers a model which accurately describes the physics and chemistry occurring in the reactor, but a compromise is always necessary between the accuracy of the description, the effort required to identify this description, and the ensuing analysis. Modelling is often described as an "art" to explain why some people succeed in modelling projects while others do not. It is in the decision of what to include and what to exclude and where to approximate that this "art" is required, where some people have an innate feel for what is reasonable, and others do not.

It is important to recognize that there are two accuracies to consider when discussing a model of a physical system:

(a) the accuracy of the description represented by the equations (i.e. the form of the equations)

(b) the accuracy by which the model predicts the result of the experiments that have been carried out on the reactor (i.e. the goodness of fit).

The discussion of empirical, semi-empirical, and mechanistic models concerns point (a) only. It could well be that the purely empirical model with enough constants would fit the available experimental results more accurately than the mechanistic model; i.e. be most accurate according to (b) but least accurate by definition (a).

2.1   WHY MODEL REACTORS?

A model of a reactor enables one to carry out a quantitative study of it. This can be very useful when a large number of results exist and some method is needed for comparing them. Which are reliable? What are the major trends in the multi-variable system? Where has there been inadequate experimentation? Where should the next experiment be made?

The investigation of a reactor system involving a number of independent variables can pose difficult problems in deciding where experiments should be carried out. By attempting to construct a model the discipline of thinking carefully about the meaning of the results available so far can show up areas that have been inadequately experimented. Once a model is available it can be used as the centre of an experimental planning procedure which predicts which next experiments to carry out to obtain useful information.

As long as the model has some mechanistic features it can throw some light on the understanding of the reaction. If there are some experimental points that cannot be predicted correctly by the model, this indicates that the mechanism contained by the model is in some way wrong or incomplete. A lack of fit indicates that we have not adequately understood the physical and chemical processes occurring. Furthermore, the nature of the discrepancies give us a hint as to the possible regions of our incomplete knowledge. Whether it is concerned with the heat balance, with a side reaction mechanism, or with a reaction order would be indicated by incorrect prediction of temperature, by-product concentration, or effect of concentration on yield, respectively.

Notice that we can never conclude from a good fitting model that we understand the mechanism occurring; there are usually many different mechanisms that produce the same mathematical form. It is only possible to conclude that from a badly fitting model we should revise our concept of what is occurring in the reactor. However, such information is still useful as it can initiate new hypotheses which in turn lead to a deeper understanding of the reaction. In the development of new processes such a deeper understanding of a reaction is invaluable in initiating new ideas on, for instance, how the catalyst should be modified or which alternative solvents should be tried. Such decisions can often produce jumps in reactor conversion unachievable by simple manipulations of temperature and concentrations.

A semi-empirical model which fits experimental data well is a great comfort when an engineer is faced with scaling-up a reactor to produce the full-scale design. There is, of course, no guarantee that the scale-up will be correct, because the model is not guaranteed to contain the correct mechanism, but the chances are very high that errors introduced by the inadequate mechanism will be tolerably small on the full-scale plant. This confidence is not given by empirical models, since the model form could give wild predictions outside the experimental region, i.e. in larger scale equipment.

From this discussion it should be clear that the preferred models are the semi-empirical, tending towards mechanistic models, movement towards empirical models being enforced to produce a manageable model, or through lack of knowledge of a reasonable mechanism.

2.2   THE MODEL EQUATIONS

The model consists of a set of equations, differential or algebraic. The first step is to collect these equations together. The equations represent the natural laws and constraints that define the behaviour of the system, and they are introduced into the model through having a good understanding of the physical nature of the system.

The following check-list can be useful in recalling that all equations defining the system are included:

(a) Each chemical reaction sets up a rate equation for each species in the reaction

(b) When two phases are present, and equilibrium between the phases is assumed, this leads to an equation for each species in the system relating the concentrations in the two phases.

(c) When equilibrium between phases is not assumed, there will be a mass-transfer equation for each component not at equilibrium.

(d) When temperatures in the system are not identical, there will be heat flow equations between the various parts not at thermal equilibrium.

(e) The system must be in heat balance, leading to a heat balance equation.

(f) The system must be in mass balance. This can lead to a number of forms:
   - the total mass-balance equation must hold.
   - a total moles balance must hold if the reaction had not involved a change in the number of moles.
   - individual elemental balances can be made on the different elements making up the molecules.

(g) Total mole fractions of any phase add up to 1.0. This gives one further equation per phase present.

(h) Various simple conversion equations are necessary when components need to be expressed in different ways, e.g. volume = weight/density or PV = nRT for a gas.

This complete list will produce redundant equations and these must be spotted before the final set is collected together. For example, to have a rate equation for every species in the reaction is making a mass-balance equation redundant since it is implicitly covered by defining every species in the reaction. As a second example, a total elemental balance makes a mass balance redundant, because the sum of the elemental-balance equations is the mass-balance equation. In cases of redundancy the least convenient equation can be discarded. Care must be taken that the computer integration errors are not unintentionally concentrated in such a selection. For example, to discard a rate equation for a low concentration component, and determine this from a system mass balance is unwise, because all errors in the system will concentrate in the predicted value of this trace component, making its calculated value unreliable.

The equations are composed of parameters and variables. The parameters have fixed values since they define the system: reaction velocity constants, densities, stoichiometric coefficients, the gas constant and so on. The variables will be the temperatures, pressures, compositions of each phase, volumes, flows, etc. There will be more of these than the number of equations, and the difference defines the number that can be taken as independent variables, i.e. those free to be chosen by the designer. The remainder are the dependent variables, and these are determined by solving the set of simulation equations forming the model.

Which variables should be chosen as independent variables is governed by two considerations. Firstly some variables present extremely difficult numerical problems if taken as dependent variables; reactor feed flows often fall into this category. These problems may be numerical convergence or multiple solution problems. Such variables had best be specified in the design as independent variables. Secondly the character of the calculation is determined by the choice of independent variable. A "simulation" model is a model with the reactor feed and its dimensions as independent variables and outlet conditions as dependent, whereas a "design" model has reactor product output as an independent variable, and reactor dimensions and feed as dependent variables.

The modeller has also to decide in which order to solve the equations, because a clever choice of order can greatly reduce the equation-solving problem.

The choice of dependent and independent variables, and the order of solution can be determined either by intuition or by the following systematic method.

Consider a set of 3 equations and 4 variables:

$$f_1(x_1, x_2, x_4) = 0 \tag{2.5}$$

$$f_2(x_2, x_4) \quad = 0 \tag{2.6}$$

$$f_3(x_2, x_3, x_4) = 0 \tag{2.7}$$

These equations can be represented by a "structural matrix":

Equation	x_1	x_2	x_3	x_4
2.5	x	x		x
2.6		x		x
2.7		x	x	x

Now 3 equations can be used to solve for 3 dependent variables, leaving one variable free to be an independent variable (in other words, the system has one degree of freedom).

We can therefore choose one variable to be independent. Let this be x_4. We can now construct a second matrix from the structural matrix with the independent variable on the left hand side, and the dependent variables forming a matrix with no entries above the diagonal, by re-ordering both the equations and the dependent variables.

	Independent	Dependent		
Equation	x_4	x_2	x_1	x_3
2.6	x	x		
2.5	x	x	x	
2.7	x	x		x

Now, if the equations are solved consecutively in the order in the above matrix, i.e. 2.6, 2.5, and 2.7, there need be no iteration because equation 2.6 can be used to give x_2, then equation 2.5 gives x_1, and finally equation 2.7 will give x_3 explicitly.

The choice of independent variables should not be random. Firstly, it alters the function of the model (from a design to a simulation model). Secondly it eases the equation-solving problem if one is chosen which otherwise would necessitate a simultaneous and not consecutive solution, and thirdly it is important to note the form of the equation itself.

If equation 2.6 were of the form

$$x_4 - A e^{-k_1 x_2} - B e^{k_2 x_2} = 0 \qquad (2.8)$$

then x_2 should have been chosen to be the independent variable and not x_4 as in our previous example. The first step of the solution, solving for x_2, cannot be done analytically, whereas if x_2 were the independent variable the solution for x_4 would be very simple.

The rearranging of the structural matrix can be done systematically in the following way.

1. Cross out all columns on the structural matrix that are independent variables. These appear in any order on the left hand side of the new matrix.
2. Take a variable that appears in the least number of rows (equations). This will be the variable on the far right hand side of the new matrix, and the equation that contains it will be the last equation in the new matrix. Cross out the variable column and equation row.
3. Find the next uncrossed out variable with the least number of entries in its column. This forms the next to last variable and next to last equation. Cross out both variable and equation.
4. Repeat step 3 until all variables and equations have been placed in the new matrix. This matrix should now have the least number of elements above the

diagonal. Each entry above the diagonal represents a need to solve simultaneously, not consecutively, which will require some sort of iteration.

Equation ordering is often done by inspection with a little flair and intuition,
and appreciation of the physical system. However, as systems become larger, or if
the engineer is lacking experience, such systematic methods as the one described
can be of value.

2.3   THE CHOICE OF COMPUTER AND SOFTWARE

2.3.a   Choice between analog and digital computers

Usually it is not possible to solve the model equations explicitly and they must
be solved by computer. If the model consists of a set of algebraic equations a
digital computer is necessary for their solution. However, if there are differential
equations involved, there is the possibility of using an analog instead of a
digital computer.

Analog computers are far less convenient to use than digital ones, because
analog computers are less readily available, and there is the tiresome task of
scaling all the variables so they do not attain the maximum voltages during the
calculation. Also, the presence of many algebraic equations or functions can cause
problems in analog programming. However, integration can be carried out on analog
machines without worrying about stability problems and step-sizes. Therefore, if
a particular reactor model consists predominantly of differential equations, and
if it is thought that these could present numerical difficulties in their solution,
the use of an analog machine is worth consideration.

Usually, however, digital machines can handle the integration adequately, and
should be considered first. Most of the past advantages of analog machines have
been eroded by the introduction of on-line simulation languages for digital
computers. These languages enable differential equations to be programmed in a very
high level language, with the engineer not having to worry about the integration, or
the output. The language looks after all these points and presents the results in
graphical form, either on a printer or on CRT graphic display equipment. CSMP,
MIMIC, GASP, BEDSOCKS, DARE, and SIMULA are a few such systems.

These simulation languages enable the engineer to get a good "feel" for the
system under study because he can conveniently try a whole host of ideas, and plot
the results out in any desired form. This can help get a qualitative understanding
of the performance of a proposed model which can be a great aid to creative thinking
in the formation and modification of the models themselves. The disadvantage is
that the simulation systems are not compatible with the later uses of the model.
They cannot be used for equation fitting by regression, or for use in process models.
Hence, at some stage of a reactor study the model must be reprogrammed for a normal
digital machine. Even the digital simulation languages should therefore only be
preferred to normal program models (in FORTRAN or ALGOL) if it is clear that the

project will benefit from this qualitative investigation to obtain a "feel" for the system under study.

2.3.b   The numerical techniques

The equations making up the reactor model are sets of algebraic or differential equations, needing suitable mathematical subroutines for their solution.

In the past it has been the practice to go into considerable detail on the choice of suitable methods and in many reactor engineering papers great emphasis is laid on the solution method used for the model. Nowadays, however, numerical packages are emerging from a number of mathematical institutions which contain large choices of professionally chosen and programmed routines together with the criteria necessary to make a reasonable choice for the particular system in question. In addition, most computer manufacturers supply standard libraries of numerical subroutines of a reasonable quality. Generally one should use the routine library supplied by the computer manufacturer and if, with this library routine, convergence or integration problems arise, then go to one of the commercially available libraries such as IMSL, NAG, or Harwell.

(i) Simultaneous algebraic equations. Problems in solutions of sets of algebraic equations show themselves in non-convergence of the algorithm, which is indicated by a message printed by the mathematical subroutine itself. During the iterative search it can be that instabilities occur in some variables in the model. This results in variations so wide that they lead to unfeasible situations within the model, which in turn result in impossible requests, e.g. log of the -ve number, before the maximum number of iterations has been reached. Hence, a sign of equation-solution problems can be calculation failures through the request for impossible operations.

When equation-solving problems occur there are a number of options open.
(a) One can try to give more accurate initial values for the iterative procedure to start from.
(b) One can rearrange the equations and choice of variables to produce a problem that is less non-linear.
(c) One can use a different (more powerful or more suitable) equation-solving routine. Guidance here is normally given by the producer of the numerical subroutine library.

Since non-linear simultaneous equation-solving routines are search routines, when they find one solution they have finished. There is no way of knowing whether this is the only solution or whether there are other solutions that have not been found. The only protection one has is to repeat the iteration with other starting values to see if other solutions are found. It is always worthwhile having good estimates of the required solution to give as initial values, and care should be taken in arranging the variables to provide a set of equations which will converge on the particular solution that represents the physical system. The problem of multiple steady-states will be discussed later when reactor stability is considered.

(ii) Differential equations. Generally, the number of differential equations representing a model is not of primary significance; 50, 100, or 200 equations can be handled without difficulty. More significant is the character of the equation set; a "stiff" or "difficult to solve" set occurs when there is a wide range in the rates of the processes occurring, when one process occurs in micro-seconds and another in seconds. Then impracticably small step sizes must be used which result in excessive computer time, and round-off error accumulation.

Particularly difficult are "two point boundary conditions" that one gets in counter-current situations such as absorption. However, most reactor simulation problems have all initial conditions known at one boundary, e.g. at time zero all input data are known and so the two-point boundary situation is not common in reactor engineering.

Although much research is carried out into advanced integration methods, the simpler methods such as Euler should not be overlooked. Euler simply states that the new value of variable y (y') after a small step h is the old value (y) plus the [slope x step length], where the slope is taken as the slope at the beginning of the step.

$$y' = y + \frac{dy}{dt} h \qquad\qquad (2.9)$$

The method is simple to program, and quite adequate when integration problems are tackled that are not "stiff".

The more sophisticated methods found in the mathematical packages can be used if the equations are stiff. However, the more sophisticated methods seem to become less efficient when the number of equations increases. There are cases reported where Euler handles a set of 200 fairly stiff equations more satisfactorily than a list of the more sophisticated methods.

It is always advisable to reformulate problems to avoid stiff systems, that is to model effects occurring only in one order of magnitude of time constant if at all possible. This can be done using the Bodenstein stationary state principle to remove very low concentrations from all differential equations because these effectively make the equations stiff. This was explained in more detail in section 1.6.a.

Problems in numerical integration show themselves by the integration routine not being able to achieve the required accuracy even by taking the lowest allowable step length. Often computation failures occur before this point if, during an integration step, impossible situations occur (e.g. a -ve concentration). This usually causes complete instability and log of negative number or accumulation overflow to occur.

2.4 EXAMPLES OF REACTOR MODELS

2.4.a Example 1: A model of a CSTR reactor

Chlorodifluoromethane (Freon 22) is produced by the reaction between HF and chloroform.

$$\begin{array}{ccccccc}
 & & & \text{catalyst} & & & \\
CHCl_3 & + & HF & \xrightarrow{\quad k_1 \quad} & CHCl_2F & + & HCl \\
A & & B & & C & & E
\end{array}$$

$$\begin{array}{ccccccc}
CHCl_2F & + & HF & \xrightarrow{\quad k_2 \quad} & CHClF_2 & + & HCl \\
C & & B & & D & & E
\end{array}$$

$$\begin{array}{ccccccc}
CHClF_2 & + & HF & \xrightarrow{\quad k_3 \quad} & CHF_3 & + & HCl \\
D & & B & & F & & E
\end{array}$$

The reaction occurs in the liquid phase, the catalyst and both reactants are liquids at the reactor pressure, but the product $CHClF_2$ and the trifluoro-impurity CHF_3 are very volatile, as is HCl. Hence the products of the reaction are removed continuously from the reactor as a gas, and the reactor is fed continuously with a liquid feed. There is no liquid outlet (see Fig. 2.1).

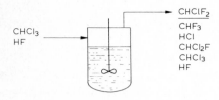

Fig. 2.1. Chlorodifluoromethane reactor. Example 1.

Step 1: Collect the equations defining the system

(a) Chemical kinetic equations

Assuming that the reactions are first-order in each component, based on the liquid concentrations, then the rate of change of each of the components, $\dot{n}$ (kmol s^{-1}) due to reaction is given by:

$$\dot{n}_A = -k_1 \, C_A \, C_B \, V_R \tag{2.10}$$

$$\dot{n}_B = -(k_1 C_A + k_2 C_C + k_3 C_D) \, C_B \, V_R \tag{2.11}$$

$$\dot{n}_C = (k_1 C_A - k_2 C_C) \, C_B \, V_R \tag{2.12}$$

$$\dot{n}_D = (k_2 C_C - k_3 C_D)\ C_B\ V_R \tag{2.13}$$

$$\dot{n}_E = -\dot{n}_B \tag{2.14}$$

$$\dot{n}_F = k_3\ V_R\ C_D\ C_B \tag{2.15}$$

(b) Mass transfer equations

Henry's Law relates liquid concentrations (C_i) to partial pressures exerted by the solution (p_i^*)

$$p_i^* = H_i'\ C_i$$

where H_i' is defined as the Henry's Law constant. The rate of mass transfer to the gas from the liquid phase ($\dot{n}'$) is given by:

$$\dot{n}_i' = K_g\ a\ V_R\ (H_i'\ C_i - p_i)$$

assuming a perfectly mixed liquid and gas phase, where K_g is the overall gas mass-transfer coefficient, a is the interfacial area per unit volume of reactor, p_i is component i gas partial pressure, which is also the exit partial pressure, because both gas and liquid are perfectly mixed. Hence the mass transfer equations are:

$$\dot{n}_A' = K_g\ a\ V_R\ (H_A'\ C_A - p_A) \tag{2.16}$$

$$\dot{n}_B' = K_g\ a\ V_R\ (H_B'\ C_B - p_B) \tag{2.17}$$

$$\dot{n}_C' = K_g\ a\ V_R\ (H_C'\ C_C - p_C) \tag{2.18}$$

$$\dot{n}_D' = K_g\ a\ V_R\ (H_D'\ C_D - p_D) \tag{2.19}$$

$$\dot{n}_E' = K_g\ a\ V_R\ (H_E'\ C_E - p_E) \tag{2.20}$$

$$\dot{n}_F' = K_g\ a\ V_R\ (H_F'\ C_F - p_F) \tag{2.21}$$

(c) Mass balance equations

1. For the overall system

For each component it is possible to write down a mass balance:
kmol s^{-1} outlet = kmol s^{-1} inlet + kmol s^{-1} produced by reaction

$$\frac{N p_i}{P} = \dot{n}_{Fi} + \dot{n}_i$$

where P is the system pressure, hence p_i/P is gas mole fraction, $\dot{n}_{Fi}$ is the kmol s^{-1} feed of i, and $\dot{N}$ is the total kmol s^{-1} fed to the reactor. Hence:

$$\frac{\dot{N}p_A}{P} = \dot{n}_{FA} + \dot{n}_A \tag{2.22}$$

$$\frac{\dot{N}p_B}{P} = \dot{n}_{FB} + \dot{n}_B \tag{2.23}$$

$$\frac{\dot{N}p_C}{P} = \dot{n}_{FC} + \dot{n}_C \tag{2.24}$$

$$\frac{\dot{N}p_D}{P} = \dot{n}_{FD} + \dot{n}_D \tag{2.25}$$

$$\frac{\dot{N}p_E}{P} = \dot{n}_{FE} + \dot{n}_E \tag{2.26}$$

$$\frac{\dot{N}p_F}{P} = \dot{n}_{FF} + \dot{n}_F \tag{2.27}$$

Since $\dot{N}$ is the total number of kmol s^{-1} entering and leaving the reactor:

$$\dot{N} = \dot{n}_{FA} + \dot{n}_{FB} + \dot{n}_{FC} + \dot{n}_{FD} + \dot{n}_{FE} + \dot{n}_{FF} \tag{2.28}$$

2. For the liquid phase

Since we have two phases, there is a second mass balance equation which can be written down, and which is necessary to define the system completely. Since the reactor feed is in the liquid phase, a balance over the liquid phase alone gives: kmol s^{-1} fed as liquid + kmol s^{-1} formed by reaction = kmol s^{-1} transferred from liquid to gas

$$\dot{n}_{Fi} + \dot{n}_i = \dot{n}_i'$$

Hence:

$$\dot{n}_{FA} + \dot{n}_A = \dot{n}_A' \tag{2.29}$$

$$\dot{n}_{FB} + \dot{n}_B = \dot{n}_B' \tag{2.30}$$

$$\dot{n}_{FC} + \dot{n}_C = \dot{n}_C' \tag{2.31}$$

$$\dot{n}_{FD} + \dot{n}_D = \dot{n}_D' \tag{2.32}$$

$$\dot{n}_{FE} + \dot{n}_E = \dot{n}_E' \qquad (2.33)$$

$$\dot{n}_{FF} + \dot{n}_F = \dot{n}_F' \qquad (2.34)$$

Assuming that we want to develop a simulation model, the independent variables are:

V_R, $\dot{n}_{FA-F}$, a, P   (9 variables).

The parameters of the system are:

k_1, k_2, k_3, K_g, H_{A-F}' (10 parameters).

The dependent variables can be divided into those of immediate interest (reactor outlet composition, p_{A-F}) and the intermediate dependent variables which are necessary for the calculation, but not of immediate interest themselves:

$\dot{n}_{A-F}$, $\dot{n}_{A-F}'$, C_{A-F}, $\dot{N}$

With 25 dependent variables and 25 equations, the system is defined and solvable.

Step 2: Simplify the equations where possible

At this stage it is probably worthwhile reducing the generality of the model to make the solution easier. HCl has a very low solubility and its liquid concentration is of no consequence in the model. HCl will also not be present in the reactor liquid feed. Hence C_E and $\dot{n}_{FE}$ can be considered to be zero, equations 2.20 and 2.33 are no longer required, and dependent variables $\dot{n}_E'$ and C_E are dropped out of the system, since $\dot{n}_E'$ is identical to $\dot{n}_E$ and C_E is zero.

Step 3: Rearrange the equations for convenient solution

The 23 equations must be solved to determine the values of the 23 dependent variables. The equations should first be ordered to enable as many as possible to be solved consecutively. Figure 2.2 lists the equations and the dependent variables involved, and Figure 2.3 shows the order of the equations to minimize the number of equations to be solved simultaneously. The rearranged structural matrix shows that all equations except 2.26 must be solved simultaneously, but if this is done in the order shown in Fig. 2.3 there are only 5 unknowns which cannot be solved consecutively.

By making initial estimates for variables C_{A-D} and C_F, equations down to 2.28 of Fig. 2.3 can be solved consecutively. Equations 2.22 - 2.25 and 2.27 can then be used to recalculate p_{A-D}' and p_F', which have already been calculated by equations 2.16 to 2.19 and 2.21. The system is solved when a set of values of C_{A-D} and C_F is found which gives acceptably low residuals (R_i) to the 5 equations, where:

$$R_1 = p_A - p_A' \qquad\qquad (2.35)$$

$$R_2 = p_B - p_B' \qquad\qquad (2.36)$$

$$R_3 = p_C - p_C' \qquad\qquad (2.37)$$

$$R_4 = p_D - p_D' \qquad\qquad (2.38)$$

$$R_5 = p_F - p_F' \qquad\qquad (2.39)$$

A standard non-linear simultaneous equation solution algorithm can be used to find the appropriate values of $C_{A-D,F}$.

Notice that we have reduced the problem from one containing 23 unknowns, to one which requires the simultaneous solution of only 5 unknowns by careful rearrangement of the equations.

Such models are often useful if they can predict the effect of temperature.

If we remain with a simulation model, and we give the operating temperature of the system as an independent variable, this can be included in the model very easily as follows.

The gas solubilities are dependent on temperature and this is given by a relation of the Henry's coefficient and temperature

$$H_i' = f(T)$$

Also the Arrhenius equation defines the effect of temperature on the reaction velocity constants.

$$k_i' = A_i \, e^{-E_i/RT}$$

Given the constants for these relationships, the appropriate k_i and H_i' can be calculated before the remainder of the model equations are solved, and there is no increase in its complexity as long as it can be assumed that adequate heat removal facilities and controls are available to maintain the required temperature.

Since HCl is acting as an inert, we can control the system total pressure by simply restricting the gas outlet flow, independent of the reactor temperature.

An example of the FORTRAN coding for a gas fed CSTR reactor is given in Appendix 3. This coded model is considerably more simple than the example here, and the equations can be arranged so that only one variable has to be located by numerical iteration.

Eq.	$\dot{n}_A$	$\dot{n}_B$	$\dot{n}_C$	$\dot{n}_D$	$\dot{n}_E$	$\dot{n}_F$	$\dot{n}'_A$	$\dot{n}'_B$	$\dot{n}'_C$	$\dot{n}'_D$	$\dot{n}'_F$	C_A	C_B	C_C	C_D	C_F	p_A	p_B	p_C	p_D	p_E	p_F	$\dot{N}$
2.10	×											×	×										
2.11		×										×	×	×	×								
2.12			×									×	×	×	×								
2.13				×									×	×	×								
2.14					×										×								
2.15						×										×							
2.16							×					×											
2.17								×					×										
2.18									×					×									
2.19										×					×								
2.21											×					×							×
2.22	×																×						×
2.23		×																×					×
2.24			×																×				×
2.25				×																×			×
2.26					×																×		×
2.27						×																×	×
2.28																						×	×
2.29	×						×																
2.30		×						×															
2.31			×						×														
2.32				×						×													
2.34						×					×												

Fig. 2.2. Structural matrix for equations in example 1.

	C_A	C_B	C_C	C_D	C_F	$\dot{n}_A$	$\dot{n}_B$	$\dot{n}_C$	$\dot{n}_D$	$\dot{n}_F$	$\dot{n}'_A$	$\dot{n}'_B$	$\dot{n}'_C$	$\dot{n}'_D$	$\dot{n}_E$	$\dot{n}'_F$	$\dot{N}$	p_A	p_B	p_C	p_D	p_F	p_E
2.10	x	x	x	x		x	x																
2.11	x	x	x	x			x	x															
2.12	x	x	x	x				x	x														
2.13		x	x	x					x														
2.15				x	x					x													
2.29						x					x												
2.30							x					x											
2.31								x					x										
2.32									x					x									
2.34										x						x							
2.16	x	x					x				x							x					
2.17		x						x				x							x				
2.18			x			x							x							x			
2.19				x										x							x		
2.21					x											x						x	
2.14															x		x						
2.28						x											x	x					
2.22							x										x	x					
2.23								x									x		x				
2.24									x								x			x			
2.25									x								x				x		
2.27										x							x					x	
2.26																x	x						x

Fig. 2.3. Rearranged structural matrix for equations in example 1.

2.4.b   Example 2: A model of a tubular catalytic reactor

Acrolein is produced by the partial oxidation of propylene over a solid catalyst. The reaction is highly exothermic, and the temperature must be carefully controlled to maintain a good selectivity. Hence the modelling of a cooled tubular reactor including both the prediction of concentration and temperature profile is necessary.

The reactions involved can be written stoichiometrically as:

$$C_3H_6 \;+\; O_2 \;\xrightarrow{k_1}\; CH_2=CH\text{-}CHO \;+\; H_2O$$
$$\;\;\;A\qquad\; B\qquad\qquad\;\; C\qquad\qquad\;\; D$$

$$C_3H_4O \;+\; 3\tfrac{1}{2}O_2 \;\xrightarrow{k_2}\; 2H_2O \;+\; 3CO_2$$
$$\;\;\;C\qquad\quad\; B\qquad\qquad D\qquad\;\; E$$

$$C_3H_6 \;+\; 4\tfrac{1}{2}O_2 \;\xrightarrow{k_3}\; 3H_2O \;+\; 3CO_2$$
$$\;\;\;A\qquad\quad\; B\qquad\qquad D\qquad\;\; E$$

$$C_3H_4O \;+\; 2O_2 \;\xrightarrow{k_4}\; 2H_2O \;+\; 3CO$$
$$\;\;\;C\qquad\quad B\qquad\qquad D\qquad\;\; F$$

$$C_3H_6 \;+\; 3O_2 \;\xrightarrow{k_5}\; 3H_2O \;+\; 3CO$$
$$\;\;\;A\qquad\;\; B\qquad\qquad D\qquad\;\; F$$

(i) The kinetic equations. Since the reaction is catalytic, fractional order kinetics are to be expected. These were found to be first-order with respect to organic and 0.05 order with respect to oxygen. The kinetic equations for the tubular reactor of cross-section A to define the change in molar flow rate with tube length can be written down as:

$$\frac{d\dot{n}_A}{dL} = -k_1 \, A \, C_A \, C_B^{0.05} - k_3 \, A \, C_A \, C_B^{0.05} - k_5 \, A \, C_A \, C_B^{0.05} \tag{2.40}$$

$$\frac{d\dot{n}_B}{dL} = -(k_1 \, C_A + 3.5 \, k_2 \, C_C + 4.5 \, k_3 \, C_A + 2 \, k_4 \, C_C + 3 \, k_5 \, C_A) \, A \, C_B^{0.05} \tag{2.41}$$

$$\frac{d\dot{n}_C}{dL} = (k_1 \, C_A - k_2 \, C_C - k_4 \, C_C) \, A \, C_B^{0.05} \tag{2.42}$$

$$\frac{d\dot{n}_D}{dL} = (k_1 \, C_A + 2 \, k_2 \, C_C + 3 \, k_3 \, C_A + 2 \, k_4 \, C_C + 3 \, k_5 \, C_A) \, A \, C_B^{0.05} \tag{2.43}$$

$$\frac{d\dot{n}_E}{dL} = 3(k_2 \, C_C + k_3 \, C_A) \, A \, C_B^{0.05} \tag{2.44}$$

$$\frac{d\dot{n}_F}{dL} = 3(k_4 \, C_C + k_5 \, C_A) \, A \, C_B^{0.05} \tag{2.45}$$

58

In all these reactions it is assumed that the Arrhenius relationship defines the effect of temperature on the reaction rate:

$$k_i = A_i \, e^{-E_i/RT} \qquad i = 1, 5 \tag{2.46}$$

(ii) Mass balance equations. Elemental mass balances on carbon or oxygen could be used to reduce the number of differential equations. For example, the H_2O and C_3H_6 rate equations above (equations 2.40 and 2.43) could be replaced by mass balance equations.

This reduced number of differential equations does not necessarily reduce the complexity of the model, because there is often little computational difference in a mass balance or a differential equation in terms of computer time (depending on the integration algorithm used). Mass balance equations can introduce errors when the components are in low concentrations because rounding and other errors (or assumptions) concentrate all inconsistencies in the mass balance determined concentrations. Unless the number of differential equations is causing problems it is therefore advisable not to introduce any mass balance equations.

(iii) Heat balance equations. The temperature change of the gas is given by a heat balance equation where

heat remaining in gas = heat generated by reactions - heat transferred through
reactor walls

i.e.

$$c_p \, \rho \, V \, \frac{dT_R}{dL} = -\Delta H_1 \, k_1 \, A \, C_A \, C_B^{0.05} - \Delta H_2 \, k_2 \, C_C \, C_B^{0.05} - \Delta H_3 \, k_3 \, A \, C_A \, C_B^{0.05}$$
$$-\Delta H_4 \, k_4 \, A \, C_C \, C_B^{0.05} - \Delta H_5 \, k_5 \, A \, C_A \, C_B^{0.05} - \pi \, d_t \, U \, (T_R - T_J) \tag{2.47}$$

where d_t is the tube diameter, T_J is the coolant temperature, V is the gas volume flow ($m^3 \, h^{-1}$), ρ is the gas density ($kg \, m^{-3}$), c_p is the gas specific heat ($kJ \, kg^{-1} \, K^{-1}$) and U is an overall heat transfer coefficient for bulk gas to coolant. It is a lumped parameter and cannot be used to predict tube diameters or particle size changes.

This equation can be rewritten as:

$$\frac{dT_R}{dL} = -\frac{1}{c_p \, \rho \, V} \left\{ A \, C_B^{0.05} \, (\Delta H_1 \, k_1 \, C_A + \Delta H_2 \, k_2 \, C_C + \Delta H_3 \, k_3 \, C_A \right.$$
$$\left. + \Delta H_4 \, k_4 \, C_C + \Delta H_5 \, k_5 \, C_A) + \pi \, d_t \, U \, (T_R - T_J) \right\} \tag{2.48}$$

Assuming the coolant flow on the outside of the tube is high, T_J can be considered to be constant, and no heat balance need be written down for the coolant.

(iv) <u>Pressure balance equations</u>. The pressure drop will be a function of the velocity (u) and density (ρ) of the gas in the tube and can be represented by the following equation:

$$\frac{dP}{dL} = - K' \rho u^2 \tag{2.49}$$

where K' is an empirical constant.

(v) <u>Linking equations</u>. The above equations are written with the variables C, V, P, T_R, and ρ. These are related by the gas laws. Assuming that the reactor feed rate is $\dot{W}$ kg s^{-1}, where

$$\dot{W} = \sum_{i=A}^{E} (\dot{n}_{i_0} M_i)$$

and the total number of kmoles s^{-1} flowing at any position in the tube ($\dot{N}$) is given by:

$$\dot{N} = \dot{n}_A + \dot{n}_B + \dot{n}_C + \dot{n}_D + \dot{n}_E + \dot{n}_F \tag{2.50}$$

and the pressure is P, then the volumetric flow is

$$V = \frac{\dot{N} R T_R}{P} \quad m^3 \, s^{-1} \tag{2.51}$$

the density is

$$\rho = \frac{\dot{W}}{V} \quad kg \, m^{-3} \tag{2.52}$$

concentrations are

$$C_i = \frac{\dot{n}_i}{V} \quad kmol \, m^{-3} \quad \text{where } i = A \text{ to } F \tag{2.53}$$

and velocity is

$$u = \frac{V}{A} \tag{2.54}$$

where

$$A = \frac{\pi}{4} d_t^2 \tag{2.55}$$

It is also possible to introduce a relationship between heat transfer coefficients and gas velocity, e.g.

$$U = K'' u^{0.8} \qquad\qquad (2.56)$$

where K" is an experimentally determined constant.

These 17 equations enable the tubular reactor to be modelled. Given defined inlet conditions as moles h^{-1} of each component, temperature, and pressure, the equations enable a profile of composition, temperature, and pressure to be calculated along the tube.

The eight differential equations can be solved using a numerical integration algorithm such as Runge-Kutta. The equations are not "stiff" (unless the temperature gets out of control) and no problems are likely to occur in their solution.

The "function" of the integration routine will, in fact, be the whole model, and the routine will take small calculation steps along the tube.

Since $\dot{n}_{A_0}-F_0$, T_R, T_J, P, and d_t are given as input data, the model must calculate $\dot{N}$, V, ρ, C_{A-F}, U, A, and k_{1-5} by equations 2.50-2.56 and 2.46, so that all the variables are known for the differential equations. The last equations in the function subroutine are the differential equations 2.40-2.45, 2.48, and 2.49. Most standard integration packages supply an outlet routine which enables the reactor compositions, temperature, and pressure against reactor tube length to be tabulated.

Two examples of FORTRAN coded tubular reactor models are given in Appendix 3.

2.4.c   Example 3: Modelling of PFR reactors when changes occur with time

There are a number of occasions when a plug-flow reactor involving a solid phase has the activity of the solid phase dependent on time and the cumulative conditions that it has been subjected to up to that time.

Catalyst deactivation is an obvious example, where the reaction velocity constants at any point in the reactor depend upon the activity of the catalyst at that point, and this is dependent on the catalyst age and its past temperature history. Modelling of the ageing of catalysts is often useful for:

(a) Modelling the reactor at all stages of its life cycle to obtain optimum process conditions for the total life and not a single point.

(b) Fitting and understanding catalyst deactivation mechanisms, as a step in devising methods of improving catalyst life times.

(c) Determining optimum operating policies and catalyst changeover procedures to maximize catalyst usage.

A second example is the regeneration of coked catalysts in fixed beds. There are a number of processes where coke is laid down on the catalyst during reaction, which reduces catalyst activity. From time to time this coke has to be burnt off by passing air through the bed. This operation is very critical in that if temperatures are too high, catalyst sintering and permanent damage result. At too low a temperature regeneration time becomes too long. To have a model of the regeneration process is of great assistance in determining the optimum regeneration procedure.

Both of these examples require models which integrate down the length of the reactor and also integrate with respect to time.

The modelling can be carried out by expressing the system in terms of partial differential equations and solving these with some computer package design for integration of partial differential equations.

Alternatively, a model can be developed which integrates down the length of the reactor at time zero, and repeats this integration at small time steps, calculating the changes that occur during the time step assuming constant conditions for that time step. This is a rough form of partial differential equation solving, but it does have the advantage over the thorough method that it is understandable to most engineers (since most self-respecting engineers have forgotten all they knew about partial differential equations before they need to use them). The process and its solution are also clearly understandable and there is no need to obtain a numerical package for partial differential equations.

Consider as an example the regeneration of a coked catalyst bed. The bed can be divided up into, say, ten segments, and each segment assumed to have a constant coke composition. The model can then integrate over each segment using a normal integration procedure to determine the gas concentration change over that segment. The gas leaving that segment enters the next, and, based on the new coke concentration, the integration continues. This procedure is repeated over all segments until the end of the reactor is reached.

Such an integration at time zero calculates the gas concentration profile and temperature over the total reactor. From the concentration profile it is possible to calculate the rate of burning of coke from each segment by mass balance and, assuming the coke is burnt off evenly in each segment, the new coke concentration in each section at the end of a small time step Δt can be determined by subtraction. Hence a new coke profile is available for time $0 + \Delta t$ and a new integration can be run for time Δt, and the process repeated until the regeneration is complete. Figure 2.4 shows the process diagrammatically. Care must be taken in the choice of time step length and number of segments in the bed to obtain a tolerable accuracy for the simulation.

When such a technique is programmed as an interactive program, the temperature, gas, and coke profiles can be printed out at specific time intervals, and gas feed compositions can be changed to keep bed temperatures within defined limits.

The regeneration of a bed of coked catalyst is a difficult operation with regeneration occurring either as a front up the bed (e.g. high temperature, limited oxygen feed) or evenly in the whole bed (low temperature, excess low concentration

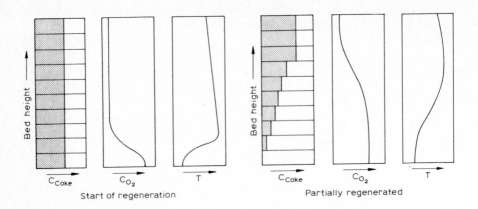

Fig. 2.4. Regeneration of a coked bed.

oxygen). Which of these procedures to choose, and whether to modify oxygen concentrations during the regeneration period can be determined using such an interactive program.

2.4 FITTING PARAMETERS TO MODELS

Besides containing dependent and independent variables, models contain physical constants that describe the system. These are the parameters of the system and they represent the reaction rate constants, apparent activation energies, densities, specific heats, and so on.

Some of these constants are well known and can be looked up in tables and entered directly in the model. Others may not be known, but could be easily measured in a laboratory - heats of reaction or viscosities of the reaction mixture for instance. A third group could only be measured with difficulty in the laboratory and only in apparatus looking very different from the reactor of interest. For example, the reaction rates for individual reactions in a reaction scheme could be measured if reactor intermediates were used in equipment design for the measurement of fast reactions. Mass-transfer coefficients could be measured in falling film absorbers, and diffusion coefficients through catalyst pellets in single pellet apparatus.

The measurement of parameters of this last class for model development can be a very questionable activity. Firstly, the effort to devise the experiments is costly. Secondly, the data obtained are not completely reliable because the "extrapolation" to a normal reactor may bring with it effects which have not been accounted for. Thirdly, the reactor model is itself an approximate model, and a parameter in the model may in reality be a lumping together of a number of effects. If this is the case, it is pointless to carefully measure a value representing only one of the effects which the parameter represents.

When measured parameters are not available, they are "fitted" to experimental results. This means values are found (usually automatically by a search procedure by the computer) which make the model predictions match experimental results as closely as possible. It is usually possible to fit up to 10 variables using computer fitting techniques, but when more than 10 variables are involved it may well be that computer costs become prohibitive.

The modeller must decide which parameters he will leave free for the program to fit, and which he will give fixed values. Contributing to this decision are:
o Are values for the parameter available?
o Can parameter values be obtained by a modest laboratory program?
o How many parameters are unspecified? If the number is high special efforts should be made to reduce it.

There are pitfalls in defining values for too many parameters. If the model is particularly sensitive to a particular parameter it may require a more precise value than that obtained from the literature or specific experiments. The work of Cresswell and Santos (1980) showed that an adiabatic batch reactor was a more sensitive device for measuring activation energies than simple thermostat kinetic experiments at three temperatures. Very poor predictions were obtained using kinetically measured parameters though good fits could be obtained by fitting the parameters. The resulting fitted parameters lay within the confidence region of the measured data. There are other examples of cases where the estimation of kinetic parameters separately brought less information than a fit of the parameter values to experimental results from the reactor itself but these are rarely published because of the convention that only success stories should be published.

If the values of a number of the parameters are left to be fitted to the experimental results, an adequate "fit" can usually be obtained with "physically reasonable" values for the parameters fitted. However, if most of the parameters are measured separately, there is often inadequate freedom to achieve a reasonable "fit". This results in a great deal of extra effort being spent to locate the reason for the discrepancy, often without result, in addition to the extra effort already invested to obtain these parameter values by independent experiments. In practice this slows the modelling work down, and the extra investigations rarely have a significant impact on the value of the model for the reactor design.

It is usually only managers who ask not to be confused by facts, but in the case of model development there is an art in not getting too deeply involved, in remembering that only a semi-empirical model is being sought.

2.6   LINEAR REGRESSION

At the beginning of this chapter there have been a few examples of equations linear in the parameters. This form of equation is rarely of use in reactor modelling, but since it is the basis of regression analysis, and regression analysis is the science of fitting equations to data, it is worth mentioning linear regression analysis as a precursor to non-linear regression analysis and experimental design, both of which are useful in the practice of reactor design.

It can be shown that the fit with the highest probability of being correct, i.e. the "best fit" of an equation to a set of experimental points which includes a normally distributed random error is the equation that minimizes the sum of squares of the discrepancies between model predictions and measured values:

$$\sum_{i=1}^{n} (y_i - y_i^*)^2$$

where there are n experiments with experimental results y_i and model predicted values y_i^*.

Consider that we want to fit a simple linear equation to a system with only one independent variable and we have n experimental points. The model equation is:

$$y^* = \theta_0 + \theta_1 x \qquad (2.57)$$

but the experimental measurements, y_i, are subject to an error e_i, i.e.

$$y_i = \theta_0 + \theta_1 x + e_i \qquad (2.58)$$

or

$$e_i^2 = (y_i - \theta_1 x - \theta_0)^2 = (y_i - y_i^*)^2 \qquad (2.59)$$

Now we are looking for values for θ_0 and θ_1 which give the total sum of squares over all experiments, E, where

$$E = \sum_{i=1}^{n} e_i{}^2$$

as a minimum. This can be found by differentiation.

$$E = \sum_{i=1}^{n} (\theta_0 + \theta_1 x_i - y_i)^2 \qquad (2.60)$$

$$\frac{dE}{d\theta_0} = 2 \Sigma (\theta_0 + \theta_1 x_i - y_i) = 0 \qquad (2.61)$$

$$\frac{dE}{d\theta_1} = 2 \Sigma (\theta_0 + \theta_1 x_i - y_i) x_i = 0 \qquad (2.62)$$

Therefore

$$n\, \theta_0 = \Sigma\, y_i - \theta_1\, \Sigma\, x_i \qquad (2.63)$$

or

$$\theta_0 = \bar{y} - \theta_1\, \bar{x} \qquad (2.64)$$

where $\bar{y}$ and $\bar{x}$ are the mean values of y and x. From equation 2.62

$$\theta_1 = \frac{\Sigma(x_i\, y_i) - \bar{y}\, \Sigma\, x_i}{\Sigma\, x_i^2 - \bar{x}\, \frac{\Sigma\, x_i}{n}} = \frac{\overline{xy} - \bar{x}\bar{y}}{\overline{x^2} - \bar{x}^2} \qquad (2.65)$$

where Σ refers to $\sum\limits_{i=1}^{}$.

Equations 2.64 and 2.65 show that the constants for equation 2.57 can be evaluated explicitly. In fact, for any equation, linear in its coefficients, it is possible to derive an analytical solution to determine the best-fitting coefficients. This is an extremely useful property of linear equations which is not possessed by non-linear equations.

For the general case matrices are used in the manipulation of the data. The completely general case, where n experiments are to be fitted by a linear equation containing m independent variables, can be written as:

$$y_1 = \theta_0 + \theta_1\, x_{1,1} + \theta_2\, x_{2,1} + \cdots \theta_m\, x_{m,1} + e_1$$

$$y_2 = \theta_0 \quad \theta_1\, x_{1,2} \quad \theta_2\, x_{2,2} \quad \cdots \theta_m\, x_{m,2} + e_2$$

$$\vdots$$

$$y_n = \theta_0 \quad \theta_1\, x_{1,n} \quad \theta_2\, x_{2,n} \quad \cdots \theta_m\, x_{m,n} + e_n$$

which can be represented in matrix form as:

$$Y = X\theta + E$$
$$\text{or} \qquad (2.66)$$
$$E = Y - X\theta$$

Hence the sum of square errors is given by

$$\sum_{i=1}^{n} e_i^2 = E^T E = (Y - X\theta)^T (Y - X\theta) \qquad (2.67)$$

To find the best fit this sum of square error should be a minimum. This is again found by differentiating and equating the partial derivation to zero. The

resulting equation can then be rearranged to give the following explicit relationship
to evaluate θ:

$$\theta = [X^T X]^{-1} X^T Y \tag{2.68}$$

This matrix equation is then used to calculate the value of $\theta_0 \ldots \theta_m$.

2.6.a   Analysis of variance

Equation 2.68 can be solved by computer for models containing very many parameters
and experiments. If the model is correct, all errors are random experimental errors
(variance σ^2) and this can be estimated from the sum of residuals

$$\sigma'^2 = \frac{1}{n-k} \sum_{i=1}^{n} e_i^2 \tag{2.69}$$

where there are k parameters and n experiments.

This value of σ^2 can also be estimated from a series of n' repeated experiments.

$$\sigma''^2 = \frac{1}{n'-1} \sum_{i=1}^{n'} (y_i - \bar{y})^2 \tag{2.70}$$

When the two estimates of σ (σ'^2 and σ''^2) are of the same order, (statistically
speaking when σ' and σ'' cannot be shown with 95% confidence to belong to different
distributions, which can be checked with an F test), the model can be considered
to be statistically "adequate". When they are not of the same order, then errors
other than experimental errors are present. These are systematic errors, sometimes
called model errors or bias errors, which arise because the model is not describing
the phenomena correctly. The treatment by which the model discrepancies are
compared with the experimental variances is generally called analysis of variance
(Daniel, 1971).

2.6.b   Parameter confidence limits

For a purely linear system it is possible to derive information regarding the
confidence limits of the parameters, related to the experimental error and number
of experiments carried out. Since

$$\theta = [X^T X]^{-1} X^T Y$$

$$Var(\theta) = [X^T X]^{-1} X^T \, Var(Y) \, X \, [X^T X]^{-1} \tag{2.71}$$

Since Var (Y) is σ^2, where σ^2 is the variance of experimental error

$$Var(\theta) = [X^T X]^{-1} \sigma^2 \tag{2.72}$$

This result is particularly useful because it enables the confidence limits of the parameters to be predicted from the experimental error (σ) and the position of the experiments in the experimental region (X).

2.6.c   Prediction of correlation between parameters

Correlation between parameters is of particular interest in reaction modelling studies, and for linear systems this can also be derived from the $[X^{T}X]^{-1}$ matrix. Complete correlation between parameters means that a change in the value of one parameter can be completely compensated for by a change in another parameter so that an equally good fit to the experimental results is achieved. Figure 2.5 shows plots of equal confidence for a two-parameter system. The ellipses show the relationship between the two parameters which give equally good fits to experimental data. This means the result can offer no information on the individual values of the parameters when there is high correlation. Degree of correlation is shown by the correlation coefficient, 1.0 (or -1.0) denoting complete positive (or negative) correlation and 0.0 complete independence. Generally correlation is not a problem unless the correlation corfficient is higher than ±0.95.

The matrix of correlation coefficients is called the correlation matrix and this is derived from the $[X^{T}X]^{-1}$ matrix by normalizing it elements.

$$\theta_i\theta_j \text{ correlation coefficient} = \frac{\text{Covariance } \theta_i\theta_j}{(\text{Var } (\theta_i) \text{ Var } (\theta_j))^{\frac{1}{2}}}$$

$$= \frac{L_{ij}}{(L_{ii} L_{jj})^{\frac{1}{2}}}$$

(2.73)

where L are elements out of the $[X^{T}X]^{-1}$ matrix.

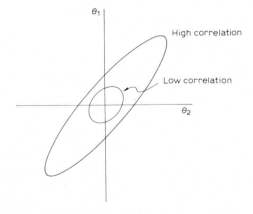

Fig. 2.5. Correlation between parameters - 95% confidence contours.

68

2.6.d   Confidence limits for a model prediction

From a linear regression analysis one can also predict the confidence limits for a model prediction using a model with uncertain parameters (Var (θ)), which have been obtained by fitting experiments with an experimental error σ^2.

The predicted point y_p is given by:

$$y_p = X_p^T \theta + e \tag{2.74}$$

Hence

$$\text{Var } (y_p) = \text{Var } (X_p^T \theta) + \text{Var } (e)$$

$$= X_p^T \text{ Var } (\theta) X_p + \sigma^2 \tag{2.75}$$

and from equation 2.72

$$\text{Var } (y_p) = (X_p^T [X^T X]^{-1} X_p + 1) \sigma^2 \tag{2.76}$$

where X_p is the set of values of the independent variables that define the point to be predicted, and $[X^T X]^{-1}$ is derived from the set of values of independent variables that were used in the experiments to find the parameter values.

To convert these variances into confidence limits (e.g. 95% confidence limits) one requires the t-value from statistical tables to determine how many standard deviations ($(\text{variance})^{\frac{1}{2}}$) to take. For most practical cases one can assume that two standard deviations gives the ±95% confidence limits.

2.6.e   Elements of the $[X^T X]^{-1}$ matrix

As the $[X^T X]^{-1}$ matrix appears frequently in all discussions over regression, it is useful to have an understanding of its physical meaning and so for the many who are unable to form a physical picture of the $[X^T X]^{-1}$ matrix let us consider a simple example in more detail.

For a linear system involving two independent variables:

$$y = \theta_0 x_0 + \theta_1 x_1 + \theta_2 x_2$$

assume we have carried out three experiments. The conditions for these experiments can be defined by:

Experiment	x_0	x_1	x_2
1	1	$x_{1,1}$	$x_{2,1}$
2	1	$x_{1,2}$	$x_{2,2}$
3	1	$x_{1,3}$	$x_{2,3}$

Now

$$X = \begin{bmatrix} 1 & x_{1,1} & x_{2,1} \\ 1 & x_{1,2} & x_{2,2} \\ 1 & x_{1,3} & x_{2,3} \end{bmatrix} \tag{2.77}$$

For the sake of clarity we can assume that the x-values have been transformed to have their mean values at the origin, i.e.

$$\Sigma\, x_{1,i} = 0$$
$$\Sigma\, x_{2,i} = 0 \tag{2.78}$$

where Σ refers to $\overset{3}{\underset{i=1}{\Sigma}}$, then the X^TX matrix is

$$[X^TX] = \begin{bmatrix} 3 & 0 & 0 \\ 0 & \Sigma x_{1,i}^2 & \Sigma(x_{1,i}\,x_{2,i}) \\ 0 & \Sigma(x_{1,i}\,x_{2,i}) & \Sigma x_{2,i}^2 \end{bmatrix} \tag{2.79}$$

and from rules of matrix inversion

$$[X^TX]^{-1} = \begin{bmatrix} \frac{1}{3} & 0 & 0 \\ 0 & \frac{\Sigma x_{2,i}^2}{D} & \frac{-\Sigma(x_{1,i}\,x_{2,i})}{D} \\ 0 & \frac{-\Sigma(x_{1,i}\,x_{2,i})}{D} & \frac{\Sigma x_{1,i}^2}{D} \end{bmatrix} \tag{2.80}$$

where D is the determinant of the $[X^TX]$ matrix:

$$D = 3\,\{\ \Sigma x_{2,i}^2\ \Sigma x_{1,i}^2 - (\Sigma(x_{1,i}\,x_{2,i}))^2\ \} \tag{2.81}$$

From these elements of the $[X^TX]^{-1}$ matrix we can now understand how the position of the experiments affects the accuracy of the predicted parameter.

To reduce Var (θ_i) the elements $L_{i,i}$ must decrease i.e. D should increase. Hence, to reduce the confidence limits of all parameters, experiments should be chosen to maximize the determinant of the $[X^T X]$ matrix (equation 2.81). This determinant is a measure of the volume of the confidence ellipse of the parameters, and so maximizing D minimizes the confidence ellipse. Alternatively, modified criteria could be used to maximize the confidence of a particular parameter, or reduce specific correlations, but most common in the literature is the minimization of the volume of the ellipse.

The expanded form of the $[X^T X]^{-1}$ matrix also shows the source of correlation and how it can be removed. Low correlation means that the off-diagonal terms are low. $\Sigma(x_{1,i} \, x_{2,i})$ is zero when every experimental point x_1, x_2 is matched by a mirror image point x_1, $-x_2$.

Hence, for a linear system, the only factor affecting correlation is the lack of symmetry of the experiments which result in the off-diagonal term not summing to zero. This is an important point when it comes to deciding where to carry out experiments to obtain most information, which will be discussed more fully in Chapter 4.

2.6.f   Improving the linear model

More important than the prediction of variances and correlation for a model is the analysis to see whether the form of the equations of the model agree with the experimental results. When this is not so, "systematic errors" become significant, and the mean residual sum of squares ($\Sigma(y_i - y_i^*)^2 \, / \, (n - k)$ for n experiments and k parameters) is significantly larger than the variance due to experimental error σ''^2 as determined by n' repeated experiments (see expression 2.70)

When these variances are significantly different, this means that predictions of the model will probably have errors at least as large as the largest residual, and the regression analysis to determine the confidence limits of parameters and predicted points is worthless, because the basis of the analysis is that systematic error is not significant.

When the model is "inadequate", the equations should be altered to give a better fit. This normally involves fitting more terms to the simple linear model to provide interaction and curvature in a series of stages, for example:

(a)   Simple linear model:

$$y = \theta_0 + x_1 \, \theta_1 + x_2 \, \theta_2 \qquad (2.82)$$

(b) Linear plus interactive term:

$$y = \theta_0 + x_1 \, \theta_1 + x_2 \, \theta_2 + x_1 \, x_2 \, \theta_3 \qquad (2.83)$$

(c) General quadratic model:

$$y = \theta_0 + x_1\,\theta_1 + x_2\,\theta_2 + x_1^2\,\theta_3 + x_2^2\,\theta_4 + x_1\,x_2\,\theta_5 \tag{2.84}$$

(d) Quadratic plus full interactive terms:

$$y = \theta_0 + x_1\,\theta_1 + x_2\,\theta_2 + x_1^2\,\theta_3 + x_2^2\,\theta_4 + x_1\,x_2\,\theta_5 + x_1\,x_2^2\,\theta_6 + x_1^2\,x_2\,\theta_7$$

$$+ x_1^2\,x_2^2\,\theta_8 + x_1\,x_2\,x_1^2\,x_2^2\,\theta_9 \tag{2.85}$$

In this way it is possible to systematically create more complex models linear in the parameters as a way of improving the model.

Although one can go to cubic and higher-order terms there is no guarantee that an "adequate" model will be found, for instance if the results represent some form of exponential decay.

It is therefore worth looking at the results and the theory of the process under discussion to try to detect any special characteristics. If an exponential term is expected, it can easily be introduced into the equation, e.g.

$$y = \theta_0 + x_1\,\theta_1 + x_2\,\theta_2 + e^{x_1}\,\theta_3 \tag{2.86}$$

A special case of this is where the equation representing the process can be linearized (for instance by taking logarithms) as described at the beginning of this chapter (equation 2.4).

The results of such model-improvement techniques are progressively more complex models with many parameters. Complex models should be avoided and models with many parameters (in comparison with the number of experiments) achieve good fits because the degrees of freedom are reduced, and not because the model is good. In the extreme case, a model with as many parameters as experimental results will fit exactly.

It is useful, therefore, to be able to reduce the number of parameters by excluding those which do not contribute to the regression. This can be the case when:

(a) the variable involved has no influence on the experimental result;

(b) there is such a high correlation between variables that one variable can satisfactorily represent the effect of two in the set of experimental results presented.

A first clue as to the significance of individual parameters is given by the confidence limits of the parameters, which are normally printed out by the regression programs. When these are more than one third of the parameter value itself, the parameter is probably not significant. The regression should therefore be repeated excluding this parameter from the model.

This approach does not identify all such redundant parameters and the only thorough method is to carry out a series of regressions leaving out one parameter at a time, comparing the "with" and "without" regression results for each parameter in turn (comparing variances by the F-test). For more details on the analysis of variance and model improvements see Daniel (1971).

For linear regression, both model improvement and determining the significance of the individual parameter are based on the systematic inclusion (or exclusion) of more terms in the regression. This procedure can be easily programmed and so some linear regression programs have the ability to automatically include terms to improve the model, or exclude terms to determine the significance of the parameters.

2.7   NON-LINEAR REGRESSION

2.7.a   Non-linear regression - theory

However many terms are included in a linear model, the linear form of the parameters is a restriction that usually prevents good fits from being obtained to experimental results from reactors. This linear constraint prevents linear regression being used for models developed from theoretical considerations.

In reactor engineering work, where a great deal is known about the process occurring in the reactor, and a mathematical description of the process is often available, what is needed is a method of fitting models that does not have a restrictive form. With this available, a model can be derived from the best present state of knowledge of the mechanism, and this model then fitted to experimental results. This is the function of non-linear regression. It is much more extensively used in reactor engineering than linear regression, but is much less convenient in terms of locating the best values of the parameters, or in defining the confidence limits of the parameters once found.

With a non-linear model, the model form is so free that there is no way of determining analytically the value of the parameters that will give the lowest least square fit, as there is with linear regression. The optimum parameter values must be obtained by a non-linear minimization procedure, using the parameters as the optimization variables and

$$\sum_{i=1}^{n} (y_i - y_i^*)^2$$

as the objective function to be minimized.

Non-linear regression packages therefore have as their first phase an optimization routine to locate the parameter values. The optimization algorithm of Marquardt (Daniel, 1971) has been specially developed for non-linear regression and is employed successfully in a number of regression packages.

In principle, this method combines the steepest ascent optimization method and Newton's method. Initially the steepest ascent search is the dominant algorithm, but as the optimum is approached, the emphasis gradually changes to Newton's method because this has fast convergence in the region of an optimum. Regression packages usually print a record of the progress of the minimization during the parameter location search, and one can see the progress of this minimization, the step size and the relative emphasis on each of the two optimization algorithms.

Non-linear optimization packages have the disadvantage that they find an optimum, but there is no guarantee that it is the global optimum for the system under study. All one can do is to repeat the optimization from another starting point to see whether the same optimum is reached. If this is so then, although it is no proof that the global optimum has been found, it at least gives more confidence in the result.

A second problem with non-linear optimization is that during the search constraints may be met, either in the values of the optimization variables or other variables calculated within the model (e.g. mole fraction which must be between 0.0 and 1.0). Constraints are a problem because they interfere with the search algorithm and so result in an ineffective search for the optimum.

Because non-linear regression contains this non-linear optimization procedure, the problems of non-linear optimization appear in non-linear regression. This can make the location of the best parameter values a considerable problem, whereas in the linear regression this never presents any difficulty.

The second stage, the non-linear regression analysis, is based on a linearization of the non-linear model about the base-fitted point. For the general non-linear model

$$y = f(\theta, x)$$

with an estimated set of parameters θ^0 it is possible to write down an equation showing the effects of changing the parameters, assuming that for small changes about the estimated set of parameter values the effect on y is linear. This will hold when the estimate is close enough to the best set of parameter values.

$$y_i = f(\theta^0, x_i) + \frac{\partial f_i}{\partial \theta_1} \Delta\theta_1 + \frac{\partial f_i}{\partial \theta_2} \Delta\theta_2 + \ldots + e \qquad (2.87)$$

or

$$y_i - f(\theta^0, x_i) = \frac{\partial f_i}{\partial \theta_1} \Delta\theta_1 + \frac{\partial f_i}{\partial \theta_2} \Delta\theta_2 + \ldots + e \qquad (2.88)$$

This model is equivalent to the simple linear model:

$$y_i' = \theta_0' + x_1 \theta_1' + x_2 \theta_2' + \ldots + e \qquad (2.89)$$

Expressed in this way, it is possible to analyze equation 2.88 to determine the variance of the parameter values and the correlation between parameters in the same way as for linear regression. Notice that in place of θ we have the deviation from a good estimate, $\Delta\theta$, and in place of x_i values we must now have the value of $\frac{\partial f_i}{\partial\theta}$, which can be derived from either analytical differentiation of the model itself, or by numerical perturbation. The X matrix becomes a matrix of the sensitivities:

$$\begin{vmatrix} \dfrac{\partial f_1}{\partial\theta_1} & \dfrac{\partial f_1}{\partial\theta_2} \\[2ex] \dfrac{\partial f_2}{\partial\theta_1} & \dfrac{\partial f_2}{\partial\theta_2} \end{vmatrix} \quad \text{in place of} \quad \begin{vmatrix} x_{1,1} & x_{1,2} \\[2ex] x_{2,1} & x_{2,2} \end{vmatrix} \tag{2.90}$$

These derivatives generally will be complex functions and not independent of θ. Hence the correlation coefficient and the parameter confidence limits for non-linear models are not only dependent on experimental conditions as with linear regression, but depend also on model form and parameter values.

The y_i of the linear analysis must be replaced by $(y_i - f(\theta^0, x_i))$, since the non-linear case is relating differences between the experimental measurement and the prediction of the model using the estimated set of parameters θ^0 due to changes in the parameter values.

In this way it is possible to get an analysis that determines the approximate variance of the parameters, and the approximate correlation matrix and approximate variance of predictions of the model. All these values are approximate because they are based on a linear approximation which holds only very close to the fitted point. Figure 2.6 shows the errors that can be expected.

This figure is a plot of the confidence regions for a two-parameter model for a first-order consecutive reaction model (Box, 1962)

$$A \xrightarrow{\;k_1\;} B \xrightarrow{\;k_2\;} C$$

and compares the linearized analysis confidence regions with the true regions.

Notice that in this case the linearized approximate analysis produces a confidence region smaller than the true region, although at 95% confidence limit the comparison is not too bad. For engineering purposes, when one needs to know if there is any correlation at all, and if so approximately how serious it is, then this linear analysis is quite adequate.

2.7.b Non-linear regression - practice

In practice, the fitting of models to experimental data is as much an art as a science; the previous paragraphs have discussed the science, we will now discuss the less tangible aspects of the subject.

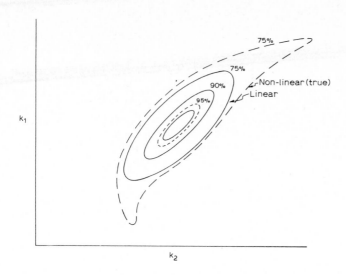

Fig. 2.6. Comparison of linearized and true confidence limits.

Obtaining the best fitting parameter set is not always simple because of the difficulties already mentioned associated with non-linear optimization.

Particularly when the model is complex, local sub-optima can be found, especially if poor starting estimates of the parameters are given. Poor starting estimates can often cause constraints to be violated, usually mole fractions being out of range, and then the regression breaks down owing to computational failure.

Correlation, which in linear regression simply means that the individual values of the parameters cannot be specified, is a much more serious problem in non-linear regression. Correlation between two parameters has the effect of making the response surface minimized into a long narrow valley instead of a circular pit, as shown by Fig. 2.7. Hence, the optimization routine has great difficulty in moving along the valley, particularly if it is curved, and there is difficulty in locating the parameter values. The least that this means is long computational times, the worst is that the minimization routine simply fails to locate the minimum. In this case more experiments have to be included, chosen so as to remove this correlation.

Non-linear models are particularly susceptible to correlation problems because it is possible to introduce correlation by the form of the model and the value of the parameters, and not solely by the selection of uneven experimental points, as is the case with the linear models.

Linear models automatically compensate for designs not being central with respect to the variables by means of the θ_0 term. Non-linear models in general have no such compensatory term, and so any non-centralized variables should be transposed by the user if it is important to reduce correlation.

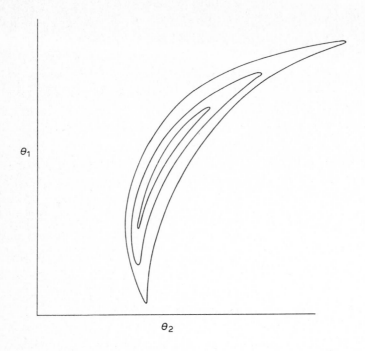

Fig. 2.7. Sum of squares surface for a non-linear, correlated system.

Particularly notorious in this respect is the Arrhenius equation

$$r_1 = A_1 \, e^{-E_1/RT} \tag{2.91}$$

Over the numerically small range of absolute temperatures where reactions occur, experiments are far from orthogonal and there is very high correlation between the two parameters A_1 and E_1. This problem can be reduced by transposing the temperature variables to be related to the difference from the mean temperature $\overline{T}$. When

$$\frac{1}{x} = \frac{1}{T} - \frac{1}{\overline{T}} \tag{2.92}$$

then

$$r = A_1 \, e^{-E_1/RT} = A_1 \, e^{-E_1/Rx} \, e^{-E_1/R\overline{T}} = A_1' \, e^{-E_1/Rx} \tag{2.93}$$

where

$$A_1' = A_1 \, e^{-E_1/R\overline{T}} \tag{2.94}$$

Hence, by using x as the temperature variable in the regression, the experiments are orthogonalized, and A_1' and E_1 have a much reduced correlation compared to the original problem of estimating A_1 and E_1. Furthermore, A_i' is the reaction velocity constant at the mean temperature, a particularly meaningful parameter.

By far the most difficult problem in non-linear parameter fitting is the choice of equations to represent the process being fitted. If the equations are an inadequate description, discrepancies occur between the model fit and the experimental results which are greater than can be expected from pure experimental error.

Usually, the laboratory conditions in which most experimental reactor research work is carried out can be so carefully controlled that results with experimental errors of less than ±5% are to be expected. Errors in fitting models are generally much greater than this because the equations describing the processes are themselves simplifications. This means that most errors in reactor model fitting are "systematic errors", i.e. errors brought about by an "inadequate" model, and not by inaccurate experiments. When systematic error has been confirmed, the next step must be to modify the model equations, unless the model is accurate enough for its intended use despite being less accurate than the experimental data.

Whereas in linear models there are systematic approaches for extending models, in non-linear regression there are not. Being non-linear, absolutely any modification is allowable, but to keep the mechanistic features of non-linear models the modifications made should have some mechanistic meaning. The source of inspiration for the modification of an ill-fitting model is the characteristics of the discrepancies. If these discrepancies have a particular characteristic, e.g. "temperature is not fitted well", model modifications to counter this tendency must be proposed, e.g. a different effect of temperature in the model.

The indications of model inadequacies from the discrepancies are usually fairly obscure because the parameter fitting has averaged the error over all experiments. For example, if a model did not describe the effect of high temperature properly, the resulting fit would not show discrepancies only at high temperatures, but both at low and high, because the parameter fitting would have distributed the error to obtain a minimum sum of squares.

The time spent modifying models and re-fitting them can be extremely informative. The fact that the proposed model is inadequate indicates that the reaction is not well understood. A discussion of possible improvements results in a new hypothesis that then results in a modified model. The results of the fit of the new model show whether the new hypothesis contains some truth. Often the fit is not significantly better, and the hypothesis must be discarded and a new one sought.

Because at this stage it is a discussion between the model and the modeller, it is as well if the discussion is not confused by less relevant information. The modeller has accurate experimental data available and he wishes to know if the model has the correct characteristics over the area of interest. Eight to ten carefully chosen experiments for the regression can often cover such an area adequately and the results are ideal for discussion, each point testing a definite characteristic of the model.

The alternative, of including all experimental data in the regression often produces a hopelessly confusing picture which stifles discussion. This advice is in direct contradiction to the statistician's rule that regressions should be done with all available information. This point will be returned to at the end of the chapter.

The procedure of fitting models to experimental data should be considered as a small research project, taking 1 - 2 months' work and needing 50 - 100 computer runs. The whole procedure consists of collecting laboratory results, sifting them for meaningful experiments, trying very many types of model, and if constraint problems make location of the parameter set difficult, then for each model a number of computer runs may be necessary. During the work much is learnt about the reactor. Hypotheses that brought model improvements teach important aspects of the reaction. Hypotheses that brought no improvements also add to our understanding by showing what effects can be neglected in our system. It is a far cry from collecting the results from the laboratory and carrying out one run with a standard regression package!

It is often the case that the model is accurate enough for the intended use before the systematic error has been removed from the fitting. It is easy to get carried away with the challenge of removing the systematic error "because it is there". This is a dangerous attitude to take when the modelling has an important part to play in a project, because it is often not possible to reduce systematic error to be less than experimental error. The modeller should always remember what the final use of the model is, and stop model development when an adequate accuracy for the use has been achieved.

2.8   MULTIPLE RESPONSE NON-LINEAR REGRESSION

Usually more than one response is measured in a single experiment. The concentration of every component is a response for the model, and temperature measurement is a further response. If the experiment is batch, every sample and every point on the temperature/time profile represents more responses.

Any model fitting study should include all responses, because these contain additional information which, though they make the fit more difficult, contribute greatly to the understanding of the reaction under consideration.

For each experiment sufficient data should be selected to transfer all the "information" given by the experiments. As with single response analysis, superfluous data should be excluded, because it confuses the general picture and only consumes computer time. For example, a temperature may be recorded 200 times during a batch experiment, but 6 selected points may be able to adequately describe the temperature/time profile. Inclusion of the six, not the 200 points is therefore to be recommended in the regression.

In principle a multi-response regression is the same as single response except for a modified objective function. This is now the sum of squares of all experiments, for all responses:

$$\sum_{i=1}^{n} \sum_{j=1}^{L} (y_{i,j} - y_{i,j}^{*})^2 \tag{2.95}$$

for n experiments and L responses, with y^* denoting the model prediction and y the experimentally measured result.

2.8.a Weighting

Once different types of response are brought together the problem of weighting responses becomes of prime importance. Expression 2.95 would try to fit temperature with the same accuracy as concentration. The result would be an excellent temperature fit with concentrations fitted probably with an error of ±0.5 mole fraction! Therefore each response has to be weighted so that they all are fitted with an accuracy proportional to their experimental accuracy. Given the variance σ_j^2 for response j (determined from repeated experiments) this can be done using the following expression for the objective function:

$$\sum_{i=1}^{n} \sum_{j=1}^{L} \frac{(y_{i,j} - y_{i,j}^{*})^2}{\sigma_j^2} \tag{2.96}$$

If one is certain that the correct model is being used in the regression, it is not necessary to give the variances as data to the regression, since this will be the residual mean sum of squares after the best parameter fit. Hence in this multi-response case it is possible to obtain a best fit without weighting and so estimate the σ_j^2 for each response for the mean sum of squares for each response. This σ_j^2 can then be used for weighting responses in a second regression. This principle has been developed to produce an algorithm which determines the weighting to employ without being supplied by experimental error. The algorithm is programmed in only the more refined regression programs and since it is conditional on the error being purely experimental, which as paragraph 2.7.b discusses, is rarely the case, it is not particularly important for engineering work.

Once the weighting facility is available in a program it can be used to help "steer" model fitting in a number of ways, both for single and multi-response studies:

(a) Inaccurate or missing responses from a data set can be weighted by 0.0, so enabling the remaining responses to be processed.

(b) Experimental results covering wide ranges often have experimental error relative to the measured value. Unweighted regression assumes that all errors are absolute. Weighting can be used to base the sum of squares on relative errors.

(c) A best fit of an "inadequate" model can be arranged so that it fits best in the region of most interest by judicious weighting.

Particularly useful is the last point, when the engineer must produce a result even when the model should be "rejected" in the statistical sense. However, by means of weighting he can make the inadequate model fit accurately in the important region giving somewhat greater, though less relevant, errors at the extremes of the experimental region.

2.9   THE ROLE OF STATISTICS IN MODEL FITTING

The first rule of statistics in regression work is to check whether the fitted model has its residuals randomly distributed and is free from systematic error. If the errors are random, then a regression analysis is meaningful. If systematic error is present, the model must be rejected. As experienced modellers know, the main problem in modelling is removing systematic error - experimental error is usually insignificant in comparison.

Statistics is the science of random errors, and random error is not the modeller's main problem. Since there is no "science of systematic errors" the modeller must take ideas from statistics, but he must do so knowing they can only be used for rough guidance as they sometimes lead to entirely false conclusions.

For example, confidence limits on parameters and predicted points are often many times too small because of model errors. As a second example, the repetition of experiments in order to produce better confidence regions is misguided when the major error is model error.

There is a great deal of work reported in the literature on improved methods of statistical analysis, e.g. maximum likelihood fits in place of least squares, complex objective functions in multi-response parameter fitting and in experimental planning, etc. Because reaction engineering is not concerned with random errors, it is unlikely that any of such work will ever find practical application in the design of reactors.

FURTHER READING

Bard, Y., 1974. Non-Linear Parameter Estimation. Academic Press. - A thorough description of non-linear regression.

Box, G.E.P., Hunter, W.G. and Hunter, J.S., 1978. Statistics for Experimenters. Wiley, New York. - Thorough description of experimental planning with linear models.

Daniel, C. and Wood, F.S., 1971. Fitting Equations to Data. Computer Analysis of Multifactor Data for Scientists and Engineers. Wiley-Interscience. - Regression explained for engineers.

Rose, L.M., 1974. The Application of Mathematical Modelling to Process Development and Design. Applied Science Publishers, London. - On uses of modelling.

REFERENCES

Box, G.E.P and Hunter, J.S., 1962. Technometrics, 4: 317.
Cresswell, D.L. and Santos, A.M., 1980. Chem. Eng. Sci., 35: 283.

QUESTIONS TO CHAPTER 2

(1) What is a reactor model and what use is it ?
(2) What role does the computer play in reactor modelling ?
(3) For the system

$$A + B \xrightarrow{k_1} P \qquad \text{products}$$

$$\left.\begin{array}{l} P + B \xrightarrow{k_2} R \\ A + A \xrightarrow{k_3} S \end{array}\right\} \quad \text{by-products}$$

what form of reactor is most suitable, and why ?
Develop the model equation for the reactor that you select.
(4) What do you understand by model fitting ?
Why is it so important ?
(5) Write down some examples of models linear in the parameters. What is the
property of these equations that makes model fitting simple ?
(6) Why are most reactor models non-linear in the parameters ?
How can fitting be carried out with non-linear models ?
(7) What is multi-response, non-linear regression, and what are the various roles
of weighting ?
(8) Why are predictions of confidence limits for parameters and predicted performance
not to be taken seriously when dealing with reactor models ?

Chapter 3

REACTOR LABORATORY STUDIES IN PROCESS DEVELOPMENT

3.1   THE OBJECTIVE OF PROCESS DEVELOPMENT

The objective of process development is the definition of a chemical process
in sufficient detail for a plant to be built which will operate safely with a
predictable output and quality at a predictable cost. Furthermore, these costs
should be as low as possible, because this, in addition to maximizing the chance
of the project making a profit, also reduces the possibility of the process being
rendered obsolete by a competitor who finds a more economic operating condition.
The process must be defined within a certain time schedule to meet marketing
requirements and engineering and construction schedules. Since the reaction is
usually the most important stage in a process, process development usually centres
around reactor development and for this reason process development becomes an
important subject in any book on practical reactor design.

The reactor development programme, being part of the process development
programme, has the same major objectives as the process development programme, and
these are discussed in more detail in the following paragraphs.

3.1.a   Scale-up

A dominant requirement is that the reactor should operate in a predictable
fashion on the industrial scale, which could be 10,000 times greater than the
laboratory scale. Changing the scale of a reaction alters the heat removal and
mixing characteristics, and these can lead to differences in temperature and
concentration profiles. For complicated chemistry these changes in temperature
and concentration change reactor selectivity, which in turn means that performance
of a full-scale reactor is difficult to predict.

One approach to predicting the effective scale-up is to carry out the reaction
at a number of scales and so gain enough confidence to make an empirical prediction
of the performance on an even bigger scale (the principle of similarity). This
method is not usually the best method in reactor engineering because it requires
many scale-up steps and the extrapolation is, despite these steps, limited.

A more fundamental approach is to understand the reactor in sufficient detail
to be able to predict the effect of concentration and temperature on the selectivity
and then to predict the temperature and mixing patterns on the full-scale equipment
(the modelling approach). Combining these two allows a prediction of the selectivity
to be made for larger scale reactor operation. This second approach is generally
preferred in industry and it has the advantage over the purely empirical scale-up

method that:

    (a) The deeper understanding that goes with this method leads to greater confidence, which in turn allows greater scale-up factors to be used.

    (b) Only two scales of reactor (laboratory and pilot plant) are needed.

    (c) The designer has flexibility in the design of his full-scale reactor, as long as he can define its heat-transfer and mixing characteristics.

The two methods are very different in philosophy and sometimes result in contradicting requirements for development work.

    The principle of geometric similarity demands that the reactors in the various stages are geometrically similar and that flow patterns are compared by operating at similar Reynolds Numbers (if necessary by changing fluid viscosities), (see Fig. 3.1).

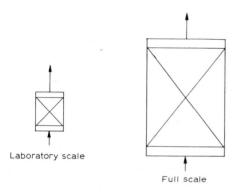

Laboratory scale

Full scale

Fig. 3.1. Scale-up stages using the principle of geometric similarity.

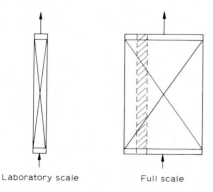

Laboratory scale   Full scale

Fig. 3.2. Scale-up using the modelling approach.

The modelling approach aims at operating a small scale reactor which represents a small section of a large scale reactor in terms of heat and mass transfer, temperatures and flows (Fig. 3.2). When it is inevitable that there must be changes compared with the full-scale (e.g. velocities or areas/unit volume cannot be held constant on both scales) the modelling predicts the differences to be expected from these inevitable changes.

3.1.b   Safety

Safe operation is an important requirement which must be clarified during the reactor development work. In addition to the usual hazards of toxicity and explosivity associated with the materials themselves, reactors generate heat which becomes more significant the larger the scale of the equipment. As temperatures rise, reaction rates increase exponentially, and in addition, new, more exothermic reactions can appear, leading to hazardous situations.

In complex organic reactions further hazards can occur. Induction periods, when no reaction occurs for a while at the beginning of the reaction, can be a safety hazard, and the complex chemistry can lead to conditions where reactors, normally considered safe, can sometimes follow a temperature composition profile, that leads to explosions. This subject is handled more thoroughly in Chapter 10.

3.1.c   Low-cost process

It is usual for the reactor to be the equipment item with the greatest cost impact on the process. The reactors consume raw materials at efficiencies less than 100%. Since raw material costs are often 80% or more of the total product cost, a change in the reactor selectivity can produce savings of an order of magnitude larger than changes on any other part of the plant.

Hence the process development work towards finding the most economic process usually results in a concentrated study of the reaction to attempt to improve yields by modifying operating conditions. This is no simple task because there may be ten or more variables associated with a reaction (feed compositions, time, temperature, pressure, catalysts composition, etc.) all strongly interdependent. The experimental location of the optimum in such a system is not easy.

In the broadest sense this search for new optimum operating conditions involves step changes such as introducing new chemical constituents into the catalyst, changing solvents, or adding small quantities of new components. It also includes changing temperatures, concentrations, and contact times. The discrete changes are normally the domain of the chemist, who, because of the nature of his problem, will probably be unsuccessful, but if he is successful handsome improvements in the process result; the continuous variables are the domain of the chemical engineer, where some success can almost be guaranteed, but the process improvements are modest, though nonetheless worthwhile.

Both the discrete and continuous changes benefit from an understanding of the reaction, both from a qualitative picture and a quantitative analysis.

An optimum reactor yield to the product does not necessarily mean that the same conditions would give the most economic process because, for example, recycling of unreacted raw materials together with the possibility of non-stoichiometric reactor feed quantities may well produce a cheaper over-all process with different operating conditions than the optimum yield conditions.

This means that in addition to knowing the reactor conditions for optimum yield, the reactor development work must supply information on how to predict reactor outlet composition for a range of inlet conditions, so that the true total process optimum can be sought in a later study.

These three requirements form the reactor development programme, i.e:
 - confident scale-up;
 - safe full-scale operation;
 - location of optimum process operating conditions,
must be fulfilled by a laboratory and pilot plant experimental programme with the reaction in question.

The type of laboratory apparatus that is available to obtain these data is discussed in this chapter. The following chapter describes the experimental design methods that are available to ensure that the experiments carried out are going to yield useful information. Chapter 5 discusses the pilot stage which is an almost inevitable last step in the reactor development work necessary to yield sufficient confidence to make a full-scale design.

The analysis of the laboratory results, and the comparison of the laboratory and pilot stages, can be most conveniently done by means of a reactor model. The model is therefore a useful tool in achieving the objective of the reactor development programme.

The model is often an aid to understanding the reaction (as a result of the parameter fitting study), it provides a basis for experimental planning to efficiently locate the optimum, and it is required when a total process is studied to locate the optimum condition for the whole process.

3.2   TYPES OF LABORATORY EXPERIMENTAL REACTOR

The development of models for reactors can follow two distinct philosophies:
 (a) The mechanistic model, built up from the elementary processes occurring in the reactor with the parameters for each of these elementary processes being determined separately in special experimental equipment.
 (b) The semi-empirical model, which uses a mechanistic form of the equations, but the parameters are fitted to the performance of a reactor quite similar in appearance to the anticipated full-scale reactor.

These two philosophies bring with them different experimental philosophies and different experimental equipment. A successful modelling project is usually a mixture of both philosophies, with some specific parameters being obtained by separate experiments, but others being determined by fitting to plant-like experimental reactor results. It is therefore possible to categorize laboratory equipment into that to obtain specific parameters, and that which simulates a full-scale plant.

The first type of equipment is usually concerned with pure reaction kinetics: the reaction rate constant, energy of activation, and orders of reaction for the various components. The object of such equipment is to remove extraneous effects due to mass-transfer or temperature changes, so that the parameters being sought can be measured as directly as possible. Some of the equipment is designed so as to minimize concentration changes so that kinetics and reaction orders can be determined without the need to integrate over concentration ranges, since this is introducing an indirect rather than a direct evaluation of the experiment.

The second type of laboratory reactor equipment is an attempt to simulate a section of a full-scale reactor, and so evaluate and optimize its performance. The principle of scaling-up has been discussed earlier, commenting that the laboratory reactor should be designed as an element of the final full-scale reactor.

The different phases and mixture of phases in the reactor result in different forms of reactor and so it is also possible to categorize laboratory reactors by phase. This is done in Table 3.1, which summarizes the various types of reactor used in the laboratory stage of process development. These various types are discussed in the following sections.

3.3   LABORATORY REACTORS FOR MEASUREMENT OF CHEMICAL KINETICS

In general, difficulties arise in measurements of chemical kinetics because temperatures are not constant, and concentrations vary. Reactors designed specifically to measure chemical kinetics attempt to overcome these problems in one way or another.

The simplest method of keeping temperatures changes small is to increase the heat-transfer area to volume ratio in the reactor and to increase the thermal capacity of the reactor compared to its contents. Both these conditions are met by having small reactors. Some of the reactors are continuous, rather than batch, since this enables a steady condition to be set up with a constant concentration. There is then no need to assume an order of reaction in order to integrate the reaction equations to determine the reaction order!

A problem in common with all chemical kinetic measurements is the stopping of the reaction in the withdrawn sample. No general method for arresting the reaction can be recommended because one attempts to use the specific chemical properties of

TABLE 3.1

Types of laboratory equipment available for reactor development studies

Reactor system	Chemical kinetic measurement	Total reactor performance
Liquid	- Ampoules in thermostat - Small scale batch - Micro reactor	- Batch reactor with cooling - Calorimetric reactor - CSTR + cooling - Autoclave - Tubular reactor
Gas	- Gas stirred reactor	- Tubular reactor
Solid	- Differential thermal analysis (DTA)	- Specifically designed equipment
Liquid/Liquid	- Small scale batch (quiescent interface) - Droplet reactor	- Batch reactor with cooling
Liquid/Solid	- Small scale batch - Differential reactor	- Batch reactor with cooling - Tubular reactor
Liquid/Gas	- Small scale batch (gas reservoir) - Wetted disc column	- Batch reactor with cooling - Packed column - Bubble column
Gas/Solid	- Stirred gas/solid - Differential reactors - Weighed pellet	- Tubular reactor - Tubular fluid bed

NB Laboratory equipment purely for safety studies are described in Chapter 10.

the mixture whenever possible. For example, if one reactant is an alkali, the reaction is "quenched" by adding an excess of acid. The back titration of this acid may then indicate the extent of reaction. If the reaction is catalyzed, the addition of a component that removes the catalyst activity can be effective as a quench. Less effective, though sometimes used, is the cooling or dilution of the reaction mixture. Reactions are effectively stopped by the start of a chromatographic analysis due to the high dilution and separation of components. This requires that the chromatograph is always available for use by the reactor, and no samples are stored.

3.3.a   Chemical kinetic determination - liquid phase

   Ampoules in a thermostat. To determine reaction rates and activation energies for slow reactions, small quantities (up to a few ml) of a reaction mixture are placed in ampoules in a thermostat and withdrawn at various times (Fig. 3.3). The reaction is then immediately quenched on withdrawal and the mixture analyzed to determine the extent of reaction.

   The very small volumes are to ensure that the mixture stays at the thermostat temperature, and sampling problems are circumvented in that the total ampoule contents are analyzed.

The method requires only small quantities of material to complete a kinetic investigation and this may be important if the reactants investigated are hypothesised reaction intermediates which may not be available in large quantities.

The method is only suitable for comparatively slow reactions.

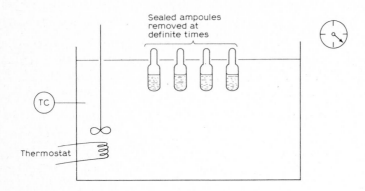

Fig. 3.3. Ampoules in a thermostat.

Small scale batch reactor. On a somewhat larger scale is the use of a small batch reactor, e.g. a 100 to 250 ml stirred flask, to obtain kinetic measurements (Fig. 3.4). In such equipment it is difficult to achieve constant temperature even though the flasks are in a thermostat because of the lower surface to volume

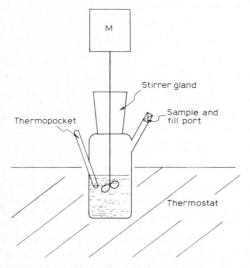

Fig. 3.4. Small scale batch reactor.

ratio. Hence it is often necessary in such equipment to work at very low concentrations to achieve an adequate approach to isothermal operation.

This method requires more reactants than the ampoule method, but if the reaction is not too fast small samples can be withdrawn from the flask during the reaction and analyzed. This gives considerable information per experiment, but is conditional on the methods of withdrawal and quenching being short in comparison to the reaction times.

For a homogeneous, adiabatic liquid reaction, after good initial mixing the continued stirring of the reaction is not important, but when this reactor is used for two phase systems (liquid/liquid, gas/liquid, or liquid/solid), continuous stirring becomes important. For kinetic measurements the stirring should be so intense as to remove all bulk mass-transfer effects, so ensuring that the experiments are only measuring kinetic effects. This can be tested by carrying out repeated experiments at progressively increasing agitation rates. A plateau should be observed (if the reaction is not too fast) which is evidence that the system is no longer influenced by mass-transfer rates. The experimental work to determine chemical kinetics should then be done at an agitation rate well onto this plateau.

When reactions involving the consumption or evolution of a gas are being investigated the batch reactor can be connected with a gas reservoir and the gas production or consumption then monitored. This change in gas quantity is an accurate measurement of the course of the reaction, and if agitation is such that mass-transfer effects have been eliminated, then the rate of change of gas volume can be used to determine the chemical kinetics. This gas quantity measurement could be done by measuring the pressure in the gas reservoir, but this is not to be recommended because the pressure determines the concentration of the gas in the liquid, and experimental equipment involving a varying gas pressure would produce a confusing concentration variation during the experimental run. Preferably, the equipment should have a variable volume and constant pressure. This can be achieved by having a comparatively large gas reservoir which has a variable gas volume in that inert liquid can be introduced or removed. The gas pressure is adjusted to its original value several times during the experiment by changing the quantity of inert liquid in the reservoir. By careful selection of the volume, pressure changes can be kept within a few percent and the liquid addition to the reservoir during the run can be measured accurately.

A plot of the liquid volume added to return the pressure to its original value provides a convenient history of the course of the reaction which can be used for the determination of the kinetics. Figure 3.5 shows the principle of such equipment.

For liquid/liquid reactions the batch reactor or the CSTR can be operated with such mild agitation that the liquid remains in two separate layers and the liquid/liquid interface between the two layers is quite smooth. Hence, the mass-transfer

90

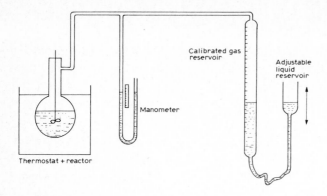

Fig. 3.5. Batch reactor with gas reservoir.

area is known, and mass-transfer data can be determined. By floating various sizes
of plastic balls at the liquid/liquid interface, the area of the interface can be
varied and its effect on the reaction rate determined. The equipment is particularly
useful for liquid/liquid reactors that are mass-transfer controlled, or that could
perhaps be mass-transfer controlled. The known and variable mass-transfer area
enables the importance of the role of mass transfer to be defined, and this affects
the final choice of process equipment for the full scale.

Clearly, such methods can also be applied to gas/liquid reactions where mass
transfer is rate-controlling and should be quantified.

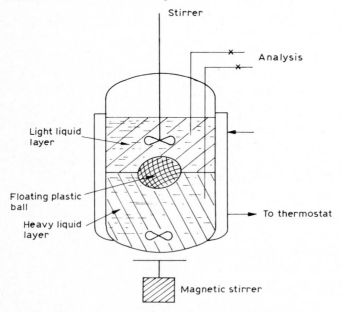

Fig. 3.6. Batch reactor with quiescent interface.

Micro-reactor. All forms of batch reactor involve changes in concentration with time. In investigating kinetics a factor of importance is the relation between concentration and reaction rates. Clearly the batch reactor is not an ideal means of measuring this relationship directly. Only indirect conclusions can be drawn after some model fitting procedure has been carried out.

The CSTR, on the other hand, works at constant concentration and temperature, which has the attraction for kinetic work that direct measurement can be made. The disadvantage of the CSTR is that large quantities of reactants are needed because the equipment must run to steady state conditions for each experimental condition investigated (compared to the batch experiments where each run covers a range of conditions).

This disadvantage is overcome by the design of the micro-reactor, which is basically a CSTR reactor on a very small scale. Hence each experiment is carried out with defined concentrations and temperatures, yet the material used is kept to a minimum because of the small scale. Figure 3.7 shows the design of the micro-reactor.

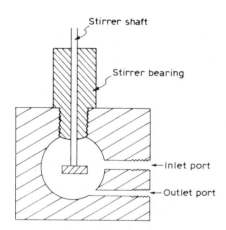

Fig. 3.7. The micro-reactor.

3.3.b   Rate measurements in two phase systems with mass transfer resistance

In both liquid/liquid and gas/liquid reacting systems there are many reactions that are so fast that the rate is always controlled by the mass-transfer rate. These systems must be identified by laboratory work because a different design procedure and full-scale equipment must be used for such systems.

If a system is mass-transfer controlled, its rate will be directly proportional to the exchange area presented and independent of the liquid volume, whereas a kinetically-controlled reaction will show the opposite characteristics.

Laboratory reactors which are used to determine the importance of mass transfers, and the evaluation of the appropriate mass-transfer coefficient are designed to present a known surface area of interface, so that K_L can be evaluated from the reaction rate results.

For liquid/liquid reactors the quiescent stirred batch or CSTR reactor can achieve this (Fig. 3.6). This can also be achieved by a droplet reactor, where droplets of the dispersed phase rise or fall through the continuous phase (Fig. 3.8). The droplets can be counted, photographed, and measured, and so the surface area can be determined and hence K_L obtained from the rate measurements.

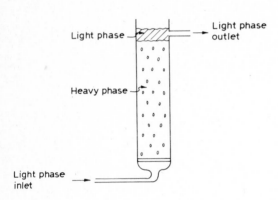

Fig. 3.8. Liquid/liquid droplet reactor.

For gas/liquid systems the wetted wall column, or wetted disc column is used (Figs. 3.9 and 3.10). This consists of a tube, which may contain a column consisting of connected discs down the surface of which the liquid flows. Gas flows up the column and so a known wetted area is presented but a very small liquid hold-up is involved. When substantial reaction occurs in such equipment it is clear that it is a very fast reaction which is mass-transfer controlled. The equipment evaluates K_g which can then be used for comparative purposes in sizing full-scale equipment.

3.3.c   Chemical kinetic determination - gas phase

Gas stirred reactor. Because of the low densities and high reaction rates involved, batch gas phase reactors are impracticable. In their place is the plug flow tubular reactor which has been shown to behave kinetically similarly to the batch reactor. For pure kinetic studies, however, the tubular reactor has the disadvantage that the temperature cannot be held constant and this puts all quantitative rate measurements in question. For accurate gas phase kinetics measurements there is little choice other than to employ a CSTR form, very similar to the micro-reactor.

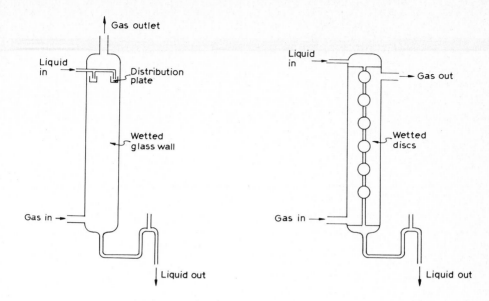

Fig. 3.9. Gas/liquid wetted wall column.   Fig. 3.10. Gas/liquid wetted disc column.

Such a reactor, however, need not have a very low volume, since usage of material is no longer critical. The size is determined by the mechanics of stirring and convenience of manufacture, and usually has a capacity of 20 to 100 ml.

Gas reactions usually have contact times of only a few seconds; many reactions occur at high temperature, and others under high pressure. These conditions require that the stirrer must

(a) stir intensely enough to mix the incoming feed in a short time compared to the residence time;

(b) have a gas-tight coupling operable over extremes of temperature and pressure. A fairly usual design is to have a magnetic drive to overcome the sealing problems and a stirrer bearing inside the reactor. To cool the bearing one of the reactants can be fed over it. Figure 3.11 shows the general arrangement for such reactors. They have become standard equipment for gas kinetic work and can be bought as a standard piece of equipment.

These reactors are comparatively small and so have a very high service area / volume ratio. Should wall effects be important (i.e. when free radicals are involved), the "scaling up" to larger reactors is not possible.

Stirred gas-solid reactors (spinning basket or SGS reactors). This is a development of the stirred gas reactor for gas reactions occurring on solid catalysts. The stirrer of the stirred gas reactor is replaced by a wire basket, usually in the form of a crucifix, (see Fig. 3.12), and this basket is filled with catalyst. This has the effect of:

94

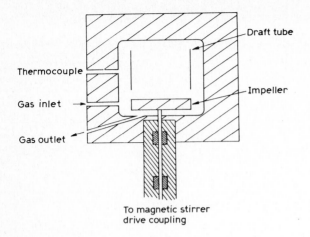

Fig. 3.11. Gas stirred reactor.

Circular wire basket   Cruciform wire basket

Fig. 3.12. Basket designs for the SGS reactor.

o mixing the inlet gas well;
o having a gas velocity round the catalyst particle high enough to remove mass-
  transfer problems associated with getting reactants from the bulk gas to the
  catalyst surface and products from the catalyst to the bulk gas; and
o increasing the heat transfer from the catalyst to gas to reduce temperature
  differences between solid and gas phases.

Solid-catalyzed gas reactions are complex processes involving mass transfer from
bulk to catalyst surface (bulk diffusion), from catalyst surface to catalyst site
(pore diffusion), and reaction on catalyst sites (chemical kinetics). The SGS
reactor is a useful tool in investigating the relative importance of these three
steps.

By repeating a series of experiments at increasing stirrer speeds, the importance
of bulk diffusion can be demonstrated. If the reaction rate is dependent on stirrer
speed, bulk diffusion is present, and kinetic measurements should be carried out
at a stirrer speed sufficiently high to be out of this region (as with the batch
reactor for two phase systems). With such a suitably high stirrer speed a further
set of experiments can be carried out with constant weight but various sizes of

catalyst particles in the basket. If the particle size has an effect on the reaction rate, then pore diffusion may be important.

If the kinetics of the surface reactions are to be investigated, then the reaction stirrer speed should be high enough to remove bulk diffusion and the catalyst particle small enough to remove pore diffusion. Only then can the experiment's results be related to chemical kinetics.

The SGS reactor is a very useful tool but its results must be used with caution. There may remain some form of bulk diffusion within the basket which is not removed by increasing the rotation of the basket so that even when plateaus of reaction rate are achieved with catalyst size, diffusion resistances are not absent.

This reactor has a high proportion of free volume and in some systems some reactions occur in the free volume (burning reactions for example) and others on the catalyst surface. When this occurs, results become very difficult to interpret and selectivities achieved on the SGS bear no relation to those of a reactor fully packed with catalyst.

Differential reactor. Although catalytic reactions are carried out in tubes or packed beds on the industrial scale, such forms of reactor are not useful for kinetic measurements because concentrations and temperatures change down the reactor and so direct measurement of kinetics is not possible. Near isothermal operation can be achieved only by working at low conversions.

The differential reactor is so designed that very low conversions occur over the catalyst layer. Because such low conversions cannot be accurately measured, the reaction gases are circulated over the catalyst bed to achieve an accurately measurable conversion. The feed of reactants is introduced into the recirculating stream, and a purge stream removed, (see Fig. 3.13).

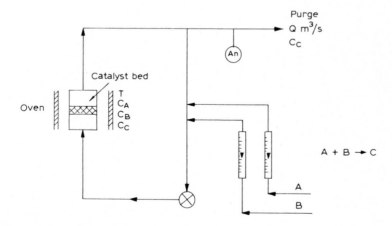

Fig. 3.13. The differential reactor.

The purge stream is analyzed and hence by mass balance the reaction rate is known:

$$\frac{dn_C}{dt} = Q\,C_C = V_R\,A_1\,e^{-E_1/RT}\,C_A\,C_B \qquad\qquad (3.1)$$

The analysis of the exit stream, and the flow rate measurement can be done accurately and so an accurate reaction rate can be measured despite the small conversion occurring across the reactor. Results from this reactor are only meaningful when homogeneous reactions are absent.

There should, in principle, be no problem in employing such a reactor for use in solid-catalyzed liquid reactions if the catalyst can be held in a thin bed.

Weighed pellet. In a solid gas reactor where the solid is not acting as a catalyst but is a reactant, the course of the reaction can be followed by suspending a pellet of the solid in the gas stream and noting its change in weight with time. Such reactions occur in oxidations, in the preparation of certain metals and in uranium extraction and are usually associated with high temperatures (Stonhill, 1959).

Sensitive thermobalances are available for continually monitoring small changes in weight of samples at high temperatures, and so such equipment can be used with a reactor and furnace to provide conversion data based on rate of change in weight together with gas analysis (Figure 3.14).

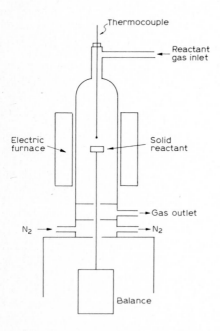

Fig. 3.14. Thermal balance for gas-solid reactions.

3.3.d Chemical kinetic measurements on solid reactions

The heating of clays, ores and coals causes chemical changes to occur, which
if followed precisely can give an understanding of the processes occurring in the
manufacture of cement, recovery of metals from their ores, and the gasification
of coal.

When the changes occurring are accompanied be evolution of gas they can be
monitored by weighted pellet experiments as previously described. However, when
the changes are of a more physical nature, such as changes of crystal structure
or complex formation, or melting of specific components, then weight measurements
will not provide any information.

Differential Thermal Analysis (DTA) is a technique whereby such changes can be
detected by differences in the apparent thermal capacity of the sample as it is
heated. A very small sample is heated steadily and compared against the similarly
heated reference material. When temperatures are reached where changes occur which
involve heat effects (melting, crystal structure changes or reaction) there is a
discrepancy between the reference behaviour and the sample. This discrepancy is
recorded on a chart, so giving as a result a series of peaks against temperature.
The temperature at which these peaks begin is characteristic of the process occurring,
and the shape of the peak gives an indication of the rates at which the process
occurs.

Very many industrially important "Reactors" involve the roasting of solids;
DTA is a method of understanding and improving the feed mixes and operating conditions
of such reactors.

3.4 LABORATORY REACTORS FOR MEASURING TOTAL REACTOR PERFORMANCE
3.4.a Batch reactor with cooling

The most usual form of laboratory reactor for reaction in the liquid phase is
the simple batch reactor, if necessary supplied with a cooling jacket (see Fig. 3.15).
This reactor provides a quick estimate of the performance of the full-scale reactor
in terms of maximum yield, the presence of heat-transfer problems, the required
contact time, and so on. An investigation with it is usually the first step to be
made when a new reaction is being studied.

It can be used for evaluation of the kinetic parameters only if the reactor is
isothermal, or if a model for the reaction is available to which the experimental
data can be fitted. It is a very useful reactor on which to develop the first
reactor model for later scale-up because it is so similar to the full-scale batch
reactor and also kinetically not dissimilar from a full-scale plug flow tubular
reactor.

When heat dissipation problems are encountered it is usual with this type of
reactor to add one of the reactants slowly, ensuring that the temperature does

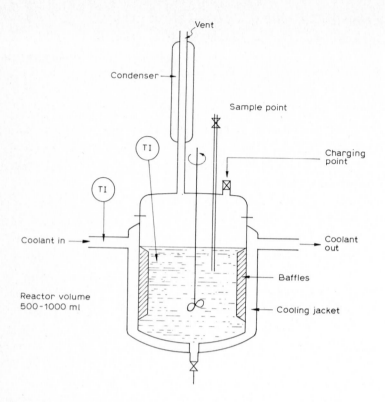

Fig. 3.15. Batch reactor with cooling.

not rise above a predetermined value. When all the second reactant has been added
the reactant then proceeds as a true batch reactor until the yield of product is
maximized (see Fig. 3.16).

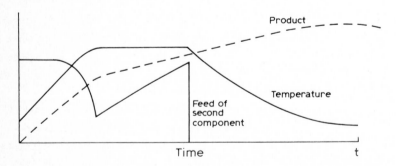

Fig. 3.16. Batch reactor concentration and temperature profiles.

This is a form of semi-batch reactor at the beginning of the batch and pure
batch reactor at the end. This is common laboratory practice but rarely appreciated

by engineers. The reaction time t on Fig. 3.16 gives no indication of the batch
time or the space time in a tubular reactor on the full scale because it is
predominantly a time determined by the ability to transfer heat from the reactor,
and not a time determined by reaction kinetics. A full scale reactor properly
designed to remove heat could result in a much smaller time than that determined
in first experimental work.

3.4.b   Calorimetric reactors

Most reactors are accompanied by heat effects and often, particularly in liquid
phase organic reactions, the analysis of the reaction mixture to determine its
composition in order to follow the course of the reaction is very difficult. This
has led to the development of the calorimeter as an advanced type of batch reactor
for process development purposes. The principle of such equipment is to follow the
reaction by measuring the heat evolved from the reactor.

The equipment consists of a jacketed batch reactor of between 0.5 and 2.5 litre
capacity with accurate measurement of inlet and outlet coolant and reactor temperatures. A simple form of the equipment then uses the temperature measurement and
coolant flow to calculate the heat evolved by the reactor, and this gives the
appropriate trace of the progress of the reaction with time.

The disadvantage of this is that the reactor temperature is not controlled,
which is a severe disadvantage if chemical kinetic data are to be derived from the
results. This disadvantage is overcome by means of the sophisticated "heatflow"
calorimeter (Regenass, 1978) whereby the reactor temperature is held practically
constant, or given a predetermined ramp function, by adjusting the temperature of
the cooling (or heating) medium in the reactor jacket (Fig. 3.17). A control system
mixes "hot" and "cold" coolant streams from two thermostats to maintain the required
temperature. After calibration, the resulting heat transfer from or to the jacket
can be used to calculate the heat evolution occurring in the reactant medium, which
can then be plotted out by the equipment (Fig. 3.18). From such results, reaction
orders, reaction rates, activation energies and heats of reaction can be determined.
The equipment is also particularly useful for examining potentially hazardous
situations. Such heat flow calorimeters have been developed for industrial use for
investigating liquid phase organic reactions and are capable of operating over
very wide temperature ranges.

3.4.c   Autoclave reactors

The Autoclave is the name for a high pressure liquid phase reactor. For volumes
less than 0.5 litres the autoclave can be a very heavy tubular body with a massive
screw cap, enabling work up to many hundred atmospheres to be carried out (Fig. 3.19).
The liquid and/or solids are charged in the autoclave, the cap is screwed down
tightly and the autoclave is heated to the appropriate temperature for the appropriate

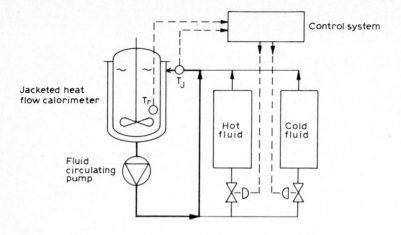

Fig. 3.17. The heat-flow calorimeter.

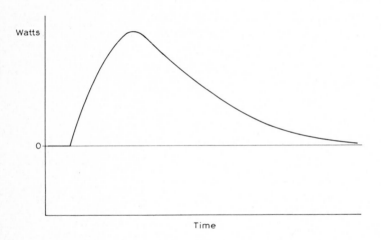

Fig. 3.18. Chart showing heat flow from a reactor at constant temperature.

time. Agitation is usually done by rotating the whole autoclave as the high pressure
put great demands on stirrer glands.

For lower pressures and higher volumes, the autoclave has more the appearance
of a sturdy batch reactor, with heavy flanges. Temperature measurement, gas
feeding and removal, and jacket heating and cooling are possible, and a conventional
stirrer is used. To overcome stirrer glanding problems magnetic coupling is employed
(see Fig. 3.20).

Glass autoclaves can be used up to pressures of 20 bars (and 3 litres volume).
Metal autoclaves of this design can be used up to 60 bars pressure.

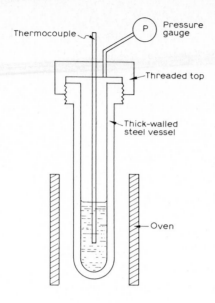

Fig. 3.19. Simple "bomb" autoclave.

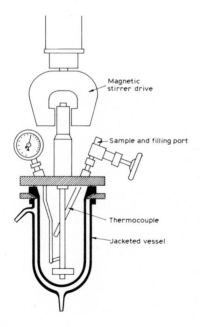

Fig. 3.20. 60 bar autoclave.

3.4.d CSTR with cooling

The disadvantage of the CSTR for laboratory work is the large quantity of raw material it consumes, hence the development of the micro-reactor.

The advantage of running a CSTR of reasonable size in a process development programme is that this is the form of reactor that will often be employed on the full scale, and to scale-up directly from micro-reactor results is too large a scale-up step; for instance heat-transfer problems will not have shown themselves in the micro-reactor.

Hence CSTR operating experience is valuable, but there has to be provision for handling the larger quantities of material involved. Hence the reactor becomes a small plant, and usually the operation of a laboratory CSTR is done by the operation of a "mini-pilot" plant. CSTRs in process development are therefore concerned with pilot plants, and this is the theme of the next chapter.

3.4.e Gas/liquid and liquid/liquid reactors

Laboratory reactors to simulate industrial reactors for these types of system again have the appearance of the proposed industrial-scale reactor. If a stirred tank is envisaged, a batch or CSTR reactor, as described in the previous section, is employed, but with the two phases present. If a bubble-tower reactor, or a liquid/liquid extraction column are envisaged full-scale designs, then the appropriate laboratory-scale reactor is a 25-50 mm diameter tube about 1 m long in which the gas is bubbled through the liquid or droplets of one phase pass through the continuous phase.

3.4.f Tubular (integral) reactors

Since many types of both liquid and gas phase reactor on the full scale are empty tubular reactors it is of interest to determine the performance of a reactor tube in the laboratory. The maximum performance of a reactor tube will be a useful estimate of the maximum performance that can be expected on the full-scale tubular reactor.

Tubular reactors are characterized by being very non-isothermal and having large concentration gradients. Hence their use for obtaining kinetic data is very questionable, but they are the useful first stage in modelling work.

Laboratory tubular reactors usually consist of a tube about 300 mm long and 25 mm in diameter. The tube is heated by being in an oven, or by electrically heated windings (Fig. 3.21).

Temperature measurements are normally taken from a movable thermocouple in a narrow tube running down the centre of the reactor. Alternatively 6-8 thermocouples may be permanently mounted in this narrow centre tube in fixed positions.

The reactor must be equipped to measure reactor feed flows and the outlet stream must be analyzed in some way. For gas phase tubular reactors this is usually done with a gas chromatograph.

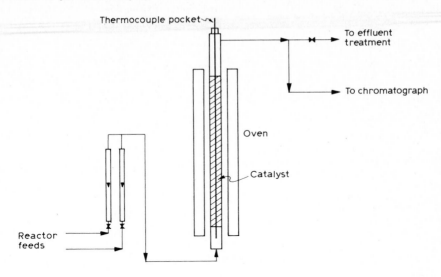

Fig. 3.21. The laboratory tubular reactor.

Full scale reactors are often about 3 metres long but still 25 mm diameter. For the same contact time this means that the gas velocity on the full scale will be 10 times that of the laboratory. This has the effect of extending the high temperature or "hot spot" region of such reactors, which can increase the peak temperature. However, the increased velocity will improve the heat transfer to the walls which will tend to reduce the peak temperature. Hence tubular reactor temperatures, concentration profiles, yield and stability regions will be changed in some not entirely obvious way in the longer tube. These factors must be borne in mind when interpreting laboratory tubular reactor data.

Laboratory tubular reactors are also used to simulate full-scale reactor vessels since a small element of the cross-section of a vessel is best represented in the laboratory by an insulated tube.

When solid-catalyzed reactions are involved, the tube contains the catalyst particles. It is often possible to use exactly the same catalyst particle size in the laboratory as in the full scale, because tube diameters are usually the same.

Reactions involving adiabatic catalyst beds are carried out in insulated tubes in the laboratory since this is an adequate simulation of a section of a packed bed.

A particularly interesting arrangement of laboratory reactors is used for catalyst screening. Because catalyst selection is only partly a science there remains a great deal of empirical experimental work in trying out new catalyst compositions. Since it is not possible to predict optima of operation conditions for new formulations a scan of temperature and concentration for each formulation is necessary.

Catalyst screening equipment consists of about six tubular reactors fed from a common mixed gas supply of reactants, and mounted in a common oven (Fig. 3.22). On the outlet to each reactor is a sample valve which can be activated to send an outlet gas sample to a chromatograph. These valves are activated in rotation so that one chromatograph can monitor the outlet concentrations of all tubes.

Different catalyst formulations are packed in different tubes and the equipment started at a low oven temperature. The oven temperature is then programmed to rise slowly over a period of 24 hours and in this time the equipment operates fully automatically to give the performance of the various formulations over the temperatur range scanned.

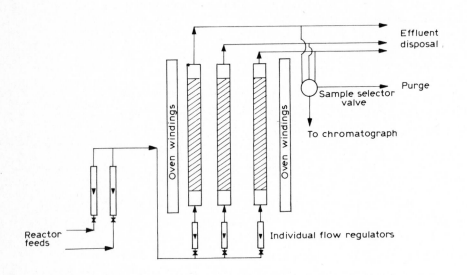

Fig. 3.22. Multitubular catalyst screening equipment.

3.4.g Tubular fluid bed

If a fluid bed reactor is being planned as the reactor to be employed on the full-scale process, then the equivalent laboratory equipment is a 25-50 mm glass tube 1-2 metres high containing the catalyst.

The operation of such a reactor gives information on achievable conversions, fluidizability of the catalyst, attrition data and regeneration data.

It will not, however, give information on the effect of bed-height or contact time changes because the narrow laboratory reactor tube has different bubbling characteristics to the full-scale reactor because of the proximity of the reactor walls. After a few centimetres the reactor operates in a slugging regime with slugs of gas passing upwards through the bed, of constant diameter equal almost to the tube diameter. On the full scale they would grow; in the laboratory they cannot.

Laboratory fluid bed reactors are usually fitted with gas distributors and support plates made of sintered glass. This differs so markedly from the full-scale equipment that again fluidization data, bed height and contact times cannot be taken from laboratory data directly for use for the full-scale design.

3.5   LABORATORY MEASUREMENT OF HEAT OF REACTION

As explained in Chapter 1 there are many reactions where the heat of reaction cannot be calculated because the appropriate thermodynamic data for all components in the necessary physical states are not available. Hence it often happens that the heat of reaction must be measured in the laboratory.

When this heat of reaction is not too high, this can most easily be done by carrying out the reaction at normal concentrations adiabatically in a vacuum-jacketed reactor to measure the temperature rise brought about by the reaction when it proceeds virtually to completion.

When the heat evolved is such that adiabatic reaction results in too high a final temperature, which leads to vaporization, hazards, or the appearance of further reactions, then dilute solution can be used for the adiabatic runs.

If dilute runs are not possible, or if the resulting thermodynamic data are thought to be unrepresentative of concentrated runs, the heat evolved by the reaction must be derived from results from calorimeter reactor type experiments. Particularly suitable for reaction heat data determination is the heat flow calorimeter described earlier in section 3.4.b.

3.6   THE CONTRIBUTION OF REACTOR MODELLING TO PROCESS DEVELOPMENT STUDIES

We have seen that reactor design must start by obtaining data and that there are various ways of doing this in the laboratory. Following this, the data will be used for pilot plant design, for comparison with pilot plant results, and finally for the full-scale design.

The link between these various stages is the mathematical model of the reactor. A model built up from data from chemical kinetic measurement experiments fitted to other laboratory total reactor performance experimental results can be used to design the pilot plant reactor and to compare pilot plant operation with laboratory operation. When comparisons are not satisfactory, model improvements and the hypothesising that preceeds them can add to the understanding of the system. When such comparisons are satisfactory the model can be used with justified confidence to design the full-scale reactor.

Hence reactor modelling can be the quantitative link between the various stages of reactor development. In addition, it can be used to plan useful experiments, as will be shown in the next chapter, and it also provides an early quantitative picture that can be used to predict the full-scale design and total manufacturing costs which can be of use in overall guidance of the project.

This chapter closes with two reactor development case studies in which reaction modelling played an important role.

3.6.a The development of an acrolein process

Acrolein can be produced by the partial oxidation of propylene over a catalyst made from a complex mixture of metal oxides. Since acrolein is a very reactive compound, it has many potential uses as a chemical intermediate if it can be produced at a low enough cost. Most published catalysts have selectivities of only around 35%. The reactor development project to be described had as its objective the development of an improved catalyst that would result in an economic process.

The main reaction is

$$C_3H_6 \; + \; O_2 \longrightarrow CH_2{=}CH{-}\underset{H}{C}{=}O \; + \; H_2O$$

but this is accompanied by a set of burning reactions in which CO and CO_2 are produced.

The model development for this reaction is given as an example in section 2.2. Figure 3.23 shows a typical concentration/time curve for the reactants and products.

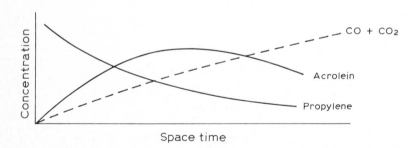

Fig. 3.23. Concentration vs contact time in the acrolein reactor.

The catalysts tested were various mixtures of four different metal oxides, and the operating conditions which would be changed were temperature, space, time and feed composition (C_3H_6, O_2, inert), giving a total of 8 independent variables. For each catalyst, the four operating condition variables had to be investigated to find the optimum catalyst performance before the different catalysts could be compared.

The inflammability characteristics of the reaction mixture meant that two operating regions came in question: fuel weak (<2 v/v % C_3H_6) or fuel rich

(>11% C_3H_6) mixtures.

There were the possibilities of operating the process under high or low pressure, and pressures up to 6 bar are considered practical. Also possible was the recycle of the reactor effluent after removal of product. This suggested two possible modes of operation; a high conversion, low selective reaction with no recycle, or a low conversion, high selectivity reaction with recycling of unreacted propylene by using pure O_2 and purging (see Fig. 3.24).

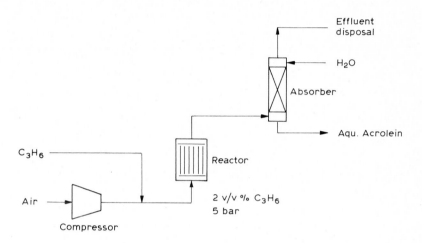

(a) Low concentration, high pressure, no recycle.

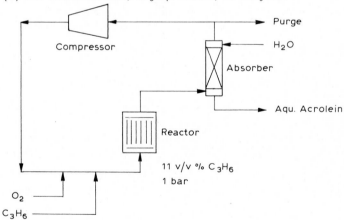

(b) High concentration, low pressure, with recycle.

Fig. 3.24. Two possible acrolein processes.

Such a wealth of possibilities on top of the large number of variables to be examined presented a problem which was virtually impossible to solve by an experimentalist, and so even at this stage of catalyst screening an engineer could be a great help in evaluating the main process alternatives, and indicating the more economically attractive alternatives on which the experimentalist could then concentrate.

An early engineering study of the various possibilities showed two attractive areas. A low fuel process (2% propylene) was only attractive at high pressures to reduce equipment size, and such a process should work without recycle; hence a catalyst with a high conversion to acrolein would be required. A high fuel process would be attractive only with recycle, but pressure operation was not necessary to give an economic process. The recycle of propylene directed the importance on to a good selectivity to acrolein and the catalyst required for this type of process demanded a high selectivity rather than high conversion. To reduce the number of alternatives by such guidelines was a great help in reducing the experimental search to tractable proportions.

The large number of variables associated with the reactor operation posed problems in experimental design and interpretation of results. To assist in this area, modelling work was initiated and those catalysts showing promise in the screening studies were taken for more detailed investigation and fitted by models. The emphasis on the more detailed investigation lay with work in a tubular laboratory reactor, but some back-up work was also carried out on a spinning basket reactor. The spinning basket reactor produced kinetic data; in particular the orders of reaction with respect to the various components and reactions. Although there was poor agreement on reaction rate comparisons between the tube and SGS reactor, the orders of reaction from the SGS reactor were successfully transferred to the tubular reactor model, which made the integral reactor modelling much simpler.

The integral reactor results were fitted by the model described in section 2.4.b. In particular, exit concentrations of acrolein, CO and CO_2 and temperature profiles down the tube were fitted. The temperature profile was particularly informative, because to obtain such a sharp initial peak as the measured data required a high initial heat liberation, which indicated the order of reaction since this has the greatest effect on the shape of the temperature peaks.

Once an approximate model was fitted to selected results, the remaining experimental results could be compared with the model, and their agreement or disagreement with the predicted results gave the necessary opportunity to draw conclusions from them. Without the model, scores of results were being obtained with no effective means of interpreting them or bringing them to a common basis.

The model showed systematic error in that temperature effects could not be

predicted by any reaction scheme obeying the Arrhenius equation. Figure 3.25
shows that either high or low temperature ranges of the curves could be fitted by
different sets of parameter values but no single set of parameter values could fit
the whole curve.

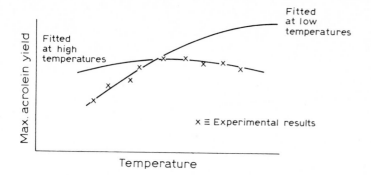

Fig.3.25. Experimental and predicted conversions to acrolein.

Low temperature fits indicated a higher activation energy and high temperature
fits a lower activation energy. This apparent change in activation energy is typical
of processes involving diffusion steps since diffusion is more controlling as the
temperature increases.

Since the model predictions gave 10% higher conversions to acrolein at high
temperatures compared with the best achieved experimental conversion, the diffusion
hypothesis was tested by reducing the catalyst size. This gave an improvement in
the direction indicated in the model, and so further experimental work was devoted
to modifying the catalyst carrier to increase its pore size, and less emphasis was
put on active metal oxide concentrations.

A thorough economic analysis during the laboratory development phase showed
that the catalyst life was economically very significant if it was less than 500
hours, so much so that the process would probably be uneconomic unless this lifetime could be achieved. This again caused a review of the experiment planning in
that lifetime experiments were given higher priority than improved conversion
experiments. The objective of the lifetime experiments was to confirm that the
catalyst would have a lifetime of at least 500 hours. In the event it was found
that this could not be achieved, which again caused an abrupt change in the process
development.

When at a later stage with a new catalyst the project moved on to a pilot plant
scale, a 40 tube pilot plant reactor 3 metres long was designed using the model
developed from the laboratory tubular reactor only 30 cm long. The simple heat-
transfer model given in section 2.4.b was thought adequate to predict the new

temperature profile as a result of the new gas flow / tube. Such a simple treatment of heat transfer was, however, thought to be too empirical to predict the behaviour of tubes with larger diameters, and changes to larger diameters were not contemplated.

3.6.b   The development of a pyridine process

Pyridine has been traditionally extracted from coal tar and this source satisfied the market until the development of pyridine herbicides caused demand to exceed supply from natural sources.

This led to efforts to find an economic synthetic route to pyridine, and one process investigated was the dimerisation of acrolein in the presence of ammonia and oxygen.

$$
\begin{array}{c}
\underset{\substack{O \\ \parallel \\ CH \\ | \\ CH \\ \parallel \\ CH_2}}{} + \underset{\substack{CH_2 \\ \parallel \\ CH \\ | \\ CH \\ \parallel \\ O}}{} + NH_3 + O_2 \xrightarrow[\text{cat.}]{400\,°C}
\end{array}
\quad
\text{pyridine} + H_2O + CO_2 + C + \text{polymer}
$$

Such a complex reaction cannot be expected to be a clean high yield reaction, and in fact the catalyst quickly was covered in "coke" which causes activity to fall after a few hours.

Laboratory work was carried out in a tubular reactor, varying acrolein, NH_3, and O_2 concentrations, temperature, and space time. After 24 hours the catalyst activity had so decayed that it had to be regenerated by burning off the coke in a gas stream with a given oxygen concentration and flow rate at a certain bed temperature. The resulting regenerated catalyst had a changed activity related to the regeneration conditions. In particular an error in regeneration could cause catalyst sintering and a sharp drop in activity. Experimentation with regenerated catalyst was considered essential as fresh catalyst showed untypical behaviour.

Such a situation put the experimentalist in a serious dilemma. With so many experimental conditions to define, modelling did provide a support on which he could compare all the reactor results he obtained. Particularly difficult was the regeneration phase, because the "coked" catalyst was the result of 24 hours of experiment, and this gave a limited number of opportunities in which to vary concentrations, flows, temperatures and their profiles with time. In addition, experiments had to be conservative to prevent the chance of sintering and so destroying important catalyst lifetime results.

In this situation modelling was helpful in that a completely theoretical fixed bed regeneration model was developed from oxidation kinetics found in the literature for petroleum cracking catalysts, and experimentally measured heat-transfer data were obtained by the temperature profiles obtained when a cold gas was passed through the hot reactor.

This model, as already mentioned in section 2.4.c, was programmed on an interactive computer with the facility to modify feeds and temperatures at frequent time intervals. The model simulated the regeneration and printed reactor temperatures bed coke concentrations, and exit gas compositions, analogous to a print out obtained in a plant control room. This enabled the experimentalist to simulate many possible regenerations, e.g. oxygen limited regeneration or oxygen excess, low temperature regeneration, and obtain a real "feel" for the problem before he carried out actual regenerations with the coked reactor.

As the project developed, a fluid bed reactor system was decided upon with separate reactor and regenerator beds. Such a system is difficult to construct and operate continuously on a small laboratory scale, and so a tubular laboratory fluid bed was operated in batch mode: a reacting period, followed by a regenerating period.

Models were fitted to these laboratory results which defined the coke lay-down and burn-off rates, and these kinetic models were then used in a model of a continuous full-scale reactor-regenerator system (see Chapter 9). This enabled the heat balance of such a system, and the effect of a catalyst recycle rate, operating temperature, and reactor size to be predicted for the full-scale system from the batch laboratory results.

Such a study does not dispense with the need for pilot plant work with a continuous system, but it does give a very early indication of the design of a full-scale plant, based on results obtained from a very different-looking type of laboratory reactor.

It is hoped that these two examples give some indication of the types of problems the engineer is confronted with in reactor development and show the need for some form of systematic experimental design, early engineering appraisal and quantitative mathematical modelling as a basis on which to correlate experimental results and predict the effect on the industrial scale.

Following a successful laboratory development study comes the pilot plant study. This is the subject of a following chapter.

FURTHER READING

Danckwerts, P.V., 1970. Gas-Liquid Reactions. McGraw-Hill. - Good survey of
    laboratory methods for these reactions.
Hill, C.G., 1977. An Introduction to Chemical Engineering Kinetics and Reactor
    Design. Wiley, New York.
Horak, J. and Pasek, J., 1978. Design of Industrial Chemical Reactors from
    Laboratory Data. Heyden, London. - Useful selection on laboratory methods.
Rase, H.F., 1977. Chemical Reactor Design for Process Plants. Wiley, New York.
    - Good description of a number of laboratory methods.

112

REFERENCES

Regenass, W., 1978. Am. Chem. Soc. Symposium Series No. 65. Chemical Reaction
   Engineering, Houston, Texas.
Stonhill, L.G., 1959. J. Inorg. Nucl. Chem., 10, 153.

QUESTIONS TO CHAPTER 3

(1) What is process development and what are the objectives of the reactor part of
    a process development programme ?
(2) What are the principal categories of laboratory reactors available for a
    process development study ?
(3) What are the advantages and disadvantages of using CSTR reactors in process
    development studies ?
    What is a micro-reactor ?
(4) What are the principal differences in using an SGS and an integral reactor for
    studies on a gas-phase solid catalyst reaction ?
(5) For the reaction

$$ \underset{\text{liquid}}{A} \quad + \quad \underset{\text{gas}}{B} \quad \xrightarrow{\quad k_1 \quad} \quad \underset{\substack{\text{liquid} \\ \text{product}}}{P} \quad + \quad \underset{\text{gas}}{R} $$

with side reaction

$$ \underset{}{P} \quad + \quad \underset{\text{gas}}{B} \quad \xrightarrow{\quad k_2 \quad} \quad \underset{\text{by-product}}{S} \quad + \quad R $$

what laboratory equipment is available to determine the reaction rate for this
reaction ?
What laboratory equipment can be used to simulate the full-scale reactor ?
How can the results from these two forms of reactor be compared ?
(6) How can reactor modelling be usefully employed in a process development study ?

Chapter 4

THE PLANNING OF EXPERIMENTS

The first important step in any reactor study is experimental. Experiments are performed to find economically attractive operating conditions, to understand the mechanism of the reaction, and to provide rate data for the full-scale design. Reactor systems are complex, with ten independent variables being not uncommon. The experimenter is often perplexed, although he may not admit it. What experiments should be performed in order to obtain the desired information as efficiently as possible ?

The experimenter also has the disadvantage of not being the end user of the data, and so he has no feel for relative priorities. How accurate the data need be for design, or when the "best" operating conditions have been found, (based on process economics, not reactor selectivity) are not the subject areas of the experimenter.

The chemical engineer can be of great assistance in such situations because he will be the end-user of the data and so he can set priorities, and decide when adequate results have been obtained.

In addition to such qualitative advice, the engineer is normally numerate enough to suggest that some experimental design is brought into the planning of the experimental work. Methods are available for the selection of sets of experiments which are best for obtaining specific information, and it is sensible for such techniques to be applied wherever possible.

4.1   GENERAL CONSIDERATIONS

The extraction of useful information from the results of a set of experiments is complicated by a number of tedious problems. Firstly, all experimental data contain experimental error, so information must be available on the magnitude of this error before one can conclude that measured differences between experiments are significant, and not just variations due to random experimental error. Secondly, drifts can occur with time; for example, catalysts decay, and such trends can mask other effects being investigated. Thirdly, step changes in experimental results can occur when new batches of raw materials are used or changes in experimental techniques are inadvertently made.

Experimental design techniques are systematic methods of overcoming effects introduced by the above problems, and at the same time they ensure that reactions are carried out at points which will yield maximum information.

As an insurance against the above three problems obscuring the interpretation of the results there are three ground rules which can be applied whatever experimental plan is employed.

(1) Before serious experimental series are attempted a number (e.g. 6-12) of repeated experiments should be carried out, which both check the functioning of the equipment and give an estimate of the experimental error that can be expected.

(2) When there is any chance of trends with time occurring, experiments should be "randomized" so that no single independent variable is changed systematically with time. This would mean that the effects of the variable and the effect of time would not be separable. When trends are suspected, periodic repeats of a standard experiment are advisable.

(3) When step changes are suspected (block effects), every step should be considered to be a pseudo-independent variable - the "blocking variable". If the following analysis shows this variable to be significant, the presence of a blocking effect is confirmed. In addition, the effect has not masked the effect of changes in other variables.

4.2   SCREENING EXPERIMENTS

4.2.a   Planning

Early in reactor development work the question to be answered is which conditions have a considerable influence on the system and which have a smaller, or negligible effect. For example, say an imaginative chemist has named eight possibly significant variables in the system. To vary each one over its range with, say, six experiments needs $6 \times 8 = 48$ experiments. Even this number of experiments would still give no information of interactions, i.e. the improvements that occur only when two or more variables are moved simultaneously. To fully cover this possibility needs a total of 6^8 experiments.

For screening experiments it is sufficient to have experiments at only two levels (denoted as +1 and -1) which are far enough apart for resulting differences to be significantly greater than experimental error. Significant interactive effects can be detected by a complete set of all possible combinations of the two levels. This is called a complete two-level experimental design and it is planned simply by combining all possibilities in the systematic way shown in Table 4.1 for two, three, and four independent variables.

However, such a design would still require 2^8 experiments to carry out an initial screening on a system with eight variables - still an unacceptably high number.

These "full" experimental designs cover all possible interactions which are first-order with respect to each variable (see equation 4.1). Hence the eight-variable system will carry out experiments to determine second-, third-, through to seventh- and eighth-order interactions. This is rather overdone for a set of screening experiments where main effects and possibly second-order interactions are the most likely to be significant.

Thus, effects can be identified with much reduced designs. In principle, only $n + 1$ experiments are necessary to identify the first-order effects of n parameters

TABLE 4.1

Full factorial designs for two, three, and four independent variables (2^2, 2^3 and 2^4 designs)

	x_1	x_2		
1	+1	+1		
2	-1	+1		
3	+1	-1		
4	-1	-1		

	x_1	x_2	x_3	
1	+1	+1	+1	
2	-1	+1	+1	
3	+1	-1	+1	
4	-1	-1	+1	
5	+1	+1	-1	
6	-1	+1	-1	
7	+1	-1	-1	
8	-1	-1	-1	

	x_1	x_2	x_3	x_4
1	+1	+1	+1	+1
2	-1	+1	+1	+1
3	+1	-1	+1	+1
4	-1	-1	+1	+1
5	+1	+1	-1	+1
6	-1	+1	-1	+1
7	+1	-1	-1	+1
8	-1	-1	-1	+1
9	+1	+1	+1	-1
10	-1	+1	+1	-1
11	+1	-1	+1	-1
12	-1	-1	+1	-1
13	+1	+1	-1	-1
14	-1	+1	-1	-1
15	+1	-1	-1	-1
16	-1	-1	-1	-1

assuming interactions are absent, and

$$1 + n + \frac{n(n-1)}{2}$$

experiments can estimate main effects and second-order interactions, assuming higher-order interactions do not exist.

Experimental plans involving incomplete interaction information are called Partial Factorial Designs, and these designs are useful because they can quickly give an indication of major effects with the minimum of experimentation.

Consider that we wish to carry out two-level screening experiments on a system containing seven variables and we believe interactions to be of less importance than the effect of changing the variables separately. Hence, a total of $n + 1 = 8$ experiments are necessary. The design is called a 2^{7-4} design and the resulting experimental plan is given by Table 4.2.

TABLE 4.2
Design and regression matrix for a 2^3 full factorial design or partial 2^{7-4} design

Experiment number	θ_1	θ_2	θ_3	θ_{12}	θ_{13}	θ_{23}	θ_{123}
	x_1	x_2	x_3	x_4	x_5	x_6	x_7
1	-1	-1	-1	+1	+1	+1	-1
2	+1	-1	-1	-1	-1	+1	+1
3	-1	+1	-1	-1	+1	-1	+1
4	+1	+1	-1	+1	-1	-1	-1
5	-1	-1	+1	+1	-1	-1	+1
6	+1	-1	+1	-1	+1	-1	-1
7	-1	+1	+1	-1	-1	+1	-1
8	+1	+1	+1	+1	+1	+1	+1

2^3 design matrix

2^3 regression matrix or a 2^{7-4} design matrix

This table is built up in the following way. Eight experiments are necessary and so they can be defined by 2^3 design. The 2^3 design has three main effects and four interactions (three second-order and one third-order). In place of estimating these interactions, these effects are to be replaced by four of the single-variable effects of the seven-variable system.

In the first experiment the interactive term x_1x_2 will be +1, since the combination of x_1 and x_2 (both negative) is positive, similarly x_1x_3 interaction is +1, x_2x_3 is +1, but $x_1x_2x_3$ is -1. These four levels are entered in Table 4.2 to the right of the main effects. When this procedure is repeated for the whole table, the result is the partial 2^{7-4} design, and the interactive terms are replaced by the variables x_4 to x_7, as shown in the table. Carrying out these eight experiments at these levels will produce results which, when analyzed, will identify the seven main variable effects, as long as interactive effects are absent or negligible.

If some interactive effects are present, the position is somewhat confusing. In the 2^{7-4} design we have shown that x_4 is taking the place of x_1x_2. If there is an x_1x_2 effect, then we are really measuring the effect of $x_4 + x_1x_2$. There is no way of really knowing how much of each is contributing to the observed effect. In fact, the story is even worse, because there are more interactions of the seven variables that produce the same pattern as x_4 and x_1x_2. The total set is called the "alias pattern" and it can be identified by the following algorithm:

(1) All the "planning generators" (I) are found from the chosen mixture of main and higher-order effects. In our example there are four:

$$12 \equiv 4 \qquad \therefore \qquad I = 124$$
$$13 \equiv 5 \qquad \therefore \qquad I = 135$$
$$23 \equiv 6 \qquad \therefore \qquad I = 236$$
$$123 \equiv 7 \qquad \therefore \qquad I = 1237$$

(2) With each planning generator the interaction of main effects can be determined by multiplying each side of the above equation by each variable in turn. When a variable is present twice it is cancelled. From the first planning generator:

$$1 = 1124 \quad = \quad 24$$
$$2 = 2124 \quad = \quad 14$$
$$3 = 3124 \quad = \quad 1234$$
$$4 = 4124 \quad = \quad 12$$
$$5 = 5124 \quad = \quad 1245$$
$$6 = 6124 \quad = \quad 1246$$
$$7 = 7124 \quad = \quad 1247$$

This must be repeated for all the planning generators in the design. In one example we have 4 and so 1 is confounded with four second-order interactive terms:

$$24 + 35 + 1236 + 237$$

and variable 2 with

$$14 + 1235 + 36 + 137$$

This treatment of "alias patterns" enables the experimental planner to choose how he presents his variables so that main effects are not confounded with possibly important second-order effects. A good design would choose less important variable interaction (i.e. negligible ones) to be confounded with more important main effects.

It is necessary to entangle alias patterns to be able to <u>completely</u> interpret the result of a set of two-level factorial experiments.

4.2.b   Evaluation of screening experiments

Looking through a complete two-level design it is possible to locate pairs of experiments that differ in only one variable. Hence, the importance of the individual effects can be indicated by finding various sets of these pairs and checking whether the difference in each pair is significantly greater than the experimental error in the system. This approach, however, is far too laborious to be considered a method of analyzing results, but at least it indicates the principles involved.

Since it is a two-level design, only straight-line relationships can be expected

118

from the model, and it is usual to analyze such designs with a model linear in the parameters.

Hence, for three variables, the model fitted is usually:

$$y = \theta_0 + \theta_1 x_1 + \theta_2 x_2 + \theta_3 x_3 + \theta_{12} x_1 x_2 + \theta_{13} x_1 x_3 + \theta_{123} x_1 x_2 x_3 + e \qquad (4.1)$$

For such a model the analysis is very straightforward. The set of experimental conditions for the variables (the 2^3 regression matrix of Table 4.2) is the X-matrix of linear regression (section 2.6), and the set of experimental results is the Y-matrix. Hence, expression 2.69 can directly evaluate the seven θ-values in the model.

The +1 and -1 levels of the independent variables give the design special properties. The design is described as being "orthogonal", because it is symmetrical in all the independent variables. In the analysis this means that the off-diagonal terms of the $[X^T X]^{-1}$ matrix are zero (see section 2.6.e) and so the correlation coefficients are zero. That is, for linear models, such a design is the best possible one for separating the effect of the individual parameters.

When the +1 and -1 levels are exact, and when the design is complete (i.e. no missing experiments), the analysis of the results takes a particularly simple form:

$$\theta_0 = \frac{\sum_{i=1}^{n} y_i}{n} \qquad (4.2)$$

and

$$\theta_j = \frac{\sum_{i=1}^{n} (x_{ji} y_i)}{n} \qquad (4.3)$$

These equations can be derived from the diagonal $[X^T X]^{-1}$ matrix, but they are also the logical extension of the simple method initially described at the beginning of this section.

In practice, the experimenter has usually not been able to keep his independent variables exactly at +1 and -1 levels for all experiments, and probably cannot present a complete set. Hence, expressions 4.2 and 4.3 cannot usually be used, and a linear regression program is normally employed to evaluate the θ-values.

Partial factorial designs are not readily accepted by experimenters because they go against the old rule "never change two variables at once" when carrying out experiments. Table 4.2 shows that in the partial 2^{7-4} design there are no pairs of experiments where only one variable has changed, as we found for the full factorial. The experimenter is worried because an inspection of the results cannot indicate major effects; these can only be done by disentangling the mixture of effects by a linear regression. The results of such a regression produce a set of θs for the main variables, but it must be remembered that these are confounded with second-

and higher-order effects, according to the alias patterns described earlier.

The object of carrying out the screening experiments is to determine which effects are significant. In our linear model (equation 4.1), this means those θ-values which are significantly greater than zero.

The linear regression will give this information because it gives the standard error of the parameters. The parameter values should be at least three times greater than the standard error for the parameters to be "significantly" greater than zero (the statistical t-test).

Alternatively, it is possible to determine the significance of the parameters by comparison with the experimental error determined from repeated experiments (conveniently done at the mid-point of the design, i.e. x_1 to x_n = 0). The normalized parameter values (i.e. with x = +1 or -1) should be at least three times greater than the measured experimental error, for the parameters to be significantly different from zero.

4.2.c   Problems associated with screening experiments

The major problems associated with two-level factorial designs arise because a linear model is being fitted to a non-linear system.

To represent the more complex systems, the interactive terms are introduced into the model. These terms take account of effects that appear only when two variables are altered simultaneously. They are unsatisfactory in that the interpretation of the meaning of the interactive effect is difficult, and their relation to a mechanistic model is almost impossible. Besides their interpretation being difficult, the determination of interactive terms causes an exponential rise in the number of parameters to fit, resulting in factorial designs requiring large numbers of experiments to be carried out.

In practice, the partial factorial designs are most useful in that they indicate that there are effects which could be main effects, or at least particular interactions. If one proceeds on the assumption that interactive effects are less likely than main effects, then the partial factorial designs yield analyzable results. If later effects cannot be properly explained, this may be because some interactive terms should not have been neglected, and the work must go back to an earlier point to analyze for the more likely interactions.

Screening experimental work is detective work; one follows clues, following the most likely paths first. It cannot be considered to be a very precise science, where the analysis produces an indisputable result as is usual with most mathematical techniques.

After all, a completely false model is being fitted, using minimum experimental data to get information on a large number of variables. As soon as the significant variables have been located, the problem becomes more tractable, and the experimental work should move on from the screening experiments to a more scientific treatment involving non-linear models aimed at describing the system in question, and the experiments planned with specific purposes in mind.

4.3   EXPERIMENTAL DESIGN TECHNIQUES AFTER THE SCREENING STAGE

The main objective in a process development laboratory is the search for "optimal" reaction conditions to maximize the yield and selectivity from the reactor, because the process costs normally are predominantly raw material costs. Secondary aims are to know the behaviour of the reaction in the region of optimum performance, and to understand the reaction well enough to be able to scale-up with confidence. This latter point can be done scientifically by defining a reactor model which fits the reactor performance over a fairly wide range of conditions.

Additionally, many specific experiments are necessary, such as catalyst life work, effects of poisons, effect of feed quality, corrosion data, and effects of recycling. Such experiments are process-dependent and are investigated by experiments specifically designed to clarify the point in question. However, the first two aims (the reactor optimization and model identification) are so general that it is worth describing generalized experimental planning methods that are available for them.

4.3.a   Experimental design methods for location of optima

Response surface methods. A reactor is a highly non-linear system, which has a yield, selectivity, or profit function which will be non-flat, and which will contain an optimum. Generally, optimum yield is a good first estimate of a profitable operating point for an industrial reactor, or, if an inexpensive recycle system is involved, the optimum selectivity may be preferred. Only at a more advanced stage of the process development need these economic indicators be replaced by an estimate of the cost of a total process as a function of the reactor operating conditions. The actual objective function used in the optimization is of secondary importance in experimental design as long as it can be calculated from the experimental results.

The problem can be summarized as the location of the position of the optimum on an unknown surface, when the responses that define the surface are subject to experimental error. The presence of this random component causes numerical optimization algorithms to fail, since such algorithms are based on the assumption that all responses are accurate.

What are necessary, are optimization procedures which handle random errors in the responses. To do this the methods developed introduce statistics and linear regression to determine which effects are significantly greater than the experimental errors, and then only these effects are used in the optimization search.

The Box-Wilson method. This method employs two-level partial factorial experimental designs to locate significant parameters and these parameters are then used to choose the direction in which to move the experimental conditions. A series of experiments is then made in the chosen direction until an optimum has been achieved.

Because the model used does not represent the whole system, but only a linearized fit about the experimental point, this new "optimum" point will have a completely new set of linear parameters which describe the surface local to it.

It is therefore necessary to repeat the partial factorial design at this new point to determine the new significant parameters, and the new direction in which to move. Experiments are carried out in this direction until a new optimum is found, and this whole procedure is then repeated.

In the neighbourhood of a true optimum the linear regression indicates that no variables are significant, and the mid-point $(0,0,...,0)$ experimental results differ from the mean of the other experiments. When this occurs the linear model is proving incapable of representing the curved non-linear surface, and then the experimental design has to be expanded to a three-level design and a quadratic model fitted.

Once a quadratic model is fitted, the optimum point on this quadratic surface can be determined by differentiation.

In more detail, the Box-Wilson method is composed of the following steps:

(1) Choose suitable +1 and -1 levels for the significant variables in the system, and carry out a partial two-level factorial design. In addition, carry out a number of mid-point experiments to determine experimental error.

(2) Fit a linear model to the results and determine the significant parameters by comparison with the experimental error. If the significant parameters are likely to be attributable to interactive effects confounded with the main variables, a more complete design must be carried out by going back to (1) to carry out a further part of the total design that will separate the important interactions. Whether interactions are important or not is purely a matter of judgement; no tests can be carried out. The best advice is probably to assume that interactive effects are negligible, unless there is strong reason for thinking otherwise.

Assuming that the significant variables can be identified from significant parameters, proceed to step (3). If no parameters appear significant, or if the model is distinctly curved, as shown by comparing the mean of the mid-point experiments with the mean of the designed experiments, a linear model is inadequate, and a quadratic design with more experiments is necessary. Go to point (4).

(3) Move from the centre-point of the design in a series of experimental steps in the direction of steepest ascent, until an optimum is found. For example, when three variables are significant with parameter values θ_1, θ_2, and θ_3, the direction given by the following steps:

x_1 length $\theta_1/\Sigma\theta_i$

x_2 length $\theta_2/\Sigma\theta_i$ where i = 1 to 3

x_3 length $\theta_3/\Sigma\theta_i$

is the direction of steepest ascent. At the optimum point, carry out a new fractional design, according to step (1).

(4) If the surface cannot be represented by a linear model, further experiments should be carried out, and a quadratic model fitted:

$$y = \theta_0 + \theta_1 x_1 + \theta_2 x_2 + \theta_3 x_3 + \theta_{11} x_1^2 + \theta_{22} x_2^2 + \theta_{33} x_3^2 \qquad (4.4)$$

The optimum point on the surface can then be located by differentiation:

$$\frac{dy}{dx_i} = 0 = \theta_i + 2\,\theta_{ii}\,x_i^* \qquad (4.5)$$

$$x_i^* = \frac{-\theta_i}{2\,\theta_{ii}} \qquad (4.6)$$

Repeated experiments at point x_i^* and an experimental sensitivity analysis about this point should be carried out to confirm the prediction of the quadratic mode The additional experimental points to be introduced in order to use the quadrati model should be the so-called "star points" to obtain maximum information from the model by keeping the design orthogonal. The star points are located on each axis at approximately +1.5 and -1.5, with all the other variables being 0.0. Further details on optimum position and number of points for quadratic designs are given by Box (1978).

The Box-Wilson method finds considerable application in process development projects for the location of optimum reaction conditions, and for this reason it is explained in some detail here. It does, however, have a number of serious defects which must be noted.

o The +1 and -1 levels of the independent variable settings must be chosen very carefully. When the step is too small, the experimental error can so dominate the results that no conclusions concerning significance of the variables can be drawn. If the step is too large, the linear model becomes unsatisfactory. For instance, a step might straddle the optimum value of a variable; the linear analysis would then show this variable to be insignificant, but, in fact, it could be the most important variable in the system!

o When to include or exclude the interactive effects is not well defined. To always carry out a full design leads to vast experimental programmes, but to assume that interactions are negligible leads to a search direction which may be quite false because an interactive term has been attributed to a completely different main variable.

o The final step with the quadratic model and the location of the optimum by differentiation appears in practice not to be without problems. With a model with so many parameters, good orthogonal designs and a considerable number of results are needed to ensure that the parameters are not correlated. Correlated parameter would predict optima in a purely random fashion. Secondly, the real system is not a quadratic model, linear in parameters. Thus, systematic error will result in a discrepancy between the real system optimum and that predicted by the quadratic model.

In general, the Box-Wilson approach needs very many experiments and the experimental results cannot be re-used because of the local nature of the fitting. Furthermore, the model uses no information on the form of equations that are known to hold approximately for the system under investigation. In many systems (e.g. biological or medical systems) it can be classed as an advantage that the method requires no prior knowledge of the form of the model, but in chemical systems a rough form of model is generally known and to work without it is not properly using all the information available.

The method has a number of attractions and this explains why it is fairly well-established in research departments.

o Firstly, the procedure always results in an improvement, even though it might not find the best operating conditions.

o Secondly, it is a systematic algorithm which is of great value in a complex situation.

o Thirdly, there is no proven alternative available.

EV-OP methods. EV-OP (Evolution Operation) is a method for optimization on pilot- and industrial-scale reactors where it can be a method of improving performance of operating plant. It is not suitable for laboratory experimental work.

The method was developed primarily for use on continuously operating production plant where experimental work in the normal sense means interruption of production, and yet there is need to modify the operating conditions as external influences (e.g. raw material quality or catalyst decay) shift the optimum operating point.

An operating plant is a source of vast quantities of operating data, but its experimental potential is limited because this means operating at sub-optimal conditions to produce information, which is defeating the object of maximizing the profitability of the plant.

Therefore, experimental results should involve as little deviation from the as yet considered "best" operating point as possible. The limit to the smallness of this "perturbation" is when the effects are completely masked by experimental error. However, by repeating small perturbations sufficiently, and fitting a localized linear model about the operating point, it becomes possible to detect trends even when the perturbations are very small. Hence, the plant is yielding information despite the smallness of the changes. When enough cycles have been done, and clear information is available, a step is taken in the direction of steepest ascent, and the process repeated.

Such a process slowly moves towards an optimum and if the optimum drifts with time, the procedure should follow such drifts, as long as the time constants for the two operations are suitably different.

There are a number of variations on the method. The original method proposed by Box suggests a factorial experimental design as the set of perturbations to be carried out. Spendley, however, proposes that this be replaced by the simplex algorithm where the best n + 1 results are taken for a system with n independent variables and the next point attempted is the reflection of the worst of these points in multi-dimensional space. If this point is an improvement on the worst point, then this worst point is discarded and the new worst point chosen from the remaining n + 1, and the process repeated. Figure 4.1 shows the process on a two-dimensional system. This is the same algorithm as used in the simplex numerical non-linear optimization procedure. Again, before any decision to drop a point is made the perturbations have to be repeated sufficiently frequently to show the trend clearly above experimental error.

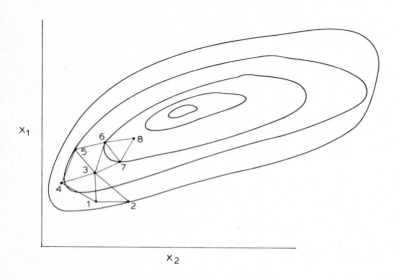

Fig. 4.1. The simplex algorithm in EV-OP.

The EV-OP method requires continuous management of operating conditions to systematically perturb the variables, the collection of a great deal of experimental results, and the continuous evaluation of these results. The method is, therefore, really only feasible for processes which have a process computer installed with capacity to carry out the control, regression, and selection of new operating points automatically. In addition, the process must be controlled, so that the independent variables can be adjusted directly by the computer and, the most difficult constraint of all, the plant performance, if possible expressed in economic terms, must be measurable and transmitted to the computer.

Because of all these constraints, the reported use of EV-OP on plants is very rare. However, with computer-controlled processes becoming common place, and with improvements in on-line analysis, it could well be that the method finds increasing application.

Mechanistic modelling methods. As mentioned as a disadvantage of the Box-Wilson method, there is a complete loss of information when the mechanistic form of the model must be replaced by a purely empirical linear or quadratic model. Usually some equation can be written down for the reaction which will approximately represent the system as a semi-empirical model, as described in Chapter 2. Once such a model has parameters fitted to it, it can be used to find an optimum by an exhaustive numerical search on the computer. Attractive regions can then be pinpointed, and laboratory work directed towards them.

When modelling work is proceeding in parallel with process development, as a model is fitted to early results it can be used to define interesting areas. As more results become available, the model improves and the optimum will be more precisely located.

In practice, this method has the disadvantage that by the time the model has been developed, and fitted to experimental results, the experimenter may have already located the optimum by laborious experimentation, together with a little help from his own intuition. Usually, the first priority is to look for the optimum, and only later is the model developed. To use a mechanistic model to find the optimum needs the model development and optimum search to be combined. This will be discussed in detail in section 4.4.

4.3.b   Experimental design techniques to define the mechanistic model

A mechanistic model grows out of a discussion with chemists on the possible reaction mechanism and sources of side reactions, and an understanding of the physical processes of heat and mass transfer occurring. The rough first model must be tested with experimental results to estimate the adequacy of the model and to locate systematic errors. A first set of experiments roughly scattered over the area of interest is a good start, and then as systematic model errors are discovered, more precise experiments are necessary to locate the source of these systematic errors.

Here precise questions should be formulated and the appropriate experiments performed:

o Does the model predict trends correctly at extreme conditions ?
   This leads to a set of experiments that the experimenter may well never make
   without its being requested by the modeller.
o Does the model predict individual effects (e.g. pressure) well ?
   This leads to a series of experiments at different settings for one variable
   holding all the others constant (a complete contradiction to the principle of
   the screening experimental design).

The location of systematic error demands a comparatively large number of experimental results to be available. Models can be fitted to small numbers of results and give a satisfactory fit, because all systematic error concentrates in regions where results are not available.

The recognition of systematic error is also not easy. One can use the scatter diagram provided by various regression programs, but the most satisfactory way is to plot the results out as experimental points and predicted curves, and then carefully study them.

Once the systematic error has been located, the model equations must be revised to try to reduce it. In this way, a better understanding of the reactor results, and a better fitting model are obtained. The procedure of proposing and checking out better models should stop as soon as a model of adequate accuracy for its purpose has been obtained. Usually too much time is wasted getting a model that is unnecessarily accurate. In particular, the challenge to get a model as accurate as the experimental results must be resisted, even though it is this model that is demanded by all statistical texts!

Hence, the experimental planning for model definition and improvement are non-systematic procedures which rely much on individual flair and intuition. This is brought about because of the nature of systematic error, for which there is no systematic treatment.

There are, however, non-linear experimental planning techniques available that can be used if the correct form of a model is known. In this case all that is necessary to define the model is to have the value of the parameters. Hence, a good experimental design technique will be one that determines the value of the parameter as accurately as possible with the minimum of experiments.

If there is no experimental error, then any n random experiments could define the parameters accurately in an n-parameter model. It is the presence of experimental error that brings in the need to choose experimental regions carefully, so that the interpretation of the results is not masked by experimental error, and the value of the individual parameters can be separated.

If we return to the theory of non-linear regression in Chapter 2, we see that the confidence limits for the parameters are given by

$$\text{Var}(\theta) = [X^T X]^{-1} \sigma^2 \tag{4.7}$$

where, because we are dealing with a linearized non-linear system

$$X = \begin{bmatrix} \dfrac{\partial f_1}{\partial \theta_1} & \dfrac{\partial f_1}{\partial \theta_2} & \cdots\cdots & \dfrac{\partial f_1}{\partial \theta_m} \\[2ex] \dfrac{\partial f_2}{\partial \theta_1} & \dfrac{\partial f_2}{\partial \theta_2} & \cdots\cdots & \vdots \\[2ex] \dfrac{\partial f_n}{\partial \theta_1} & \dfrac{\partial f_n}{\partial \theta_2} & \cdots\cdots & \dfrac{\partial f_n}{\partial \theta_m} \end{bmatrix}$$

Hence, an orthogonal set of experiments to produce a good set of parameter confidence limits and low correlation requires a good coverage of the $\frac{\partial f}{\partial \theta}$ values. That is, experiments where the dependent variable is sensitive to the individual parameter values. Whereas in linear systems, a good experimental plan covers all combinations of the x values, in non-linear designs all combinations of $\frac{\partial f}{\partial \theta}$ should be planned.

The location of these best experimental points involves a great deal of computing, but it is possible once a correct model and a set of approximate parameter values are available.

If the model can be easily differentiated, the expressions for the elements of the X-matrix are available as equations:

$$\frac{df}{d\theta_i} = f'_i(x_1, \ldots, x_n, \theta_1, \ldots, \theta_n) \qquad (4.9)$$

The best next experiment is that set of x_1 to x_n which gives the lowest values of some criterion derived from the $[X^T X]^{-1}$ matrix when a new matrix is constructed comprised of the existing experiments plus the proposed experiment.

Different elements of the $[X^T X]^{-1}$ matrix estimate the different parameter confidence levels. To combine the confidence levels for all the parameters in one function, the determinant of the $[X^T X]$ matrix is usually chosen because this function appears as the denominator in every parameter confidence limit expression (see section 2.6.e). Hence, these experimental planning techniques generally choose a next experiment which maximizes the value of the determinant of the $[X^T X]$ matrix.

The only way of locating the optimal set of x_1 to x_n is by an n-dimensional non-linear optimization, with the determinant of the $[X^T X]$ matrix as the function to be maximized.

The computational problem is made worse if the model cannot be analytically differentiated to define $\frac{\partial f}{\partial \theta_i}$. In this case, these values must be estimated by numerical differentiation, which means taking differences between pairs of evaluations of the model for each derivative.

The method is one of _sequential_ planning. It plans only one experiment at a time, otherwise the optimization problem would have too great a dimension to be soluble. The surface of the function to be optimized is complex, leading frequently to the identification of local optima and not global optima (see Fig. 4.2). This problem can be overcome at the expense of increased computing by repeated optimizations from different starting points.

The method is impressive; out of a system too complex for the average modeller to comprehend and a set of random experiments already carried out, the method will predict quite reliably where to carry out the next experiment to get most information.

The most sensitive points to experiment are generally at extremes of conditions, well away from experimental regions of interest. So the method directs the

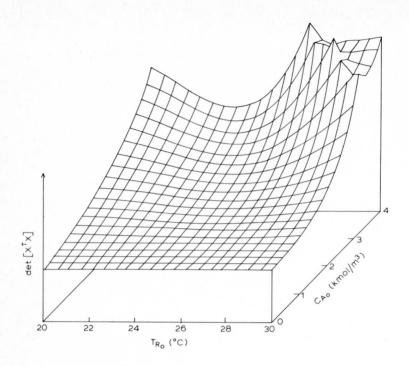

Fig. 4.2. Surface from which the next experiment must be located.

experimental work to these points. There is often one point per parameter, which is understandable when one considers the physical system. After these n experiments have been performed, it is usual that more information is retrieved from repeating these points than by trying new points, because this improves the accuracy at these sensitive points.

Hence these techniques perform a limited number of experiments at points of no operational interest so that they best determine the parameter values in order to result in the most accurate model.

The only flaw in this cool logic is that in reactor modelling one rarely has the correct model, and so this procedure with an incorrect model is concentrating the systematic error in the middle of the experiment area, hence producing a model with minimum reliability in the region of maximum interest.

The method is, however, so promising that it must somehow be adapted for use in reactor engineering. In particular, it always predicts areas where the experiment provides much information, and this is probably true both for models with and without systematic error. It could at least be combined with other experiments to provide part of a reliable technique. This discussion will be continued in the next section, which is a case study describing experience with various non-linear experimental planning algorithms.

The experimental design method just described had the objective of minimizing the confidence region of the parameters. Minor changes allow for other objective functions to be used.

One variant of particular interest is a method of designing experiments with the objective to most accurately predict the performance (y_p) at a particular design point $x_{1p}, \ldots, x_{np}$ (X_p).

As shown by equation 2.77 in Chapter 2:

$$\text{Var } (y_p) = \{X_p^T \ [X^TX]^{-1} \ X_p + 1\} \ \sigma^2 \qquad (4.10)$$

Hence, when a model and some experiments are available, new experiments can be planned to minimize the variance of any particular point by choosing a set of X which minimizes expression 4.10. This is only a minor variation of the principle of minimization of the confidence limits of the parameters by minimizing a criterion based on expression 4.7, but it is simpler in that $\text{var}(y_p)$ is neither a matrix nor a vector.

The method has the potential of carrying out experiments in the area of interest (X_p), and so could potentially overcome the disadvantage of the first method. In practice, it is often dominated by the need to obtain accurate parameters for the model, and so it again carries out experiments at the boundaries. At other times, however, it carries out repeated experiments at the point of interest.

An extension of these methods are the experimental design procedures to predict the best experiment to carry out to distinguish between a set of models, one of which is the correct model. There are a number of methods of doing this which vary in the complexity of the objective function taken.

The simplest method finds the experiment which produces the greatest difference between the model predictions

$$\sum_{j=1}^{m} \sum_{i=1}^{m} (y_i - y_j)^2 \quad \text{for m models}$$

More refined methods use more complex objective functions which enable the probability of each model being correct to be calculated.

As the sequence of experiments proceeds, one model is shown to be more and more probable as the probabilities of the other models being correct diminish.

A problem no workers have yet considered is what exactly is this procedure doing when no model is correct, when all have varying degrees of systematic error.

The model discrimination methods also tend to request experiments to be carried out at specific points on the boundaries (often one point per model pair). Having carried out these experiments once, it is usually requested that they be repeated rather than new areas be entered.

Summarizing this section on statistical methods of non-linear experimental planning, it can be said that:

o The methods are only appropriate when there is no systematic error in the models.

o There are a very limited number of experiments planned, and these give no indication as to the adequacy of the model used.

o The methods achieve their objectives when correct models are available and, probably, predict experiments that are at least useful to carry out even when the model is only approximate.

What is wanted is some experimental technique that takes the best from these statistical methods and modifies it in some way so that it can be used in practical situations for optimization as a replacement for the scarcely satisfactory Box-Wilson method, and for model identification and development work.

4.4 NON-LINEAR EXPERIMENTAL PLANNING - A CASE STUDY

The general message in this chapter is that, after the initial screening stage, there are no completely satisfactory experimental design procedures available to enable systematic experimental programmes to be planned.

The Box-Wilson method requires much experimentation and does not use all the information available, and the non-linear statistical methods demand as a condition of their use that the correct model is known.

In practice, only a roughly correct model is known at the beginning of the work, which may be refined after the experiments have been performed. It is, however, often never sufficiently accurately refined to be an "adequate" model in the statistical sense.

A method robust enough to give assistance even when the model is only approximately correct is necessary.

This problem has been tackled as a research project at E.T.H. Zürich (Schifferli, 1979) and a technique has been developed which attempts to meet these requirements.

Though the results are far from exhaustive, it is at least an attempt at a practical solution, and the work has demonstrated its application to a real experiment. The results of this research work are reported here in some detail; it serves as a good example of the application of the various non-linear design methods and the pitfalls that can result.

The object of the research project was to test various non-linear experimental design procedures experimentally, and to determine their suitability for use in a process development environment. In practice, this means the suitability of them for locating the "optimum" performance of a reactor.

Further to the testing of existing methods, the work was aimed at providing an improved method, should the existing reported methods be shown to be inadequate.

Experimental. The reaction chosen for this study was the hydrolysis of propylene oxide to propylene glycol and diglycol.

$$H_2O \;+\; H_3C-CH\overset{\displaystyle O}{\overbrace{\qquad}}CH_2 \;\xrightarrow{\;H^+\;}\; H_3C-\overset{OH}{\underset{}{C}}H-\overset{OH}{\underset{}{C}}H_2$$

$$\quad W \qquad\qquad\qquad P \qquad\qquad\qquad\qquad\qquad G$$

$$H_3C-CH-CH_2 \;+\; H_2C\overset{\displaystyle}{\underset{O}{\overbrace{\qquad}}}CH-CH_3 \;\xrightarrow{\;H^+\;}\; H_3C-CH-CH_2-O-CH_2-CH-CH_3$$
$$\quad\;\; OH\;\; OH \qquad\qquad\qquad\qquad\qquad\qquad\qquad\;\; OH \qquad\qquad\qquad OH$$

$$\qquad G \qquad\qquad\qquad\qquad P \qquad\qquad\qquad\qquad\qquad\qquad D$$

The reaction was carried out batchwise in a 500 ml glass, stirred reactor. The reactor was computer-controlled with a PDP 11/45 computer with 64 K memory, which controlled the filling, starting, emptying, analysis, and coolant temperature, and recorded the weights of each component, temperatures, and results of the chromato-graphic analysis. The same computer also carried out the non-linear regression on the experimental results, and the non-linear experimental planning optimization to locate the next experimental point.

Hence, in principle, the equipment was capable of carrying out the whole sequential experimental planning procedure automatically. A potential of such an on-line experimental design technique is to create a system which can operate continuously until it automatically drives the reactor so that it is operating at its optimum point.

A comparison of existing algorithms. The equipment was firstly used to predict an optimum operating point using a single criterion. Three cases were considered:
(a) Experiments were planned to most accurately identify the model parameters.
(b) Experiments were planned to most accurately predict the performance of the reactor at the optimum operating point (as predicted by the model).
(c) Experiments were carried out at the optimum operating point as predicted by the model. This is not a reported algorithm, but it is conceivable that such an algorithm would converge at the optimum as the increasing number of experiments gave a more accurate model. At the same time it would deliver most information in the area of interest.

The optimum operating point was defined as the initial oxide concentration, temperature, and reaction time which determine the maximum profit from the plant defined by:

$$F = a(\text{glycol produced}) - b(\text{propylene oxide}) - c(\text{maximum rate of heat}) \qquad (4.11)$$
$$\text{consumed} \qquad\qquad\qquad \text{generated}$$

For demonstration purposes constants a, b, and c were chosen to bring the optimum point inside the permitted experimental region.

The model used in this work took the form:

$$\frac{dC_P}{dt} = - k_1 \, C_P \, C_{cat} \, e^{-E_1/RT} \tag{4.12}$$

$$\frac{dC_G}{dt} = -\frac{dC_P}{dt} \tag{4.13}$$

$$c_p \, \rho \, V_R \frac{dT_R}{dt} = - \Delta H \frac{dC_P}{dt} - U \, A' \, (T_R - T_J) \tag{4.14}$$

Within the range of operating conditions it was possible for the reactor contents to boil. When this happened, the temperature calculated from the above heat-balance equation was replaced by the boiling temperature of the reaction mixture. This required the model to have a boiling point/composition relationship built into it. The discovery and incorporation of this "non-ideality" into the model took a considerable time to accomplish.

The result of the three sequences of experiments are shown in Table 4.3. This shows that methods (a) and (b) converge on the same point (within experimental error), and they also do so from different starting points (series 1 and series 2). Method (c), on the other hand, converges on a different point, and yet a different point in the second series.

Clearly, method (c) alone is an ineffective algorithm, which can be explained by presuming that the first experiment produced a poor estimate of the parameters and this, in turn, produced a poor estmate of the location of the optimum. Experiments at this point presumably did not contain adequate information to correct the initially wrong parameters and so the model repeated experiments at the wrong point, never receiving any new information to correct its error.

Methods (a) and (b) both achieved their respective objectives best, and both produced a convincing result. However, within these runs, neither method designed an experiment to operate at the predicted optimum and so the experiments in themselves would be inadequate in a process development environment, unless backed up by experimental work at the optimum itself.

A development from these results is to combine methods (a) or (b) with (c) so that more emphasis is brought to the point of interest. Method (d) was therefore tried, in which experiments were carried out in pairs; one experiment at the predicted optimum (method (c)), and one at the point to best reduce the variance of the predicted optimum (method (b)).

In process development the experimental systems under investigation are normally fairly complex, and not usually represented by a single reaction, as was the case in the series shown in Table 4.3. The objective function and experimental conditions were therefore modified to produce significant quantities of diglycol, and this was assumed to be the product of the reaction. The objective function was taken as:

TABLE 4.3
Experimental design with a single criterion
(a) To minimize the variance of the parameters
(b) To minimize the variance of the predicted optimum
(c) By experimenting only at the predicted optimum

Expt. planned	Method (a)					Method (b)					Method (c)			
	C_{p_0} kmol m^{-3} exptl. settings	T_0 {°C}	opt. time {h} predictions	opt. profit predictions	det. $[X^T X]$	C_{p_0} kmol m^{-3} exptl. settings	T_0 {°C}	opt. time {h} predictions	opt. profit predictions	var (y_p)	C_{p_0} kmol m^{-3} exptl. settings	T_0 {°C}	opt. time {h} predictions	opt. profit predictions
Series 1														
Start	2.68	20.6	4.2	.346	.20	2.68	20.6	4.2	.346	17.3	2.7	20.6	4.22	.346
1	2.62	21.7	4.3	.314	.14	2.58	21.8	4.3	.306	6.3	2.8	20.1	4.21	.353
2	2.48	23.4	4.3	.288	.03	2.61	22.0	4.3	.306	3.6	2.9	20.0	4.10	.360
3	2.45	23.7	4.3	.293	.028	2.49	22.4	4.3	.296	2.0	2.9	20.0	4.21	.361
4	2.45	23.6	4.2	.296	.014	2.39	22.7	4.2	.289	1.7	2.9	20.0	4.10	.359
5	2.42	23.7	4.2	.289	.013	2.47	25.6	4.4	.290	1.3	2.9	20.0	4.10	.355
Series 2														
Start	1.20	25.4	4.4	.131	.018	1.2	25.4	4.4	.131	670.	1.2	25.4	4.0	.131
1	3.10	20.0	4.3	.360	.071	2.6	22.7	4.3	.320	131.	1.7	20.0	4.0	.233
2	2.30	23.7	4.2	.279	.056	2.1	25.0	4.3	.249	16.2	1.7	20.0	4.0	.232
3	2.20	25.0	4.2	.276	.048	2.2	24.2	4.3	.271	2.4	1.7	20.0	3.9	.231
4	2.20	24.9	4.3	.276	.040	2.1	24.6	4.3	.262	1.7	1.7	20.0	3.9	.233
5	2.20	24.8	4.4	.277	.033	2.2	24.4	4.3	.266	1.5	1.7	20.0	3.9	.233

F = (diglycol produced per batch) x (number of batches per year)

A fixed preparation and cleaning time is assumed between batches.

Optimization variables: reaction time, initial concentration, initial monoglycol concentration, initial temperature.

The model was:

$$\frac{dC_W}{dt} = -k_1 \, C_W \, C_P \, C_{CAT} \, e^{-E_1/RT} \tag{4.15}$$

$$\frac{dC_D}{dt} = k_2 \, C_G \, C_P \, C_{CAT} \, e^{-E_2/RT} \tag{4.16}$$

$$\frac{dC_G}{dt} = -\frac{dC_W}{dt} - \frac{dC_D}{dt} \tag{4.17}$$

$$\frac{dC_P}{dt} = \frac{dC_W}{dt} - \frac{dC_D}{dt} \tag{4.18}$$

$$\rho \, c_p \, V_R \, \frac{dT_R}{dt} = + \frac{dC_W}{dt} \, \Delta H_1 \, V_R - \frac{dC_D}{dt} \, \Delta H_2 \, V_R - U A' \, (T_R - T_J) - (U A')' \, (T_R - T_{amb}) \tag{4.19}$$

where $(U A')'$ is a correction term for heat loss.

To be sure that this was a reasonable model took many weeks of work and a considerable number of preliminary experiments.

The results of this second set of experiments is given in Table 4.4. Three series of experiments were carried out, each with a different initial experiment as starting point.

All three series identified the same optimum within ±15% but the pattern of designed experiments repeated itself after the first five experiments. Later experiments did not search new areas, but only reinforced old experiments. As shown by the results of Table 4.3 method (c), any algorithm that repeats itself can lead to incorrect results, if the limited number of points do not discriminate properly between right and wrong sets of parameters. Because of this limited set of experimental points, there is no information given over possible model errors. Clearly, if the experiments beyond experiment 5 had involved new areas, then valuabl model inadequacy information would have been obtained at the cost of slightly reduced parameter information.

A proposed algorithm. Experience with the results shown in Tables 4.3 and 4.4 shows the number of weaknesses that must be overcome before a serious planning algorithm for process development purposes can be proposed. Such an algorithm should:

(a) not be bound rigidly to having an exact model,

TABLE 4.4

Experimental design using method (d) - pairs of experiments: at optimum and to reduce variance of predicted optimum

Experimental condition			Predicted optimum				
C_0	C_g	T	Time	C_g	C_0	T	Objective function = production/year
Series 1							
1 4.7	0.	30.6	4004	3.0	4.0	30	1077
2 4.0	3.0	30.0					
3 4.0	4.4	30.0	3758	3.6	4.0	30	1490
4 4.0	3.6	30.0					
5 4.0	4.9	30.0	3882	3.9	4.0	30	1540
6 4.0	3.9	30.0					
7 4.0	4.8	30.0	3960	3.9	4.0	30	1507
8 4.0	3.9	30.0					
9 4.0	4.8	30.0	3917	4.0	4.0	30	1526
10 4.0	4.0	30.0					
11 4.9	4.9	30.0	3819	3.9	4.0	30	1483
Series 2							
1 1.1	3.8	24.9	1914	5.6	4.0	30	5914
2 4.0	5.6	30.0					
3 4.0	0.3	30.0	3102	3.3	4.0	30	1631
4 4.0	3.3	30.0					
5 4.0	2.9	30.0	3266	3.3	4.0	30	1553
6 4.0	3.3	30.0					
7 4.0	3.1	30.0	3290	3.4	4.0	30	1574
8 4.0	3.4	30.0					
9 4.0	3.1	30.0	3270	3.4	4.0	30	1591
Series 3							
1 3.9	0.	19.9	1001	3.7	4.0	30	22000
2 4.0	3.7	30.0					
3 4.0	3.3	30.0	3694	4.4	4.0	30	1820
4 4.0	4.4	30.0					
5 4.0	4.1	30.0	4643	4.4	4.0	30	1753
6 4.0	4.4	30.0					
7 4.0	4.3	30.0	3619	4.4	4.0	30	1773
8 4.0	4.4	30.0					
9 4.0	4.4	30.0	3652	4.5	4.0	30	1779

136

(b) include experiments at the points of maximum interest,

(c) be a mixture of criteria so that there are different sets of information available,

(d) provide some indication of model error over the range of the model. This means that a repeated experiment has probably less value than an experiment in a new area.

From these requirements, the algorithm method (e), as shown in Fig. 4.3, was built up.

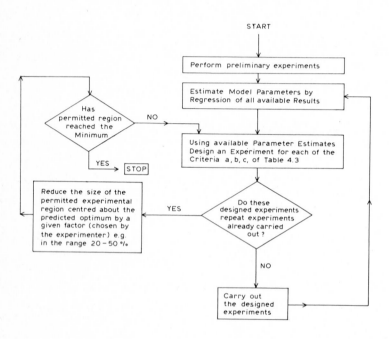

Fig. 4.3. Design procedure for use with approximate models (method (e)).

This algorithm was tested experimentally, using three different models. Model 1 was the best developed reactor/heat-transfer model for the reaction. Model 2 was an assumed "isothermal" reaction where temperature profile during the reaction was ignored, and model 3 was isothermal and had an inadequate representation of the chemical kinetics involved.

The objective function for this set of experiments was taken as the annual profit arising from diglycol production.

$$F = \left\{ \begin{array}{c} a(\text{diglycol produced}) - b(\text{glycol fed}) - c(T_0 - T_{amb})^{k'} \\ \text{per batch} \qquad \text{to batch} \end{array} \right\} \times \begin{array}{c} (\text{number of batches}) \\ \text{per year} \end{array}$$

<u>Optimization variables</u>: Initial oxide concentration, initial monoglycol concentration, initial temperature T_0.

The third term of the objective function represents the cost of providing heating from atmospheric temperature T_{amb} to the initial reaction temperature. In this case the reaction time is no longer an optimization variable, but it is assumed that the reaction must be terminated when the oxide concentration falls to 0.5 kmol m^{-3}. The values of a, b, c, and k' were again chosen to keep the optimum point within the experimental region.

Table 4.5 shows the results from these three experimental series. It can be assumed that model 1 has accurately predicted the optimum after the 14 experiments (based on the known accuracy with which model 1 fitted the experimental results). It can be seen that model 3 also predicts the same optimum and has an acceptably low error, even though it is, mechanistically speaking, a "poor model".

Model 2, however, predicts a different optimum, 15% below that of the model 1 optimum. It is possible to see that this model is not a good overall fit by plotting out the individual concentration and temperature profiles of each run and comparing them with its predicted values.

This discrepancy in the optimum may have been because the form of equation of model 2 does not allow it to follow the real reaction. In this case, the optimum given by model 2 would not correspond to the real optimum, but the algorithm has no means of knowing this and concentrates on accurately predicting a point which is not the optimum.

In non-linear experimental design the model is performing two distinct functions: as a function to fit to experimental results, and as a device for predicting the optimum. When the model is mechanistically incorrect, it will predict an optimum somewhat different from the true optimum. There is no way of knowing what this error will be. However, we can use its first formulation (the fitting of the experimental data) to guide us to its possible accuracy in predicting the optimum. When there are experiments which fit badly, we must suspect the "optimum point", even though the model predicts very well at this point. The only recourse is to carry out a sensitivity perturbation at this point or to revise the structure of the model in the light of the discrepancies and repeat the analysis.

4.5 CONCLUDING REMARKS

Satisfactory experimental design methods exist for initial screening experiments, but as soon as these have been done, a model must be available to enable any further experimental planning (other than pure intuition) to be practised.

The simplest method is to use the model in a predictive way to suggest interesting areas but this is not systematic and depends much on the intuition of the modeller and experimenter.

The use of the more systematic and theoretically-based non-linear method for

TABLE 4.5
Experimental design using proposed criterion for approximate models (method (e))

3 Models: model 1 best available developed model
model 2 isothermal model to non-isothermal reaction
model 3 isothermal model with inaccurate chemical kinetics

criterion	Model 1					Model 2					Model 3				
	C_{oxide}	C_{glycol}	Temp.	Time	predicted optimum profit $\times 10^{-5}$	C_{oxide}	C_{glycol}	Temp.	Time	predicted optimum profit $\times 10^{-5}$	C_{oxide}	C_{glycol}	Temp.	Time	predicted optimum profit $\times 10^{-5}$
1	2.1	2.0	20.3			2.1	2.0	20.3			2.1	2.0	20.3		
2	2.0	2.0	30.3			2.0	2.0	30.3			2.0	2.0	30.3		
3 c	4.0	5.1	38.9	13134	3.45	4.0	6.0	29.9	13480	4.37	4.0	2.5	25.7	4023	1.73
4 a	4.0	5.7	39.2			4.0	5.7	29.0			4.0	1.1	27.5		
5 b	4.0	2.4	35.0			4.0	5.7	35.0			4.0	0.4	35.0		
6 c	4.0	5.3	29.0	14964	2.81	4.0	5.2	27.5	16936	2.75	4.0	2.9	26.6	3000	3.68
7 a	4.0	6.4	32.5			4.0	5.0	19.0			4.0	4.6	25.0		
8 b	4.0	7.0	35.0			4.0	3.1	15.0			4.0	4.6	35.0		
9 c	4.0	4.3	28.2	11979	2.67	4.0	5.2	27.8	19429	2.60	4.0	3.7	27.6	6690	3.24
10 a	4.0	2.8	25.9			4.0	3.1	21.6			4.0	4.5	26.9		
11 b	4.0	2.6	34.0			4.0	4.9	32.4			4.0	4.5	33.0		
12 c	4.0	3.6	28.0	8989	2.78	4.0	5.2	28.2	20929	2.52	4.0	3.7	28.4	7398	3.15
13 a	4.0	3.5	28.4			4.0	5.4	27.4			4.0	4.0	29.2		
14 b	4.0	3.5	31.0			4.0	4.1	31.0			4.0	4.7	31.0		
**	4.0	3.8	28.2	10531	2.95	4.0	5.1	28.5	21083	2.50	4.0	3.9	28.5	7269	3.03
**				10200	2.90				22000	2.50				10500	2.75

* Final predicted optimum from the series of experiments.

** Observed performance at predicted optimal initial conditions of experiment 12.

process development studies is as yet still far from routine, and would only be recommended in a laboratory that was interested in being involved in the development of new techniques.

Experimental designs using these methods in reactor studies in process development are most likely to succeed if a mixture of criteria is employed and provision is made for ensuring that the whole experimental area is thoroughly investigated by preventing new experiments repeating old ones. Because correct models are generally not known, a mechanism has to be written down which is plausible, and this used in the design. The model predictions of the individual experimental results show the degree of systematic error to be expected from the model. If this is intolerable, then a new mechanism must be proposed in the light of the nature of the discrepancies before the model can be used for optimization studies.

FURTHER READING

Box, G.E.P. and Draper, N.R., 1969. Evolutionary Operation. Wiley. - EV-OP and all its related methods.
Box, G.E.P., Hunter, W.G. and Hunter J.J., 1978. Statistics for Experimenters. Wiley, New York. - Thorough study of linear models.
Himmelblau, D.M., 1970. Process Analysis by Statistical Methods. Wiley. - Survey of available methods.

REFERENCES

Box, G.E.P. and Wilson, K.B., 1951. Journ. Royal Statistical Society, B 13: 1.
Schifferli, C.C., 1979. Ein integriertes rechnergesteuertes System zur Datenerfassung, Versuchsplanung und -Durchführung für die chemische Prozessentwicklung. Ph.D. thesis, E.T.H Zürich, No. 6459.

QUESTIONS TO CHAPTER 4

(1) What is the meaning of "screening experiments" and when are they useful ?
(2) For the system

$$A + B \longrightarrow P + R + S$$

liquid   gas   Catalyst C_{cat} byproducts
pH=2

what variables would you incorporate in an initial screening study ?
Design a full, and a partial factorial design for this screening study ?
(3) What are the advantages and disadvantages of partial designs ?
(4) How can linear models be used to find optima in reactor studies ? What disadvantages have to be accepted ?
(5) What is the difference between the Box-Wilson and the EV-OP methods ?
(6) What is the principle of true non-linear methods of experimental planning ? What are their disadvantages ?
(7) How can models be useful in experimental guidance in practice ?

Chapter 5

THE PILOT STAGE

   Having carried out laboratory experiments on a reactor simulating full-scale,
and having fitted the results from it with a mathematical model, it might be thought
that one is now in a position to design the full-scale reactor directly.
   In practice, this is hardly ever done; normally the next stage is the "pilot"
stage, where the whole process is tested on an intermediate scale, where a number
of kilos to some tons per day of product are made.

5.1   THE FUNCTIONS OF THE PILOT PLANT
   The pilot-scale serves a number of functions besides that of purely checking
the technological feasibility of the design. The functions of a pilot plant can
be summarized as:
   (a) to produce material for evaluation and introduction into the market;
   (b) to define effluent problems and test methods of their solution;
   (c) to check the feasibility of continuous operation;
   (d) to check the effect of build-up of impurities and other long-term effects;
   (e) to check that the process can be "scaled-up";
   (f) to check materials of construction;
   (g) to obtain design information for difficult unit operations;
   (h) to check that no important factors have been overlooked or misinterpreted;
   (i) to obtain the confidence of those allocating the capital by showing that
       the process is technically and economically feasible.

5.1.a   The production of material for evaluation and market development
   There is a difference between being able to produce a certain chemical, and
producing a "product" which is suitable and competitive for its intended use.
Certain impurities in a product can have very adverse effects on following production
stages, and in a competitive situation a competitive material would then be bought.
Polymerization can be affected by trace impurities which reduce the quality of the
resulting product, and therefore any new process for the production of a monomer
must supply enough product to enable the following stages to be carried out on a
large enough scale to obtain test samples of the final polymer for quality analysis.
   Products destined for the consumer market are particularly susceptible to
rejection by properties that cannot be defined by a chemical analysis. A soap from
a new process may have a slightly foreign odour which is simply not accepted by the
customer. Hence again it is necessary to make sufficient quantity of the product
to be able to fully examine it and test consumer reaction to it before the final

decision is made to invest very large sums of money in a full-scale plant.

The market for a new product often builds up slowly which could make the economics of a full-scale plant unattractive. To have some product to market before the full-scale plant comes on stream is a good way of ensuring an early demand for the new product and to improving its economics. This product can be produced in the pilot plant.

5.1.b The definition of effluent problems and the testing of methods of their solution

As with item 5.1.a, it is difficult to predict simply from laboratory experiments what the effluent problems will be, and if they can be satisfactorily solved. Sufficient quantity of material leaving the reactor must be collected, and the proposed plant work-up carried out to isolate the product. The resulting effluents must be thoroughly analyzed to identify potential effluent problems. Methods of solving these effluent problems must then be tested and be shown to be satisfactory. A full-scale plant may have to be shut down if effluent problems remain unsolved, and so the consequences can be just as bad as if an inoperable reactor were built.

Effluent problems are difficult to quantify. To be able to guarantee that a full-scale plant involving a mercaptan will not produce such an objectionable smell as to be shut down by the local community is a more difficult problem than guaranteeing that the reactor will function satisfactorily. It can therefore be a stronger reason for the pilot plant than the reactor development programme.

5.1.c To check the feasibility of continuous operation

Most industrial processes are continuous processes whereas many reactor investigations are made in batch reactions. When no continuous reactor experiments have been performed, they can be carried out on the pilot plant. Continuous reactors present problems in laboratories because of the large quantities of raw material they consume, and reactor effluents they produce. These problems are solved in the pilot plant, because the reactor outlet can be worked up into product and sold. Hence, the pilot plant is a suitable piece of equipment for an extended continuous operation investigation.

Continuous behaviour is different to batch behaviour in practically all unit operations, not only reactors. Reactors produce different yields and selectivities, distillations produce different compositions and so on. A continuous full-scale plant requires considerable investigation to have been made with continuous operation before it can be built with confidence.

5.1.d Checking the effects of concentration build-ups of impurities and other
long-term effects

Processes involve a maximum of recycling of solvents, mother liquor, process
water, and unreacted raw materials, both on economic and ecological grounds.

Recycling of streams involves not only the recycling of the major component,
but also that of any component not specifically removed from the stream. When a
process runs for many days, small impurities, undetected or unrecognized in a
single experiment can build up in these streams to such a level as to adversely
affect the reaction or cause corrosion or other problems.

To be sure such problems do not arise, it is advisable to operate the proposed
process, including all the proposed recycles, for an extended period of time.
This can be very conveniently done by the operation of a pilot plant.

If such problems do arise, then purges or recycle purification stages must be
introduced and then new tests made.

All long-term effects are difficult to establish in a laboratory because the
normal laboratory is not prepared, either physically or psychologically, to carry
out experiments 24 hours a day for weeks at a time at constant conditions. Such
work is much more suitably done on the pilot plant. Most important in this respect
are catalyst-life work and corrosion testing.

5.1.e Checking that the process can be "scaled-up"

When a reactor operates on the one litre scale, but does not operate on the
100 litres scale, the reason for the failure is a "scale-up problem". The term
"scale-up" is a synonym for "unexplainable" - because unexplainable is a somewhat
unprofessional term. Scientifically speaking, all effects can be explained in terms
of the differences that occur when scale changes, and one should try to explain
effects in these terms and not hide behind the phrase "scale-up".

On changing the scale of a reactor, a number of factors inevitably change; these
are described in detail in the following sections:

(1) Surface to volume ratios. This has an important effect on heat transfer
and leads to different temperatures in different scale reactors, unless specific
care is taken to see that the heat removal facilities are expanded in the same
proportion as the reactor volume.

For some reactions, the reactor walls take part in the reaction, hence changing
scale-up changes this effect, which can lead to different reaction conditions. The
reaction may be catalyzed by the reactor wall, or quenched. Free radical reactions,
for example, have reduced rates in the presence of high surface areas because of
the radical quenching reactions that occur on the walls.

When the reactor geometry remains the same, the reactor liquid surface area/unit
volume decreases with scale and hence, if a gas or vapour is involved, the superficial gas velocity is altered. This change can also bring about problems of

vapour/liquid disengagement, resulting in flooding or foaming appearing at different scales.

(2) Mixing in tank reactors. On the small scale it is possible to have a much greater degree of agitation per unit volume than on the large scale. A normal laboratory stirrer will produce more agitation than can be installed on a full-scale reactor. When the gas phase is involved, for example when one of the reactants is a gas, then the different levels of agitation on the different scales produce different mass-transfer rates and different standing concentrations of the gas in the liquid phase. This can affect not only the reaction rate, but also the selectivity in a complex reaction situation. The larger reactor will have the lower concentrations, lower rates, and possibly lower selectivity.

Mixing is also important in single-phase reactions to ensure that the incoming raw materials are dispersed as quickly as possible into the bulk of the reactor fluid. Whenever they remain as two streams, there are local high concentrations and this can lead to unwanted side-reactions and a lowering of the selectivity.

Besides the macro-mixing which breaks up the entering streams into a homogeneous mixture, there is micro-mixing, where this homogeneous mixture of packets of the raw materials breaks down further so that homogeneity is present at the molecular level.

During the time taken for this second mixing step to be completed there is further opportunity for local high concentrations developing and unwanted reactions occurring. The different degrees of mixing on the two scales can thus produce different yields if the reactions involved are both very fast and concentration-dependent.

(3) Tubular reactors - diameter changes. Clearly, any scale-up of a laboratory tube-reactor by increasing the diameter is producing a very different set of reaction conditions because the heat transfer will be so much reduced. The matter becomes even worse when the tube is packed with catalyst because the lack of radial mixing will then produce much higher tube centre temperatures which can produce side reactions (and increase heat generation!) and catalyst deactivation. A tube diameter change is a very radical change which must be very carefully investigated, both theoretically and experimentally, before it is introduced.

(4) Tubular reactors - length changes. As previously explained in section 3.4.f, plant reactors are often longer than laboratory reactors, and this results in higher gas velocities with changed reactor temperature profiles. One can expect some change in stability, conversion, and selectivity as a result of the different temperature profiles. Diffusional resistances and hence selectivities are also affected by velocity changes.

(5) Tubular reactors - multiple tubes. Laboratory work is usually carried out with a single tube which has a central thermocouple pocket. A full-scale reactor may have many thousand tubes with only a few containing thermocouple pockets. In

such a change there are many possibilities for "scale-up" effects to be present. The even distribution of gas over all the tubes is always a matter for discussion. If there were maldistribution of flow through the tubes, this would lead to sub-optimal conversions for the reactor as a whole. If the monitor tubes with the thermocouple pockets were no longer "typical" because of the themocouple pockets, this again would cause apparent differences due to scale. The coolant characteristic and temperature profiles differ from those of the laboratory, and the possibilities of reaction after leaving the tubes are different on the different scales.

5.1.f   Checking materials of construction

Laboratory work is usually carried out in glass apparatus, and it is not until the design of the pilot plant that much effort is put into selecting the most suitable materials of construction.

The selection of materials of construction can often pose problems because information concerning the suitability is given for pure components, whereas in processes mixtures are present for which data are not available.

Reactors pose a particular problem in this respect because they involve novel mixtures of components under extreme conditions of temperature, and so generally experimental work has to be done on the suitability of materials of construction before the full-scale plant is built. This can conveniently be done on the pilot plant.

5.1.g   Design information for unit operations

The difficult unit operations such as crystallization and the following filtration need to be carried out on a large enough scale for pilot equipment to be tested, and the work done with the feed liquor that arises from the process itself. Impurities can change crystal shape and crystal shape affects filterability, and so pilot plant trials become necessary.

Distillation is often calculable, requiring no pilot trials, but when potential problems occur (thermal instability in the boiler, reaction in the column, or the distribution of unidentified components in the product streams), piloting may be necessary.

5.1.h   The overlooking or misinterpretation of important factors

The development of a new process is an extremely complicated procedure, and although throughout the development every attempt is made to understand the physical and chemical processes occurring, it can and does happen that effects are overlooked or misinterpreted so that the full-scale design is inoperable.

The pilot stage serves as a final check that all the thinking that went into the reactor, the work-up, the recycles, and materials of construction, is in fact correct, and all will function when put together as a process.

It is not possible to discuss this topic rationally, because once the mistake is pointed out it is difficult to believe that it could have been overlooked. It is only possible to say that "experience" shows that in the development of new processes points are overlooked and the pilot scale helps to minimize the number of such points. The last section of this chapter tries to summarize such "experience" by describing some actual problems that have arisen in process development that were far from obvious at the time.

5.1.i Developing confidence in those in control of finance

A new process represents an enormous investment and a considerable risk. The risks are both technical (will the plant work; will it work economically?) and commercial (will we find a market for our product; will the market remain long enough for us to recover our money?).

Any method of reducing the risks by providing information on the operability of the process or the sales potential of the product are welcomed by company management.

The pilot plant is an excellent tool in both these areas. It is evidence that there are no technical problems and provides further evidence that customers will buy the product. Though a pilot plant may cost many millions of dollars, it is a good investment when it gives information regarding the viability of an investment in the order of hundreds of millions of dollars. This common sense conclusion can be illustrated using probabilities as follows.

Say a pilot plant would cost $10 million for a process whose full-scale capital investment would be $100 million. Let us assume that the project has been subjectively estimated, based on past process development performance, of having an 80% probability of being technically successful and that the building of a pilot plant would remove all doubt as to the success of the process. If the process is a technical success, it will bring in a total of $200 million income discounted to the time of the investment.

Proposal A: Build no pilot plant

Capital cost:	$100 million
Income:	80% probability $200 million
	20% probability $0
Expected return:	0.80 (200 - 100) + 0.20 (-100)
	= + $60 million
Maximum loss:	$100 million

Proposal B: Build a pilot plant first

Capital cost:	80% probability	$(10 + 100)$ million
	20% probability	$10 million
Income:	80% probability	$200 million
	20% probability	$0
Expected return:	$0.80 \ (200 - 110) + 0.20 \ (0 - 10)$	
	= + $70 million	
Maximum loss:	$10 million	

Hence proposal B (to build a pilot plant) is the preferred route, since it has
a much lower risk (maximum loss) and is the correct policy when the firm meets
a number of these decisions each year (the expected return is higher). Only the
compulsive gambler would choose proposal A.

5.2   TYPES OF PILOT PLANT

Having discussed the function of a pilot plant, we should now consider what
such a plant should look like to perform these functions satisfactorily. The major
decision variables are how big should the plant be, how much of the final process
should it contain, and what should be the criterion for the equipment design.

If it is important that substantial quantities of product are available for
product and market trials, the plant must be large, and its output defined by the
requirement for the trials. Furthermore, the whole process must be piloted in
order to get a proper sample of material of identical quality as that to be
expected from the full-scale.

Such a plant is going to look quite similar to a small version of the full-scale
plant, rather in accordance with what one normally expects by the term "pilot-
plant". The cost will be high, $1 - 10 million being not unusual, and the plant
will probably take a year to build.

If the process were extraordinarily complex, very novel, or the full-scale
plant very expensive, then again only a complete, quite large plant would satisfy
functions (h) to (i), that nothing is overlooked and to give confidence to those
allocating the money for the full-scale.

There are, however, many instances where these conditions do not apply; market
trials will not be necessary if it is to be a firm-internal intermediate, and most
new processes are not highly novel to the firm developing them, because they have
other processes that they operate which are similar. Under these conditions a
smaller plant can be decided upon. The new limit may be an adequate size to obtain
reliable data for certain unit operations, or to involve a reasonable scale-up
factor between the laboratory and the pilot and the full-scale.

Neither need the plant pilot all stages all the way to produce the final product.

It may be unnecessary to pilot the process beyond the last recycle, because the final work-up may be known and calculable. Hence the pilot plant should be situated on a site where the pilot plant outlet streams can be fed to another plant for work-up or destruction. Clearly, such smaller, partial pilot plants would cost less to pilot and be done more quickly than a larger complete version.

If no unit operations needing pilot results are included, and the process is judged to be fairly straightforward, then even smaller plants can be suggested, with reactor volumes in the 1 - 10 litre range. Such plants are called mini-pilot-plants, and they are more like overgrown laboratory equipment than chemical plants. However, they normally operate continuously with all recycles to be expected in the final process, and they operate 24 hours a day, so obtaining all necessary information on recycle build-up, lifetimes, reactor performance, etc.

Only those stages essential to the continuous operation of the process are included in the mini-pilot-plant. Final make-up and product isolation can be carried out separately in other standard equipment in the development department. To include a distillation section unnecessarily in a pilot plant is not advisable because distillation problems may delay the operation of the plant. Work-up done separately allows parallel development to take place. Mini-pilot-plants are built up from standard pieces of equipment. In fact, a development department may try to have multi-purpose mini-pilot-plants, particularly if all its processes are broadly similar.

Mini-pilot-plants are sometimes made of glass, being then composed of standard laboratory equipment and being very easy to construct and alter. However, this means that the plant does not automatically deliver materials of construction information. Material samples, however, can be hung in the glass apparatus and the corrosion measured.

In some rather rare cases it is possible to go to full-scale design without going through the pilot stage. However, "dropping the pilot" is a dream that rarely comes true. Only if the new process is very similar to an existing one, and it is very well understood theoretically would this approach be advised.

A sensible development project is a mixture of laboratory investigation, modelling, and pilot experimental work. The pilot experimental work should probably be smaller than the "ideal" size because this results in a saving in time and money, but to make up for this being off-optimal, a back-up laboratory programme and modelling programme can replace the confidence loss by the too-small pilot plant.

The advantage of pilot plants is that in one piece of equipment very many problems are answered, and that when the plant runs successfully, this bestows great confidence; technically the process will be a success.

The disadvantage is that very little is learnt of the fundamentals of the process. A laboratory programme, on the other hand, requires very many individual experiments, but this results in a wealth of knowledge about the process that may be useful in

the future if unexpected process troubles arise. It is worth repeating that an
intelligent process development programme contains a proportion of both types of
work.

5.3   PILOT PLANT DESIGN

The pilot plant is a piece of experimental equipment, being built to obtain data.
It should not just be a miniature model of the intended full-scale process.

The important decisions (how big should it be, and what stages should be
included?) have already been treated. The design can now start, and the designer
should firstly decide what data will be required from the operation of the plant,
and then design it to supply this data.

The plant clearly needs many more sample points than the full-scale. Measurement
of flows, temperatures and pressures should be more extensive, more accurate, and
variable over a greater range than for a full-scale plant. Yield information should
be accurately measurable, and this preferably means weighing, which for a continuous
process with recycles involves sets of weighed feed, product and recycle tanks.

Facilities should be provided for mass and heat balances to be performed over
the various units, and provision should be made for calibration of all instruments.

5.4   PILOT PLANT INSTRUMENTATION

The larger type of pilot plant has 25 mm piping and large enough flows to be
able to use the instrumentation used on full-scale processes. Hence, well developed
accurate (when clean and calibrated!) flow meters, control valves and pressure
measurement are available, capable of remote control, digital output, etc.

The picture is not so satisfactory when it comes to the smaller pilot plants and
the mini-pilots. Here industrial scale equipment is too large and less well developed
laboratory measurement devices must be sought. In fact, it may be that the size of
the mini-plant is determined by the smallest available instrument to control it.
The following sections give some indication of suitable types of instruments for the
small-scale pilot plant.

5.4.a   Flowcontrol

Much preferred to control valves for flows of the order of $1 \text{ m}^3 \text{ h}^{-1}$ are positive
displacement pumps. Piston pumps with a variable stroke, peristaltic pumps with a
variable speed and diaphragm pumps with a variable frequency are available which
are all remote controlled, for instance by a -5 to +5 voltage.

These pumps have flows reproducible to within 1 or 2%, and so after calibration
they can be used to provide fixed flow rates independent of up or downstream pressure
variations. That is, they replace the flow meter and control valve of the feed
flow controller (Fig. 5.1a). Set up as Fig. 5.1b enables the flow to be controlled
by a further measurement such as a reactor temperature.

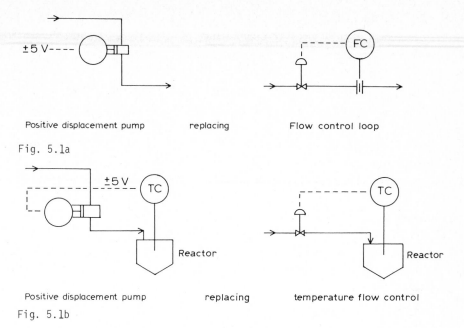

Positive displacement pump   replacing   Flow control loop

Fig. 5.1a

Positive displacement pump   replacing   temperature flow control

Fig. 5.1b

Fig. 5.1. Reactor feed control system.

One problem with such pumps is that they have a pulsating flow which is
disturbing in that one is never sure of the effect of the pulsations on the process.
Careful choice of stroke, and pressure damper (see Fig. 5.2) can minimize such
pulsations.

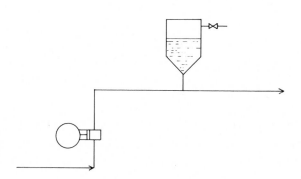

Fig. 5.2. Pulsation damping for displacement pumps.

150

5.4.b   Flow measurement

Flow measurement should be avoided by setting as many flows as possible constant and providing the constant flow by means of a pump. There will inevitably be some flows that are "dependent variables" that have to be measured, and this can be done by small laboratory rotameters. Small rotameters have the disadvantage that they cannot transmit their reading for data loggers or control purposes, and so for such duties the rather expensive, but highly satisfactory constant temperature anemometer must be used. This instrument contains an electrically heated wire and the gas or liquid flowing over the wire abstracts heat from it and the power required to maintai the temperature of the wire is a measurement of the flow rate. These instruments are very accurate, but require calibration because each fluid composition has different characteristics.

For calibration purposes, and also for accurate spot mass balance checks, the rate can be measured volumetrically by switching in a calibrated reservoir and measuring the time required to fill it (Fig. 5.3). Such measurements can, however, upset a process if the corresponding intermediate storage is not present.

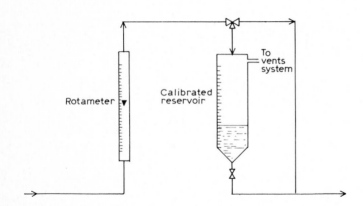

Fig. 5.3. Rotameter with installed calibration vessel.

Accurate liquid flow measurement can be done by using a reservoir mounted on a balance. Since modern electronic balances have analog or digital signals which can be read directly by computer, this set-up can be very accurate and very convenient when the balance is used in conjunction with a computer.

5.4.c   Temperature measurement

It is surprising how inaccurate most industrial temperature measurement is, with ±1°C being a usual accuracy supplied. For reaction rate measurement work, and particularly for heat balance work, an accuracy of ±0.1°C is to be preferred and thi usually requires platinum resistance thermometers rather than thermo-couples, and electronics associated with laboratory instruments rather than with industrial instruments.

Temperature control presents a problem in that there are so many types of cheap temperature controllers available for small scale equipment that one is tempted to use them rather than the much more expensive standard-type of 2-term-controller.

The mercury thermostat is robust and very accurate, but its on/off features make it a suitable controller only with large thermal capacities such as water tanks. A second disadvantage is that there are no facilities for remotely modifying the set point.

Various combined cheap dial one- and two-term controllers are also available for oven temperature control which are tempting to try for process control, but again they are so limited in scope that they are best avoided.

Temperature control is best done using industrial two- or three-term controllers; they have a full range of settings possible and have output and set-point both remotely controllable, and computer compatible. They are, however, expensive.

5.4.d Analysis

The more compositions that can be automatically analyzed, the better, since this leads to better control of the process. However, the analyses that can be done automatically are rather limited and sometimes the equipment needed is very delicate and requires a high degree of maintenance.

On the pilot plant are staff who are research and laboratory oriented and this does mean that the pilot plant does give the opportunity to use methods of continuous analysis that would not be considered robust enough for normal industrial use.

Probably the most universal method of analysis is Gas Chromatography which is now robust enough to be considered standard equipment also for industrial plants. Gas chromatography is discussed in more detail in Chapter 10.

5.5 USE OF COMPUTERS ON PILOT PLANTS

With a pilot plant being a piece of experimental apparatus operating 24 hours a day, it is reasonable to consider incorporating a computer with a pilot plant to be an advanced form of data logger, also capable of carrying out mass balance and yield calculations. Seen as a temporary development stage, however, the use of a computer does not seem so attractive because of the inevitable software and interface work required before a system is operational.

As computer systems improve, the software associated with their application reduces, and as there is a move towards electronic instrumentation, interface problems will become less important.

Comparing the advantages and disadvantages it would appear that when extensive use of the pilot plant is anticipated, for example if it is intended as a general-purpose mini-pilot plant, the installation of a computer would seem worthwhile. In general, the computer chosen should be supported with a high level process-control language, so that the user does not have to write his own procedures in FORTRAN.

A highly sophisticated application would be one with a computer large enough to include modelling, regression and optimization routines so that the pilot plant could locate optimum performance itself using EV-OP or one of the more sophisticated automatic experimental design methods described in Chapter 4. This is mentioned more as a challenge to an innovative development group than a serious recommendation for normal practice, but in time, one would hope to be able to use such techniques on a routine basis.

5.6   PILOT PLANT OPERATION

The first achievement is for the pilot plant to manufacture product at an economic yield. Once the plant is operating, however, there is opportunity to obtain far more data than simply this.

Particularly for reactors, the pilot plant gives the opportunity to check the reactor model against another reactor on another scale. If the model predicts the pilot reactor performance over a range of conditions, then it is a good sign that we are competent to design a full-scale reactor. When the fit is not so good, some form of re-fitting and explanation-seeking is called for.

Since the pilot plant is operating continuously, there is ample opportunity for optimization of the operating conditions. Also of interest is the maximum capacity of the various units, what happens when the reactor feed is doubled for instance. The operation of equipment below its capacity gives very little information because of experimental error becoming overriding as performance asymptotically approaches equilibrium values. Data obtained at conditions of overload are far more informative.

Very important data for economics considerations, though rarely available from the laboratory experiments, are catalyst lifetime and recycle build-up data, and particularly the various factors that influence these. The pilot plant is usually the only opportunity to obtain such data before the full-scale plant is operating.

5.7   SOME CASE STUDIES

In section 5.1.h it is claimed that experience shows that important factors are overlooked; by means of these case studies it is hoped to impart a little of this experience.

It is hoped that these examples show that it is difficult to foresee all the strange effects that can occur in new processes. Every example shows that unexpected problems arose and these were all reactor problems, evidence enough to show that process development and pilot plants are essential parts of practical reactor engineering.

Example 1: Acrolein reactor development

In the acrolein process development already cited in two previous examples, the time was reached when a tubular laboratory reactor was operating with a conversion and catalyst lifetime which were shown to be adequate for the economic operation of a full-scale process. The next stage in the process development was therefore the pilot stage. An impressive pilot plant was built with a reactor containing 40 tubes, five times as long as laboratory tubes (the full-scale plant was expected to need 4000 tubes). Quench towers, acrolein absorption from the effluent gas stream with water, and distillation to recover the acrolein were all part of the pilot plant. A data logger was also installed.

When the plant operated it was found to be impossible to achieve the conversions obtained in the laboratory. Despite all modelling work which one expected to adequately predict the effect of the tube length changes, 30% lower yields of acrolein were obtained than predicted, with correspondingly higher CO_2 and CO outlet concentrations. With these conversions the full-scale process would be uneconomic.

The changes from the laboratory reactor were:
(1) A multi-tubular reactor was involved and not a single-tube. This brought in queries of inlet gas distribution and uniformity of catalyst filling.
(2) Only three tubes had thermopockets, meaning that they were different from the remaining 37 which determined the performance. This brought queries concerning the operation of two types of tubes together; did a tube with a thermopocket represent the condition of a tube without?
(3) The reactor was cooled with a molten inorganic salt, circulating in the reactor shell, compared with the laboratory reactor heated by electrical windings. This brought queries of coolant temperature gradients, coolant distribution and "dead spots" in shell coolant flows.

All these hypotheses were proved not to be the explanation of the reduced yield. An observation of the gas temperature leaving the reactor body showed it to be considerably higher than the gas temperature at the end of the tubes containing thermocouples.

After-burning in the empty volume of the reactor exit header was suggested, and a literature search showed this to have a reported problem in the acrylic acid reactor, where propylene is partially oxidized to acrylic acid using a different type of catalyst. After-burning appeared in larger-scale equipment because the chain burning reaction was no longer adequately "quenched" by the reactor walls. The smaller equipment had a large enough ratio of surface area to reactor outlet empty volume to prevent this chain reaction establishing itself.

The reactor header, up to the quench-cooler was then packed with Raschig rings and this provided an adequate surface / volume ratio to prevent the burning reaction becoming established and acrolein yields in line with predictions were then obtained

Example 2: Chlorination of ethylene

With the development of the petrochemical industry in the early 1960s, steam cracking led to the availability of ethylene as a cheap organic raw material. Previously much of the organic chemical industry was based on acetylene as a raw material from carbide furnaces. This change in raw material brought with it the need to develop a series of new processes to replace those based on acetylene. In the chlorinated solvent area a dichlorethylene from ethylene process had to be developed to operate on a very large scale, since this would be the starting material for a whole range of solvents and chlorinated monomers.

The reaction

$$CH_2=CH_2 \ + \ Cl_2 \longrightarrow CH_2Cl-CH_2Cl$$

is very exothermic. Too high a temperature causes degradation to carbon and HCl, and dilution of the reactants or only partial chlorination and recycling on the scale required would be uneconomic.

A fluidized reactor was therefore chosen because the backmixing of the solid particles produced an even bed temperature and minimized hot spot formation. The fluidized solid was to be no more than a heat transfer device and so sand was chosen for this duty.

Laboratory experiments in a glass tubular reactor had shown that the reactor was operable and an economic process could be envisaged on the basis of the experimental laboratory results.

Coming to the pilot stage, consideration had to be given to the materials of construction to withstand dry chlorine at 350°C. Nickel is known to be suitable for such conditions and corrosion tests showed that nickel was in fact perfectly satisfactory at 350°C in the presence of chlorine and organic vapour.

A pilot nickel fluid bed reactor was constructed and operated but after only two weeks, reactant gases were found in the thermocouple, and within a few more days the body of the reactor started leaking in a number of places.

Contrary to all tests and literature information, the nickel was corroded by the gas mixture. A closer analysis of the reasons why nickel was not corroded under the conditions of the corrosion test was that a nickel chloride layer was very quickly formed, and it was in fact this nickel chloride that was passive, and protecting the nickel itself. However, the presence of the fluidized sand caused abrasion which continuously scoured away the protective layer and exposed a nickel surface to chlorination.

Since the only other suitable materials of construction were very expensive, the process became uneconomic and so had to be abandoned. The consternation this caused in a chemical company planning to change its source raw material, and then finding that the key process step has to be abandoned has to be experienced to be believed.

Example 3: Methyl chloride reactor

Methyl chloride can be produced by the reaction between methanol and hydrogen chloride:

$$CH_3OH + HCl \longrightarrow CH_3Cl + H_2O$$

The reaction occurs in the liquid phase using zinc chloride as a catalyst, but the reactors are large and the medium very corrosive and inconvenient because of its tendency to foam.

A far neater process was developed from some information available in the literature which reported that reaction had been observed to occur in the gas phase on alumina catalysts. The process promised better conversion, smaller reactors, more convenient operation and a solution to the corrosion problems.

The tubular reactor stage was investigated and conditions were found that produced a very attractive looking full-scale plant design. Modelling was carried out and the development team felt very confident to move on to the pilot stage. A pilot plant consisting of a single tube of similar diameter but much longer than the laboratory reactor was constructed. This pilot plant was very simple because it was located on the existing liquid phase reactor plant from which it obtained its vaporized methanol and where there was a stream into which the reactor effluent could be sent. HCl was taken from liquefied HCl in cylinders.

This simple pilot plant gave results comparable to the laboratory results and a full-scale reactor section was designed to replace the existing liquid-phase reactor section.

The 4000 tube full-scale reactor operated for about 7 days and then a rise was noted in the pressure drop across it. The pressure drop continued to rise until after about 14 days throughput had so reduced that it was necessary to shut the reactor down, cool-off, empty and repack the 4000 tubes.

During this operation it was noted that the catalyst was packed densely in about the top 20 cm of the tubes, with the catalyst below this being as new. This enabled simpler shut-down procedures to be employed, whereby only the top 20 cm of the catalyst were sucked out and replaced by fresh. This, of course, could only be a temporary solution and the reason for the repeated blockage had to be found.

The first 20 cm of the catalyst, when analyzed, contained some iron, whereas the catalyst was originally iron-free alumina. This naturally led to the supposition that iron chloride from the pipework in the existing plant was filtered out by the reactor and so was causing the problem. Whether the iron chloride was always there and the reactor was now acting as a filter, or whether it had arisen because of corrosion in the modified pipework because the pipework was not adequately heated above the dew point of the gas, had to be established by a detailed survey of pipework temperatures around the plant.

This study drew negative results; such a constant corrosion rate would have been noted on the pipe thicknesses, and no low temperature piping could be found.

A physical examination of the alumina catalyst spheres showed that the spheres had to some extent been crushed and a new theory that the blocking was due to the catalyst disintegrating was developed and then shown to be correct. In fact, the HCl supply on the plant had pressure fluctuations induced from an upstream compressor, and these fluctuations caused the inlet catalyst particles to disintegrate and block the bed. The pilot work was carried out with HCl from cylinders which gave none of the vibrations found under industrial conditions, hence the pilot plant in this case had not adequately simulated full-scale operating conditions.

This particular project had a happy ending. It was, in fact, possible to develop a catalyst sphere which was physically stronger but had the same activity, hence a change to this new catalyst grade gave a successful process which completely replaced the liquid phase reactors.

Example 4: Pyridine process development

A method for the synthesis of pyridine which involved the ring opening of a cyclic ketone, its amination, ring closure and a partial oxidation to remove the side methyl groups was under development. The processes occurred together in a single reaction stage, where the ketone was heated with ammonia in the presence of oxygen, using concentrated acetic acid as the solvent and copper acetate as the catalyst.

Although the starting material was rather expensive, the fact that the reaction occurred in only one stage made the process economically attractive.

Kinetic measurement was made in ampoules and also in batch equipment; a model was developed and the conversion was optimized. The next development step was to operate the reactor continuously. It was decided to use a mini-pilot plant, and a continuous process with a one litre reactor was built in the laboratory.

To properly simulate the process, the reaction mixture was distilled, giving a pyridine water azeotrope with various picolines as top product, leaving the acetic

acid and ammonium and copper acetates as bottom product which were then recycled
to the reactor after addition of ketone and ammonia (Fig. 5.4).

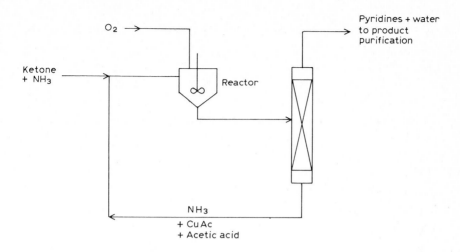

Fig. 5.4. Solvent recycling in proposed pyridine process.

The process ran well, and satisfactory conversions were obtained, but when
long-term experiments were attempted, a fall in reactor conversion was noted after
a few days' operation and the conversion became progressively worse with time.
The reason for this was shown to be a formation of small quantities of a by-product
which chelated strongly with the copper catalyst. This meant that the process
would have to have a further separation stage to separate the copper salts from the
acetic acid and solvent which could then be recycled. Because no convenient method
could be found of breaking this complex, recycle of the copper acetate was not
possible.

The economics of this new process, with a solvent recovery stage and the cost
of new copper acetate catalyst, made the process no longer economic, and the
project had to be abandoned.

Example 5: Process integration by the use of hydrogen chloride

Within a single works was a plant for producing fluorocarbons, and a second
plant for the production of methyl chloride, according to the two following reaction
schemes:

$$CCl_4 + HF \xrightarrow{\text{5 bar}} CCl_3F + HCl$$

and

$$CH_3OH + HCl \xrightarrow{\text{1 bar}} CH_3Cl + H_2O$$

158

The fluorination plant produced a gaseous HCl stream which was absorbed in water and sold cheaply as pickling acid in an already saturated market, and the methyl chloride plant burnt Cl_2 with H_2 to produce its HCl, with chlorine being a valuable material in short supply.

A project was therefore initiated to use the HCl from the fluorocarbon plant on the methyl chloride plant.

The HCl from the fluorocarbon plant contained small quantities of HF ($\approx 1\%$) and this had to be removed before the gas could be used on the methyl chloride plant because this plant contained much glass and silica equipment which would be quickly corroded by the HF. Pickling acid, however, was preferred by customers when it contained some HF because it had extra "bite". The proposed solution was to firstly counter-current wash the HCl gas with an aqueous effluent stream of 20% HCl from the methyl chloride plant. This would simultaneously upgrade this stream to a 36% pickling acid (containing HF), and give an HF-free HCl gas for use on the methyl chloride plant, with an overall reduction in chlorine usage and pickling acid production. Figure 5.5 shows the before and after situations.

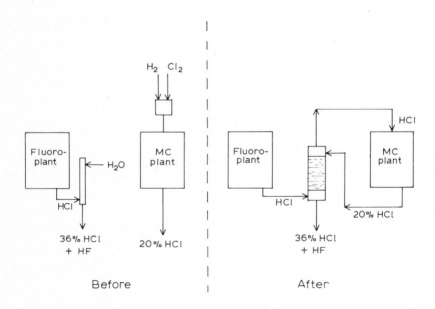

Fig. 5.5. Integration of HCl production on two plants.

The laboratory carried out a successful short-duration test to show that a methyl chloride reactor operated satisfactorily with HCl directly from the fluorocarbon plant (which nobody doubted) and sufficient HCl/HF solubility data were available to design an absorber system which could wash out HF to a known safe level.

Since no difficulties could be anticipated, it was decided to carry the project through without a pilot stage. The $300,000 was allocated and the equipment designed and installed. The plant started-up satisfactorily; no control problems were met, and the methyl chloride process ran satisfactorily. However, the gas leaving the absorber still contained considerably more HF than expected from the absorber design calculations. After two weeks, noticeable thinning of the reactor glass piping on the methyl chloride plant was detected.

Absorption tests using aqueous HCl in a simple Wolfe gas absorption bottle showed that equilibrium concentrations were achieved very slowly. This suggested that a reaction was involved, and this was confirmed by discussions with the chemical analysts. Throughout the years the fluorocarbon plant had developed a method of measuring "total HF", meaning HF and compounds that hydrolyse to HF. A major portion of this is carbonyl fluoride which is only slightly soluble in water but which hydrolyses to HF once in solution. Carbonyl fluoride when passing through an absorber with aqueous HCl would not be washed out because of its low solubility. However, wet surfaces in the gas stream, e.g. the reactor inlet pipes, would with time attain high HF concentrations as the carbonyl fluoride absorbed and hydrolysed, and so be subject to high corrosion rates.

No satisfactory way of removing the carbonyl fluoride from the HF was found, and the plant had to be shut down and the capital written off.

Example 6: Development of a carbon tetrachloride process

Methylene chloride and chloroform can be produced by the chlorination of methyl chloride

$$CH_3Cl + Cl_2 \longrightarrow CH_2Cl_2 + HCl$$

$$CH_2Cl_2 + Cl_2 \longrightarrow CHCl_3 + HCl$$

Small quantities of carbon tetrachloride are also present.

$$CHCl_3 + Cl_2 \longrightarrow CCl_4 + HCl$$

Within the company concerned there was a considerable demand for carbon tetrachloride which was supplied by an outside supplier. To produce this carbon tetrachloride on the methyl chloride plant had the advantage that chlorine was used without the production of HCl because this was integrated with the methyl chloride plant, and a higher degree of chlorination produced a lower methyl chloride recycle and so reduced methyl chloride production costs. To convert the plant to a carbon tetrachloride / methylene chloride plant, however, required operation of the plant in a very different region, in particular the reactors had to have very different operating conditions.

Modelling of the reactors over the limited operation range of the existing plant enabled a prediction to be made of various ways of producing the new product ratio. The best possible way appeared to be the installation of a further adiabatic reactor in series with the existing two, and the introduction of a high recycle of chlorofor

This combination was tested on the full-scale plant. Having installed a third reactor and a chloroform recycle system, the product ratios from the test showed remarkably good agreement with the reactor model predictions.

The test, however, showed up two problems that had not been considered. Firstly, the reactor effluent contained soot due to a burning reaction which gave HCl + carbon. Secondly, the quantity of chlorinated ethylenes rose considerably due to the higher chlorination level.

The soot gave concern regarding the lifetime of the following packed distillation columns and their boilers, and it was considered that some development programme to improve the feed distribution of the recycled chloroform would have to be embarked upon to solve the problem.

The chlorinated ethylenes caused a more serious problem because one particular component, trichloroethylene, was difficult to separate from carbon tetrachloride, and so it was difficult to achieve the required product quality. Distillation tests showed that the separation became very difficult at the low concentrations demanded by the carbon tetrachloride consuming plant. This difficulty was attributed either to peculiar VLE-data at low concentrations, or due to continuous development of trichloroethylene from the cracking of other chlorinated ethylenes in the boiler.

Before claims that a reliable process had been developed could be made, these two problems, which originally had never been anticipated, had to be satisfactorily solved.

Example 7: Scale-up of kinetic data for a gas / solid catalyzed reaction

A particular catalyst was developed for a process which was troubled by a consecutive reaction which seriously affected the selectivity of the process:

$$A + B \xrightarrow{k_1} P$$

$$P + B \xrightarrow{k_2} R \quad \text{(unwanted)}$$

Experimental work was carried out on an SGS reactor, on a single tube integral reactor in the laboratory and on a pilot plant. A model was developed for the reaction, and fitted to the results from these various reactor types. In all cases satisfactory fits to the experimental data could be obtained with the same model formulation. Table 5.1 summarizes the parameters obtained.

TABLE 5.1

Comparison of kinetic data from different reactor scales

Scale	A_1 A_2	E_1 E_2	$r_1 \times 10^3$ $r_2 \times 10^5$		Ratio $\dfrac{r_1}{r_2}$		OHTC Constant
			370°C	430°C	370°C	430°C	
S.G.S. kinetic data	3.62×10^1 4.56×10^{-3}	20208 15824	0.83 8.02	2.89 22.00	10.3	13	---
1.5 m tube (lab)	1.108×10^4 3.935×10^6	16290 26237	1.70 18.9	4.60 99.2	9	4.6	4.4
3 m tube (lab)	9.13×10^2 4.85×10^3	14000 25000	0.91 9.0	2.13 43.6	10.1	4.9	2.25
pilot plant (3 m)	5.5×10^2 5.7×10^5	12812 24244	1.38 13.1	3.0 60	10.6	5.0	2.6

This table shows that the agreement between the different scales is not good, with an error of 50% in the rate of the main reaction being recorded. More serious is the discrepancy of activation energies which results in very different selectivities at different temperatures, the SGS reactor producing a result very much out of line for the relative reaction rates at the upper temperature.

The explanation for these discrepancies was never sought, because the objective in the project was the design of a full-scale plant, and so only the results of the pilot plant were used since the full-scale plant was to have the same tube-size.

However, the table is very interesting in that it shows an example of errors that are obtained in different equipment and different scales. No doubt a deeper study could explain the discrepancies; most probably the model used was too simple and product retardation, or a form of competition for sites existed which was not detected by the relatively narrow range of experiments carried out. Additionally, parameter correlation during fitting may have been a significant problem.

However, these results were obtained in a normal process development environment, and they show that something is not fully understood. They indicated the danger of being too clever. To have designed a full-scale plant from the SGS results alone would have resulted in a small high temperature reactor, whereas a large, low temperature reactor was required to obtain an acceptable selectivity.

It is hoped that the above examples, besides presenting a little light reading, have left the impression that such obscure problems arise in process development that one cannot expect to solve them by cool deep thinking alone; the pilot plant is an essential stage in the development of almost all processes.

FURTHER READING

Jordan, D.G., 1968. Chemical Process Development, Wiley, New York. - Concerned with scale-up and pilot plants for all types of equipment.

QUESTIONS TO CHAPTER 5

(1) Under what circumstances, during the development of a new product, would it be necessary to have:
  (a) a large complete pilot plant ?
  (b) no pilot plant ?
  (c) a small complete pilot plant ?
  (d) a large partial pilot plant ?
(2) What are "Scale-up Problems"?
(3) Why is instrumentation important on pilot plants and what particular instrumentation difficulties are encountered ?
(4) We are all forgetful. What methods are employed by companies to check for forgotten factors in new processes ?
(5) What chemical engineering information can be usefully obtained from a pilot plant ?
(6) What is the role of the modelling in handling pilot plant performance results ?

Chapter 6

THE LOCATION OF OPTIMUM CONDITIONS

Any company owning or designing a chemical plant wants it to incur the lowest
total costs, whilst satisfying the market requirements. Whether the reason for this
is to maximize profits for shareholders, minimize risks from competition, or reduce
the product price to the customer, is of secondary importance to the design engineer,
since they all involve searching for process conditions that will result in maximizing some form of economic criteria.

Chemical processes are highly non-linear systems with optimum settings existing
for most of the independent variables in the system. There are strong interactions
between some variables, and very weak interactions between others. Variables
associated with the same piece of equipment, or associated with recycle flows, have
strong interactions; variables associated with different pieces of equipment not
directly connected usually have only very weak interactions.

The reactor is usually the most important item of equipment for which to select
the best values for the independent variables, because its operation has a direct
effect on raw material costs, and the composition of the reactor outlet stream
determines the cost of the product work-up section of the plant. Hence the optimum
operating conditions for a chemical process are closely associated with the reactor
technology. Because one item is predominantly more important than the others, there
exists the possibility of two degrees of thoroughness for the optimization study:
a study on the most significant equipment item (the reactor) alone, or a more
complete study of the total process.

Precisely what are these process conditions that have to be optimized, particularly
with reference to the reactor? Most important are the reactor conversion and
selectivity to be achieved, and these are defined by the reactor temperature,
compositions, and volume chosen. Secondly, if raw material costs (or recycling
potential) of all raw materials are not identical, optimum reactor feed ratios of
raw materials may not be the stoichiometric ratios. If more than one phase is
involved, mass transfer, and hence operating pressure and degree of agitation,
should be optimized.

These six, namely:
o temperature
o compositions
o volume
o feed ratios
o pressure
o degree of agitation

are the primary reactor process variables to be optimized at the stage of process definition.

In the detailed reactor design stage where the reactor is to be fully dimensioned these six variables are the basis for further sets of variables which must be decided upon before the reactor can be specified and built. These are

heat-transfer area
coolant flow
in and out temperatures
coolant medium
stirrer type
stirrer dimension
stirrer power
stirrer speed
baffle arrangement
reactor geometry.

6.1   THE OBJECTIVE FUNCTION

As already pointed out, there is the possibility of carrying out useful optimiz-ation studies on the reactor alone, on part of the total process, or on the complete process, depending on the accuracy required for the location of the optimum point.

This enables simplified optimization procedures to be used at the beginning of a project, developing into more complex procedures as the project progresses through the design stage.

For each of the stages of thoroughness of the optimization it is necessary to define an appropriate objective function.

6.1.a   The reactor alone

Raw materials are usually the most significant cost item and this leads to reactor optimization conditions being near to the optimum conversion of raw material to product.

In the very early stages of process development the objective chosen is usually to find the conditions that give the highest single-pass conversion of reactants to products.

If the raw materials can be recovered and recycled, then the selectivity, rather than yield, becomes important. If the costs incurred by recycling are negligible, then the point of optimum selectivity will be a good estimate of the best reactor conditions. As costs of recycling increase, the best operating point will move from the best selectivity towards the best conversion.

In cases where there is no "optimum" conversion, but it increases asymptotically to a plateau, the question is how far towards completion is it worth letting the reaction proceed.

The best conversion in this case is a balance between reactor cost and raw material savings (having brought the capital costs onto a yearly cost basis), as will be described in the next section.

For the batch reactor this problem is equivalent to determining the optimum reaction time. This time is again a balance between raw material savings and reactor volume, calculated via the number of batches per year.

For a yearly output of T' tons per year, given 8000 operating hours per year, and a batch cycle time of t hours with a reactor product rate of W tons per m^3 of reactor volume:

no. of batches per year $\qquad\qquad\qquad\qquad$ $8000/t$ $\qquad\qquad\qquad$ (6.1)

production required per batch $\qquad\qquad\qquad$ $\dfrac{T' t}{8000}$ tons $\qquad\qquad$ (6.2)

and so reactor volume V_R is given by $\qquad$ $V_R = \dfrac{T' t}{8000\ W}\ m^3$ $\qquad\quad$ (6.3)

Given a relationship between V_R and reactor costs $C_R(V_R)$, the yearly cost of reactor volume V_R is $f\ C_R(V_R)$ (see section 6.2.a). Hence a rough estimate of the optimum batch time can be obtained from a plot of "yearly cost of reactor volume V_R" plus "yearly raw material costs" vs. "batch time".

These calculations can be used to obtain general guidance on the probable best operating conditions for a plant, which can then be used to guide research into areas of interest, and also be used for an early economic assessment of the process.

6.1.b   Reactor plus an estimate of other plant costs

Methods based on the reactor alone are soon found to be inadequate for process development purposes, particularly if recycles are involved, or work-up costs are considered important. The costs of the remainder of the process can be approximately accounted for in the optimization in a simple way, once a rough cost estimate is available for the individual sections of the plant, in the following way. The reactor influences the design of the rest of the plant because the volumetric flow and concentration of the reactor outlet stream determine the size of all the downstream equipment. For each section following the reactor, the critical component for its cost can be defined, and it can be assumed that the cost of each section is proportional only to this factor.

Examples

(1) The size of a compression section of a process may be assumed to depend only on total flow, Q. For a standard case worked out in detail, a flow of Q_S gave a compressor section investment cost of $\$I_S$. Then for a reactor outlet flow Q', the cost $\$I'$ is given by

$$I' = I_S\ (Q'/Q_S)^{0.6} \qquad\qquad\qquad\qquad\qquad\qquad (6.4)$$

(assuming that investment costs are approximately proportional to the 0.6 power of the size).

(2) A distillation section investment cost may be considered to be proportional to the molar flow of an impurity (C), F_C moles per hour. The distillation section cost for a standard case has been shown to be $\$I_S$ for a flow of F_{CS} moles per hour impurity. For any other impurity flow leaving the reactor, F_C', the investment cost will be approximately

$$I' = I_S (F_C'/F_{CS})^{0.6} \tag{6.5}$$

This method can be extended to provide an estimate of utility costs. If, in the second example the total utility costs per year were $\$I_{US}$ for the standard case, then for any other reactor outlet condition

$$I_U' = I_{US} F_C'/F_{CS} \tag{6.6}$$

Clearly, this is only an approximate method, but it is useful in early development work once a first complete cost estimate is available for a base-case design.

6.1.c Total process optimization

More detailed optimization requires a redesign and re-costing of the whole process for a number of different points until the optimum is found. In this case the objective of the optimum can be an accurate economic evaluation by any economic criteria preferred. This, of course, cannot be done very thoroughly because of the enormous calculation effort required, unless the total process and economic calculation are carried out by a computer.

Hence, for a thorough total process optimization, a process model is essential.

6.2 ECONOMIC CRITERIA

Economic analysis of projects is complicated by the two types of costs involved. Early single payment capital costs must be combined with annual cash flows in some way before a profitability can be determined.

Various economic criteria have been defined which combine these two costs in different ways and the selection of the most appropriate criterion is always a point for discussion, but never one of agreement.

Net Present Worth (NPW) assumes all money flowing shall be credited with a fixed interest rate (r), and if any project has overall a positive sum after payment of the interest, the project is worth embarking on.

$$NPW = \sum_{n=1}^{L} \frac{C_n}{(1 + r)^n} \tag{6.7}$$

where C_n is the total cash flow from year n, and L is the life of the project.

The discounted cash flow (DCF) or IRR method calculates the interest rate r that will make the NPW equal to zero, i.e. when

$$\sum_{n=1}^{L} \frac{C_n}{(1 + r)^n} = 0 \tag{6.8}$$

then

r = DCF.

This interest rate is then compared with some standard interest rate to see whether the project is economically attractive or not.

The annual cost method is a much less thorough method than the above two discounting methods, but it can be applied when yearly costs are approximately constant.

The analysis is made for a typical year, and the method tests that the cash flow C_n in the typical year minus the depreciation gives an acceptable return on the capital employed.

For a capital cost I and an L year life project, the annual cash available after putting aside the depreciation is

$$C_n - I/L \tag{6.9}$$

This must be an acceptable interest rate on the investment I, i.e. where

$$r = (C_n - I/L) / I \tag{6.10}$$

r must be attractive when compared to currently available interest rates.

The two discounting methods are generally to be preferred to the annual cost method because they correspond more exactly to economic theory. Unfortunately, the NPW and DCF method do not produce identical results and this is particularly evident when they are used as the economic objective in optimization studies.

Much process optimization is balancing capital and annual costs, and it is here where the two methods show their major differences. If the DCF is used for the objective in choosing between two alternatives, the alternative with the greater DCF rate is chosen. As the discount rate increases, early cash flows (particularly, therefore, the investment) become relatively more important than late cash flows. Hence this method will tend to choose optima with a low investment and a high early cash income.

The NPW method discounts at a fixed minimum discount rate and so has much less emphasis on the early cash flows. Hence, optimization with NPW will give a different-looking plant, with higher capital and lower running costs than a DCF optimized plant would give, unless the DCF rate is below the NPW discount rate, in which case the process is simply uneconomic and will not be built anyway.

For proper economic investment studies the firm's management must decide which is the most appropriate criterion to use and this should, in theory, be used at the detailed optimization level.

For engineering purposes, DCF is far less convenient to use than NPW, and so, in practice, unless there are very strong grounds for objection, the NPW is generall used as an economic objective function for optimization studies. The differences between the two methods can be minimal if the discount rate used in the NPW calculation is raised artificially to the region that the DCF rate is expected to attain at the optimum condition.

6.2.a   Modified economic criteria for optimization studies

The full economic analysis requires all sales income and expenses to be included in the calculation. At the process development and design stages this information is often not available, and it is also an unnecessary burden to obtain it for the technical design.

The NPW method is simply a discounted summation of all flow occurring in each year.

$$NPW = \sum_{n=1}^{L} \sum_{m=1}^{K} \frac{C_{m,n}}{(1 + r)^n} \qquad (6.11)$$

where m are the individual cost centres.

These cost centres could be divided into two:

$C_{1,n}$ being cost associated with the technical design - capital cost, raw material cost, utility cost, maintenance cost, labour cost, etc., and

$C_{2,n}$ being purely commercial cash flows - sales income, packaging and transport costs, company overheads, etc.

Hence,

$$NPW = \sum_{n=1}^{L} \frac{C_{1,n}}{(1 + r)^n} + \sum_{n=1}^{L} \frac{C_{2,n}}{(1 + r)^n} \qquad (6.12)$$

or

$$NPW' = NPW - \sum_{n=1}^{L} \frac{C_{2,n}}{(1 + r)^n} = \sum_{n=1}^{L} \frac{C_{1,n}}{(1 + r)^n} \qquad (6.13)$$

For the optimization of the technical details of a process, the commercial costs ($\sum \frac{C_{2,n}}{(1+r)^n}$) remain constant. Hence the optimization of the discounted technical costs alone (NPW') will result in the same optimum point as the optimization of the complete NPW.

This is a very convenient condition which makes NPW very attractive as an

economic objective function. This is not the case for the DCF method, and full
economic cost information must be available for this method to be used.

The annual cost method is also very convenient for engineering work because it
is possible to relate capital and yearly cost together very simply as follows.

For an amount of capital spent (I), the income in an average year should result
in sufficient income to cover annual charges, such as maintenance, associated
with it, should make good the steady loss in value through depreciation, and should
still yield an acceptable income from the capital involved. Table 6.1 shows the
magnitude of these items and possible ranges.

TABLE 6.1
Annual capital-incurred costs (% of capital)

	Mean	Min.	Max.
Depreciation	7	4	20
Interest on capital	15	5	25
Maintenance	5	2	10
Insurance, capital incurred costs	1	0	2
f =	28	11	57

By discussion with accountants and Works' personnel, the appropriate f can be
determined for the type of process under consideration.

Process optimization can now be carried out to minimize total annual costs,
where these are given by:

$$\frac{f}{100}I + C$$

where C are the total annual costs for the process (excluding capital associated
costs), i.e. raw materials, utilities, and labour costs in a typical year.

This method is excellent for quick studies. In general, for an investment to
be worthwhile, it must produce annual cost savings of between 1/3 and 1/4 of the
investment involved.

6.2.b   The inclusion of tax

The economic criteria so far discussed have not included taxation. Taxation
complicates the whole economic analysis somewhat, and the effect is difficult to
predict;   it   sometimes improves the economics, and sometimes makes the investment
worse.

Investments are allowed against tax, and so profits equal to the investment made,
spread over a number of years, are exempt from tax. The number of years over which

this exemption can be spread is the "depreciation life-time of the plant for taxation purposes". This can be different from the expected life-time of a process.

This has the effect of making investment less costly and so the inclusion of tax in the economic objective function tends to give preference to capital investment at the expense of running costs. In geographical areas where a government might feel investment should be stimulated, particularly speedy tax allowance depreciation rates, or even investment grants, may be in force. These again have the effect of preferring capital to running costs, and so they will affect the position of the optimum in a detailed optimization study, as well as the overall profitability of the project.

It is therefore advisable that, in a serious process optimization study, the economic objective function should include the tax calculation.

For details of such calculations, a standard engineering economic text should be consulted. For a real study, company accountants and local authorities should also be consulted to determine the exact rules that are in force in the area under consideration.

For engineering purposes, the effect of tax relief on investment capital can be lumped entirely into a tax factor (F') which is applied to all the investments in the process. In place of investment I, F'I is used in the calculation of the objective function.

It can be shown that F' is related to the straight-line depreciation lifetime (for taxation purposes) m, discount rate r, and fractional taxation rate r_t, by the formula (Rose, 1976)

$$F' = 1 - \frac{r_t}{(1 + r)^{mr}} \left(1 - \frac{1}{(1 + r)^m}\right) \tag{6.14}$$

The use of this tax factor is analogous to the accountant's use of tax tables. It dispenses with the need to calculate annual depreciation to be removed before tax is deducted from the yearly profit. Its use is demonstrated in the following example.

Example. A process has an investment capital I and a yearly sales income of S_n, with yearly total costs of C_n, where n = 1 to L, since the project has an L year life.

Hence, the yearly "profit" for taxation purposes is:

$$S_n - C_n$$

With a tax rate r_t, this means the firm keeps

$$(S_n - C_n)(1 - r_t) \quad \text{in year n.}$$

No allowance for depreciation is necessary because this is allowed for in the tax factor.

Hence, the NPW after tax is given by:

$$NPW = F'I + \sum_{n=1}^{L} \frac{(S_n - C_n)(1 \div r_t)}{(1 + r)^n} \tag{6.15}$$

assuming investment is paid out one year before income is received.

This overcomes the tedious business of allowing for yearly depreciation, which is particularly messy when investment in more than one year is involved.

It also enables one to see the immediate impact of investment changes on the NPW after tax without having to recalculate the whole NPW over m years.

6.2.c   The effect of inflation

The discussion of any optimization method which is balancing capital costs to be spent now to save future annually occurring costs, must define its position on inflation before the results have any worth.

Since prices rise as costs rise, the effects of inflation are minimal when all inflation rates are the same, and prices can be adjusted without delay to account for inflation.

When this is the case, the economic analysis can also be simplified by using the concept of constant value money, as opposed to current money, (the money in our pockets), whose real value falls with time.

Hence the objective function for process optimization is based on constant value money, and changes with time occur only if there are expected to be relative differences in inflation rates between, say, labour and raw materials.

We are all aware that inflation stimulates investment; if we do not buy our car this year, we will not be able to afford it next year.

This tendency does not show itself in our optimization using constant value money, and this is because the discount rate used in the analysis is the average long-term rate to be expected through the process life, whereas our decision with the car is a very short-term study.

Long periods of inflation which cause a lowering of the constant value money interest rates should be treated by using the lower interest rate in the objective function. This will then push the balance towards capital investment rather than annual costs. It is usual to use too high values of the discount rate in economic analyses; 12-15% is a usual range employed, whereas surveys suggest that 7% is the average achieved by industrial companies, and in inflationary periods the constant value interest rate quoted by banks is of the order of 2% or even less. However, it is not the job of a reaction engineer to point out these errors to the company accountant!

6.3 OPTIMIZATION METHODS

6.3.a The case study approach

The most usual way in which engineering designs are "optimized" is by a series
of case studies. One complete design is carried out, and then alternative cases
are repeated with changes in those variables thought to be significant. The
repeated calculations show which are the more important variables, and these are
then further changed until the best values are found.

When this is done by hand it is very time consuming and is usually therefore
limited to about ten cases. Before the work starts, the engineer, from his experienc
and his own initiative, will have already suggested a good process as the base
case, and these further calculations of other cases are really perturbations to
check that he has made a reasonable choice.

The method is not very thorough and not very accurate, but full use of engineerin
judgement is made, and the process becomes well-understood as the man carries out
his series of hand calculations. It requires no special training and there is no
long period of inactivity as there is when computer programmes are under development
When optima are fairly flat, and they usually are, not very much is gained from an
accurate location of the optimum, and so such approximate methods can be quite
satisfactory.

6.3.b Sensitivity studies

This, in principle, is similar to the case study approach, where a systematic
variation of all important variables is made to determine in which direction to
move to find the optimum.

Starting from a reasonable standard operating point, every variable is moved
forward and backward one step, and the trend on each variable determined. After
this round it is possible to generalize, and say that certain variables should be
increased, others decreased, and some seem to have no significant effect. A new
standard is then defined taking these trends into account, and a new sensitivity
study performed about this point. This is continued until no further improvement
can be found. Figure 6.1 shows one round of such a procedure.

The method can be applied without the need for specialized knowledge, and after
the procedure has been carried out, a good deal is known about the reaction. Highly
correlated variables, such as reactor temperature/contact time, cause problems, and
choice of step length can be difficult.

Developments of this method are the experimental planning algorithms such as the
Box-Wilson response surface method. Experimental design methods, however, have also
to cope with random experimental error, which is not the case in process optimizatio

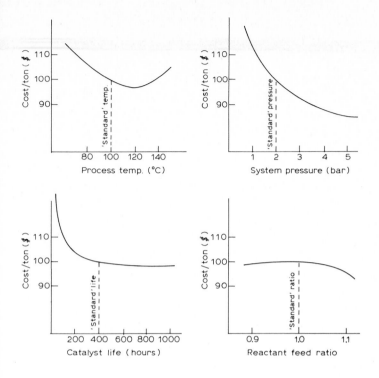

Fig. 6.1. Sensitivity curves for a new process.

6.3.c   The use of non-linear optimization methods

The most sophisticated approach is to use one of the standard non-linear optimization packages to enable the optimum values of the parameters under study to be determined.

Such methods require a complete description of the process to be available in a programmed mathematical form. They then use this model to carry out repeated evaluation with different values of the parameters in a systematic way until an optimum is located which cannot be bettered.

The different optimization algorithms use different systematic ways of searching. The most common is the direct search method, (e.g. Rosenbrock, Powell), which moves along different axes until the optimum in each direction is found. Then compound directions are used and the cycle repeated until no further improvements can be made (see Fig. 6.2).

Within the general framework there are a number of different details possible and this gives rise to the different direct search algorithms.

Rather less useful in a process optimization context are the steepest ascent methods (e.g. Davidson, Fletcher and Powell). These require the derivatives of the objective function with respect to every optimization variable to be analytically

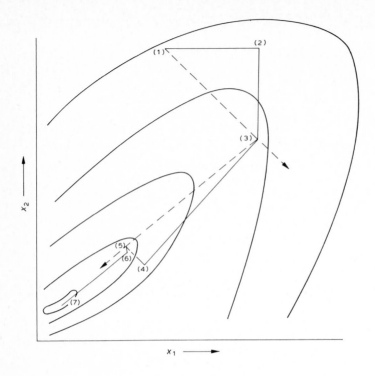

Fig. 6.2. Optimization as a hill-climbing procedure.

or numerically available. For a complex model this often introduces difficulties.

Adaptive random search (ARS) is an interesting method in that it consists of searching randomly, simply remembering the best-yet result, and adapting the distribution from which the random variables are chosen to centre on this best result. As the optimization proceeds, the search area follows the best point and also narrows down on it. Although not very sophisticated mathematically, the method does have certain advantages in an engineering environment, as will be discussed later.

Constraints. Reactor and process models have severe constraints: temperatures, pressures, mole fraction, etc. must all be constrained to physically meaningful values. Constraints are very difficult to handle satisfactorily with mathematical optimization procedures. The routines must always have a numerical response with which to determine the best direction for the search. To return with "value not calculable - constraints violated" is not allowed. The value returned by the model must also belong to a smooth response surface; any discontinuities cause most optimization routines to fail. Even mathematical routines specially developed to handle explicit or implicit constraints have their efficiency of search reduced by the presence of constraints. The algorithms spend much time developing a definition

of the surface, so that they can finally make progress in the best direction. If
they hit constraints in this direction, not only is that single evaluation wasted,
but also the whole previous set which determined that best direction.

Excluded from these generalizations is the adaptive random search, because it
makes no use of previous information and can handle "no function value - constraint
violated" returns simply as unattractive function values. In a highly constrained
system, there comes a point where more progress is made in random function evaluations
than in sophisticated algorithms which have a high chance of failing due to constraint
problems. This makes ARS more attractive for process modelling work than would at
first seem likely. ARS is highly unlikely to locate an optimum accurately, but it
will always provide some improvement on the starting value, as the result of many
hundreds of random experiments.

A general problem in non-linear optimization is the presence of more than one
optimum. Algorithms locate one optimum, but this is no guarantee that it is the
only, or the best, in the allowable field. Repeated evaluations from different
starting points is the only solution to this problem. This increases the chance of
finding other optima, and then the best can be selected.

Optimization algorithms are very attractive methods of locating optima in
reactor and process models. There should, however, be some words of warning, since
they are not always the best approach. The following restrictions and disadvantages
can be listed:
(1) The process must be completely describable by a robust continuous mathematical
    model.
(2) The objective function must be mathematically definable and singular; i.e. a
    subjective balance between economics and operability cannot be handled.
(3) Constraints must not cause problems.
(4) A good understanding of the response surface of the model is not obtained. An
    optimum point may be found, but the process understanding given by a case
    study or sensitivity approach is lacking.

An interesting combination is the use of a case study or sensitivity approach
but each time using an optimization routine to locate, say, three important
interactive variables. In this way, the problem associated with the sensitivity
approach caused by strong interaction between variables is solved, the "feel"
given by the sensitivity approach is not lost, and the optimization problem is
simplified, since only a limited number of variables are involved in the search.

6.4   TOTAL PROCESS MODELS
To achieve a thorough optimization study one must study the effect of the
reactor conditions on the whole process. One therefore needs a model of the total
process, somewhat more detailed than the rough method described in section 6.1.b.

6.4.a   Individual equipment models

As usual, there are a number of different levels of complexity that can be
chosen for the models of the individual items in the process, and the key to good
modelling is to choose the right degree of detail to employ to obtain a satisfactory
result without excessive model development work.

In principle, the equipment model must calculate the flows leaving the unit, as
a function of the flows entering the unit. If the model is then to be used for
economic optimization, it must also give the capital cost of the unit and the
yearly operating costs.

The relationship between leaving flows and entering flows can be based upon
simple split fraction rules built in the model, or based upon short-cut design
methods (or idealized design methods), or based on detailed design methods that
would be used to specify the dimensions of the equipment. As an example, consider
a distillation column:

o The split fraction model would specify always x_i% recovery of compound i in the
   exit distillate stream, the remainder leaving in the bottom stream.
o The short-cut method would use a Fenske-Underwood model to determine the componen
   split.
o The idealized design method may have a constant molar overflow procedure.
o The detailed design method may include enthalpy balances and plate efficiency.

The determination of the costs could also be carried out at different levels. Fro
the process flows it may be possible to relate the costs to a cost for a separately
calculated standard condition, as described in section 6.1.b. In more detail, one
could dimension the equipment and accurately determine the capital cost from the
equipment dimensions. The degree of detail employed in this calculation leads to
further degrees of sophistication of the model and accuracy of the result.

6.4.b   Specific process models

If computer models are prepared for all the significant items in the process then
these can be linked together in a master program, the individual output from one
model being used as the inlet to the next. This can be done conveniently by having
the main program as a series of subroutine calls which call each equipment model in
turn. When recycles are involved, some thought to determine the best simultaneous
equation-solving method must be given.

The simplest approach is to use repeated substitution in which the whole process
model is repeated, using the last calculated value of the recycle stream. Such a
technique is used widely in process modelling and is generally stable. More
sophisticated is the use of a non-linear simultaneous equation-solving algorithm to
accelerate the convergence.

The fact that all the models have been individually written and compiled together
into a specific model provides the opportunity to include a specific solution

method for the process in question. It is often possible to use some algebra to determine an explicit solution, which does away with the need to iterate, or at least derive a system containing fewer unknowns to be solved simultaneously. This is particularly important if the model is to be used in conjunction with an optimization procedure.

To illustrate this point let us consider the following example.

Example. Consider a simple process containing a reactor and distillation column, with the first-order reaction

$$A \to B$$

occurring in the reactor with a fixed conversion X_A. The distillation column produces a stream of unreacted A containing 5 mole % B, which is recycled to the reactor, and the bottom stream is the product B containing 10 mole % of A (see Fig. 6.3). The plant output required is T' kmol s^{-1} of B.

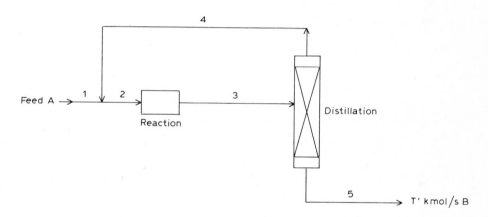

Fig. 6.3. Simple process stream connections with recycle.

For the modelling of the individual units it is necessary to know the flows in all streams required to produce an output of T' kmol s^{-1} product.

We can write down the flow in stream 5 as:

$$B_5 = T' \tag{6.16}$$

$$A_5 = T'/0.9 \tag{6.17}$$

and by an overall mass balance

$$A_1 = T' + T'/0.9 \tag{6.18}$$

Let the flow of A in recycle stream 4 be A_4, as yet unknown. Hence, by the definition of the performance of the distillation column

$$B_4 = 0.05 \ A_4/0.95 \tag{6.19}$$

Hence,

$$A_2 = A_1 + A_4 = A_4 + T'(1 + 1/0.9) \tag{6.20}$$

$$B_2 = B_4 \qquad = 0.05 \ A_4/0.95 \tag{6.21}$$

The reactor converts this to stream 3, which is the sum of streams 4 and 5; hence, a balance on A about stream 3 gives:

$$(1 - X_A) \ A_2 = (1 - X_A)(T'(1 + 1/0.9) + A_4) = A_4 + T'/0.9 \tag{6.22}$$

which by rearrangement gives

$$A_4 = \frac{T'}{X_A} \ [(1 + \frac{1}{0.9})(1 - X_A) - \frac{1}{0.9}] \tag{6.23}$$

Hence all stream flows are known in terms of X_A and T', and there is no need to involve iterative techniques to determine the recycle flows.

Such manipulations are specific to each process and so can only be carried out when the process model is specifically written for that process.

This algebraic manipulation holds for the significant components in the process. Trace components, or components which have no significant effect on the process can be ignored in the convergence, and also ignored in many parts of the mass balance.

The preparation of the equipment models for all the significant items in the process is an extensive task, taking months of programming work. The result, however, is a very satisfactory process model, tailored exactly to the needs of the problem, providing mass balance, investment cost and running cost data, taking a minimum of computer time because insignificant component flows have been ignored, and recycle convergence problems have been solved by selecting the best method for each individual case. The model is therefore ideally suited for optimization using a standard optimization algorithm.

The big disadvantage of the specific process model is the time and effort required to produce the model. The programming costs may be higher than the project development budget allows, the project may be moving faster than the programming work, so that the job is finished using "subjective optimization" before the model is ready, or the project may have been abandoned before the modelling can make any contribution whatsoever.

These disadvantages are overcome using standard models and program structure offered by generalized flowsheet programs.

6.4.c   Generalized flowsheet programs

A generalized flowsheet program is a set of computer programs and subroutines that allow any process to be modelled by data input alone. This means the user is no longer troubled by having to do any computer programming. All unit operations are present in a library of standard unit operation subroutines. The flowsheet is fed in as data, by numbering all streams in the process, and defining which items they come from, and which items they go to. The "executive" part of the flowsheet program manages the whole calculation, and ensures that the output from one unit is the input to another, as defined by the data defining the process streams.

These programs deal also with the equation-solving problem caused by recycles. Because the programs are completely general, a general form of equation-solving has to be used, and this is usually repeated substitution or an accelerated form of iteration method such as Newton-Raphson or multi-dimensional secant. All cases require the repeated evaluation of the flowsheet until convergence is achieved. Although algebraic manipulation methods akin to the matrix manipulation methods of Chapter 2, section 2.3.b, and the specific model example (section 6.4.b) have been developed, they are not available in well-established programs.

The library of standard unit operation models usually consists of between 10 and 30 fairly robust general-purpose models for reaction, distillation, adsorption, compression, heat exchange, flash, etc. Because these models are general, they must be rigorously correct and treat every component thoroughly; e.g. a trace of an inert gas must enter every plate calculation of a distillate column calculation, although in practice we know it can be simply placed completely in the distillate.

The advantage of the general flowsheet program is that within a matter of hours one can have a process model available which gives a complete mass and heat balance, and with which one can alter conversions and feed ratios to see the effect on the overall process.

The disadvantage is that the models are all very general, so specific reactor problems cannot be considered without writing new unit subroutines, and the general nature of all the subroutines makes them pedantic and slow. Similarly the convergence of recycles cannot be enhanced by explicit solution.

The result is that flowsheet programs take probably ten times as much computer time as the specific process models to produce a process mass balance. This is of significance if a numerical optimization technique is being considered to locate the optimum values for the design and operating variables.

Types of standard reactor model. Some progress has been made in representing reactor models by standard programs, and there are a number of levels of complexity that have been attempted. A good flowsheet program will have a selection to represent the different levels of detail of study.

(a) The simplest level is the model which takes a general reaction:

$$\nu_A A + \nu_B B + \nu_C C \ldots . \longrightarrow \nu_G G + \nu_H H + \nu_I I$$

where A to I are components in the process, and ν_A to ν_I are the stoichiometric ratios of each component.

   The reactor conversion X is stated, and the reactor routine has to identify which component is limiting, relate the conversion to this component, then by stoichiometry determine the quantity of the remaining components reacted, and so calculate the total outlet gas flows which are then returned to the executive part of the program. For data this model needs to know

(1) which components are involved in the reaction

(2) what are their stoichiometric coefficients (negative for raw materials, positive for products)

(3) what is the reactor conversion.

(b) The next level of reactor model is the one that considers an equilibrium reaction

$$\nu_A A + \nu_B B + \nu_C C \ldots \underset{K}{\rightleftharpoons} \nu_G G + \nu_H H + \nu_I I$$

with an equilibrium coefficient

$$K = \prod_{i=A}^{I} C_i \tag{6.24}$$

and conversion on the limiting component is X.

(c) A stage more complex is a model to represent multiple reaction occurring:

$$\nu_{A_1} A + \nu_{B_1} B \xrightarrow{k_1} \nu_{G_1} G + \nu_{H_1} H + \nu_{I_1} I$$

$$\nu_{A_2} A + \nu_{B_2} B \xrightarrow{k_2} \nu_{G_2} G + \nu_{J_2} J + \nu_{K_2} K$$

$$\vdots$$

   etc.

Hence, one must define how many simultaneous reactions are occurring, and for each reaction what are the components involved, what are their stoichiometric coefficients, their equilibrium constants and their conversions.

   Because of the multiple reactions it is not possible to easily identify the limiting components for each reaction on which to apply the conversion criterion and special techniques have to be incorporated to do this.

Generalized models of this complexity are becoming a contradiction in themselves, because they are very complex algorithms based on very rough suppositions that reactors can be categorized by quoting a conversion on their limiting components.

(d) It is therefore not surprising that rather than even more complex general models it is usual to write specific reactor models which are compatible with the generalized flowsheet program subroutine interface, and then use the specific reactor models with general models of all other items of the plant. Such models would be virtually identical to those derived in Chapter 2, with a standard beginning and end so that they accept the inlet flows from the executive program and deliver the exit flows in a format acceptable to the executive. Examples of this are given in Appendix 3.

6.4.d Optimization with process models

Having mounted the reactor model in a process model one is then prepared to really determine the "optimum" conditions for the reaction which best benefit the total process as a whole. Unfortunately, the step from the material balance to a process model that also carries out the economics, i.e. calculates utility usages for all items and capital costs, is still considerable.

With a specific process model this can be done if there is time in the project for the engineer to carry out the extra programming necessary, but with generalized flowsheet models most available programs do not go so far as costing and the process model must be used without an economic criterion.

In practice, therefore, process models are often used to determine optimum operating conditions for the reactor and the rest of the process without an objective function!

This is possible because:

(1) First of all the model allows a large number of alternative mass balances to be investigated, and one can use simple criteria such as minimizing recycle flows as a first estimate of good operating conditions.

(2) Much emphasis is put on the engineer's previous experience, intuition and simple rules-of-thumb. Approach temperatures of approximately 5°C, and operating points a little before steep rises in costs can be expected are chosen. When two counteracting factors are the major contribution to an optimum, then a simplified hand calculation to find the approximate location of the optimum point for these two factors is carried out independent of the model.

(3) Thirdly, the case study approach is used, whereby, after considerable testing of alternatives with the model has been carried out, a few (e.g. three) specific cases are taken and worked up by hand through the capital and running costs, to an economic analysis. This enables the best of the three to be chosen, but also the work-up by hand provides further insight, and this suggests new cases that

may have further advantages (i.e. the engineer's experience required for (2) is deepened).

It is always a source of confusion for students to be confronted with a process model with which to set process conditions without having an economic objective. Their usual comment is "Having gone so far, why not go all the way to the economics?" It is all a factor of time, and it is an art that must be learned to use these tools with understanding and not to hunt randomly for the highest economic result.

With the specific process model, and given time, however, it is possible to include some form of economic objective in the process model, so that in addition to the total process mass balance, the model delivers the economics of the case under study. There remain various levels of detail by which the cost and economics can be included in the model. A useful quick method is that of scaling all capital and operating costs for each item, according to a significant flow for that equipment item.

This method is described under section 6.1.b. Before this method can be used there must have been one complete costing exercise carried out on each item (by hand) in order to get the basic figures which are then scaled according to the new conditions. Given more time, one can calculate an accurate cost for each equipment item (section 6.4.a).

Having obtained the costs for the process, the economics can be calculated in a simplified form (section 6.2.a); given more time, with forecasted sales figures available and a need to impress management, a full NPW or DCF analysis can be programmed (section 6.2). In all cases the contribution of tax relief on investment capital can be accounted for by using the tax factor approach, as described in section 6.2.b.

Once the model calculates a single objective function, the way is open for optimization as described in section 6.3. The case study approach, or sensitivity analysis and following manual optimization, are very useful methods in that they achieve guaranteed improvements and deliver an understanding of the process.

Given still more time and a sense of adventure, the total process model can be used as a subroutine of a non-linear optimization algorithm and the optimum sought automatically, as described in section 6.3.c. Note that this is usually only possible with specific process models because

o they can be programmed to evaluate a single objective function;
o they can be programmed as a subroutine of an optimization algorithm;
o they have comparatively low calculation times which allow optimization to be carried out within reasonable computer costs.

The only reported work on the use of optimization routines with generalized process flowsheets uses the adaptive random search method to overcome constraint problems. With computer runs of 1-2 hours' duration handsome improvements over initial conditions are achieved (Gaddy, 1979), though there is no evidence that the "mathematical" optimum had been achieved.

Optima are often very flat, and so there is no great loss in being slightly off the optimum settings for the variables. It is this fact that enables the engineer to achieve good designs by intuitive use of a flowsheet program without having a single economic objective. The few critical variables that there are he is able to recognize and treat separately.

A question that has never been satisfactorily resolved is the worth of very accurately locating an optimum when the models used in the search are themselves not accurate. To precisely locate an optimum within 1% when the model used has a ±5% error seems questionable. The situation is not as bad as it looks at first sight, as there is a little evidence to show that rough models indicate optimum settings more accurately than the model accuracy might suggest (because optima are flat), but it does suggest that in practice the extra benefit gained from the use of optimization algorithms is not going to be more than a few percent.

6.5   TIME PROFILES
6.5.a   Batch and semi-batch reactors

Continuous reactors have constant operating conditions, and the determination of the optimum operating conditions requires a single value to be determined each for temperature, feed flow, feed ratio, etc. Batch reactors have conditions that vary with time, and this means that the optimum condition of each operating variable should be defined as a function, varying with respect to time. Hence, all variables require an optimum "profile" to be determined.

Our objective function F is an integrated function of the initial condition in the reactor x_{i_0} and the operating variables settings U_j, where U is a function of time.

$$F = \int_0^t f(x_{i_0} U_j(t)) \, dt \tag{6.25}$$

The optimum control profile for each operating variable $U_j(t)$ is that which maximizes the objective function F.

The objective could be maximum yield, selectivity, or in a production limited situation, maximum output per unit time from the reactor.

In practice, this optimization is severely constrained in that all the operating conditions have limited allowable ranges. The strict mathematical solution to this type of problem was developed by Pontryagin in 1956. The method demands a high degree of mathematical skill, as each case must be developed analytically afresh. The result of the method is often proof that a variable must stay on its minimum or maximum constraint, or change from its minimum to its maximum at some point, but sometimes the method results in a mathematical function describing the optimum profile to be followed by a specific control variable.

The advantages of the method are primarily that it provides a mathematical proof that a certain control policy is optimal, and secondly it defines that control polic It finds application in control engineering, particularly in space travel, where the systems can be easily described mathematically.

The disadvantages of the method for use in reactor engineering are:

o the systems are far from easy to describe by a mathematical function;

o the mathematics can be formidable, requiring the attention of a specialist;

o the resulting profile cannot easily be installed on a plant because of the complexity of the function;

o and finally there is a need to relate the control function to the state in the reactor. The control function can be related to time and temperature, but conversion or composition usually cannot be continuously measured.

Hence the main use of the Pontryagin maximum principle in reactor engineering has been to study various idealized cases and published "optimal" profiles for specific examples (see Tables 6.2 and 6.3). When we have a real, complex system, we can compare it with published idealized results and propose simplified functions of similar shape to the optimum one, but that are capable of being installed on a plant.

It is often possible to use simpler methods to define reasonably good profiles which are easier to employ than Pontryagin, but still achieve adequate accuracy. For example, consider a batch chlorination where chlorine is fed to a reactor containing the material to be chlorinated (A). Clearly, for acceptable low chlorination times and chlorine usage, a high chlorine feed should be used at first, reducing as the reaction rate reduces due to the lower concentration of A.

For the chlorine feed F_{Cl}, one could choose a feed linearly decreasing with time:

$$F_{Cl} = a - bt \tag{6.26}$$

This equation could be introduced into the reactor model and the values of the parameters a and b determined which gave optimum performance, if necessary by means of an optimization routine.

In a somewhat more refined way, we could see from Table 6.3, case F3, that a Pontryagin study of such a system showed the optimum function to be:

$$F_{Cl} = a + b \, C_A^{\frac{1}{2}} \tag{6.27}$$

Again, entering this in the model and using an optimization routine would give an optimum value of a and b, so defining the feed profile.

However, such a profile has no practical value if C_A cannot be continuously monitored. Hence, a function

$$F_{C1} = a\ e^{-bt} \qquad\qquad (6.28)$$

may be more nearly optimal than the linear profile and also capable of being installed on the plant.

If very complex profiles are expected, the reaction time could be divided into, say, eight equal sections and eight feed constants used:

$$F_{C1} = a_i \quad i=1,8 \qquad\qquad (6.29)$$

An optimization routine could then determine optimum values for a_1 to a_8, and, depending on their values, a more convenient function might be devised.

Profiles are important to the feeds to semi-batch reactions. Also often effective are temperature profiles when competing reactions are present. When a parallel reaction has a higher activation energy than the main reaction, high temperatures should be avoided, but this leads to long reaction times. Economic operation, therefore, would suggest low temperatures at first, increasing at the end, where concentrations and reaction rates are low and side reactions less significant.

An opposite profile is preferred by an exothermic equilibrium reaction. An initial high temperature provides a quick reaction, and a low temperature at the end improves the equilibrium concentration.

Tables 6.2 and 6.3 summarize profile studies available in the literature (Rippin, 1979) for both feed profiles and temperature profiles in batch reactors, most of which are the result of Pontryagin analyses of simple reaction systems. Improvements in performance reported over non-profiled operation are entirely dependent on the specific case under study. Savings between 3 and 50% are reported.

TABLE 6.2

Performance improvement by profile of temperature in batch reactors

Key: P = product; R,S = waste products.

Kinetics	Conditions of reaction parameters	Form of best profile
T1 $A \underset{2}{\overset{1}{\rightleftharpoons}} P$	1. $E_1 > E_2$	$T = T_{max}$
	2. $E_1 < E_2$	
T2 $A \underset{2}{\overset{1}{<}} \begin{array}{c} P \\ R \end{array}$	1. $E_1 > E_2$	$T = T_{max}$
	2. $E_1 < E_2$	
T3 $A \begin{array}{c} {}^{1} \nearrow P \\ \underset{3}{\overset{2}{\rightarrow}} R \\ \searrow S \end{array}$	1. $E_1 > E_2, E_3$ and $E_1 < E_2, E_3$	can be treated as in case T2.
	2. E_1 between E_2 and E_3	1. where reaction time is not significant, maximum yield of product is obtained with a very long operating time at a calculated constant temperature. 2. when available reaction time is limited:
T4 $A \overset{1}{\rightarrow} P \overset{2}{\rightarrow} R$	1. $E_1 > E_2$	Operate at $T = T_{max}$ for a calculated time.
	2. $E_1 < E_2$	

TABLE 6.2

Performance improvement by profile of temperature in batch reactors (contd.)

Kinetics	Conditions of reaction parameters	Form of best profile
T5 $A \underset{2}{\overset{1}{\rightleftharpoons}} P \overset{3}{\rightarrow} R$	1. $E_1 < E_2, E_3$	If reaction time is not significant use very long operating time at $T = T_{min}$. Otherwise
	2. $E_1 > E_2, E_3$	Treat as case T4.1
	3. E_1 between E_2 and E_3	$E_1 > E_2$ gives rising T profile. $E_1 < E_2$ gives falling T profile.
T6 $A \overset{1}{\underset{3}{\underset{2}{\rightrightarrows}}} \begin{smallmatrix}P\\R\end{smallmatrix}$	$E_1 < E_2, E_3$	Similar to T5.1
T7 $A \overset{1}{\rightarrow} P \overset{2}{\underset{3}{\rightrightarrows}} \begin{smallmatrix}R\\S\end{smallmatrix}$	E_1 between E_2 and E_3	Maximum yield of P obtained by operating at constant temperature equal to that calculated for case T3.1, but for a calculated time.
T8 $A \overset{1}{\underset{2}{\rightrightarrows}} \begin{smallmatrix}P \overset{3}{\rightarrow} R\\S\end{smallmatrix}$	$E_3 < E_1 < E_2$	
T9 $A \overset{1}{\rightarrow} I \overset{3}{\rightarrow} P$ $\,{}_{2\downarrow}\quad{}_{4\downarrow}$ $\,R\quad\ S$	1. $E_1 < E_2$ and $E_3 > E_4$	Rising T profile
	2. $E_1 > E_2$ and $E_3 < E_4$	Falling T profile

TABLE 6.2

Performance improvement by profile of temperature in batch reactors (contd.)

Kinetics	Conditions of reaction parameters	Form of best profile
T10 $A + B \xrightarrow{1} P + R$ $P + A \xrightarrow{2} R + S$	1. $E_1 > E_2$	Operate at $T = T_{max}$ for a calculated time.
	2. $E_1 < E_2$	Possible forms of profile.

TABLE 6.3

Performance improvement by profile of feed addition rate

Key: A,B = reactants; P = product; R,S = waste products

Kinetics	System studied	Form of best profile
F.1 $A + B \overset{1}{\underset{2}{\rightleftharpoons}} P + R$	PFR Gas phase reaction in tube (i.e. multiple feed points along the tube)	Rate of addition of A_2 approximately constant after initial charge.
F.2 $A + B \rightarrow P$ $A + 2B \rightarrow R$	Semi-batch	
F.3 $A + B \rightarrow P + R$	Continuous bubbling of gas B through charge of liquid A to produce liquid product P and gaseous waste product which carries B with it. (Semi-batch)	$F \propto C_A^{\frac{1}{2}}$

6.5.b Time profiles in continuous reactors

This usually implies some form of catalyst deactivation, resulting in the reactor performance dropping with time. The question arises, how do the operating conditions change to still achieve optimum performance with the deactivating catalyst ?

This is again an optimum profile with constraints problem that has been tackled by a few research workers using Pontryagin, but such techniques are not yet applied in practical situations.

The problem is, in practice, further complicated by the following common characteristics.

○ Precise data on deactivation are generally not available except for very few well-established processes.

○ The catalyst deactivity is often a function of operating conditions, and so the very change of operating conditions to follow an optimum performance may accelerate the deactivation. Precise data on such deactivation are generally not available, and so quantitative treatment is not possible.

○ There are often requirements of product quality which demand reactor deactivation to be countered by a temperature rise to achieve the same reactor conversion as before, so limiting the changes to the reactor itself.

When deactivation occurs mainly in the front part of the reaction zone, deactivation moves progressively down the reactor with a fairly sharp boundary. In this case it is usual to employ excessive catalyst; the conversion remains fairly constant until the bed is used up and break-through occurs, see Fig. 6.4. This point is fairly sharp and choice of optimum catalyst life and operating conditions is therefore not difficult.

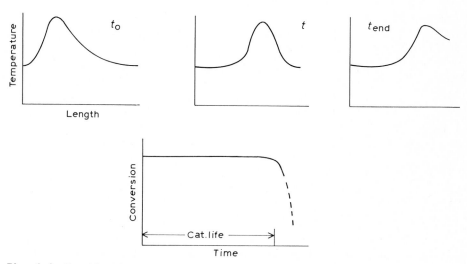

Fig. 6.4. Deactivation progress through a catalyst bed.

However, when deactivation is steady over the whole reactor bed, one cannot rely on a sharp break-through and the overall reactor performance steadily decreases until a point is reached when it is not considered economic to proceed further (see Fig. 6.5).

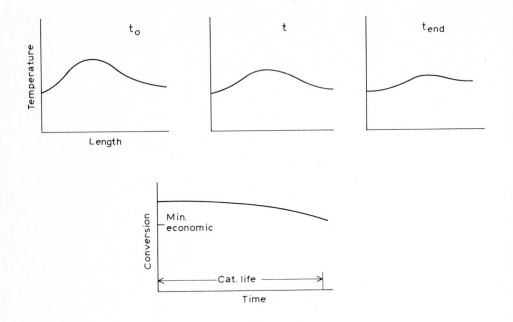

Fig. 6.5. Steady catalyst decay through the whole reactor bed.

Over the life of the catalyst there is a steadily declining conversion in the reactor. To choose the optimum conditions during a cycle it is necessary to repeat the process optimization at a number of time points to determine both the optimum profile for the operating conditions and the economic catalyst life.

When it is necessary to ensure a minimum variation in reactor conversion then temperature must be increased to maintain conversion, or the throughput must be progressively reduced, or the reaction bed broken into a number of segments, and each segment changed in rotation, so that at steady state the reactor contains a spectrum of catalyst activities, and when one section is exhausted and replaced by fresh catalyst the change in conversion is only a fraction of that of changing the whole bed (see Fig. 6.6). If the policy is to maintain conversion by reducing throughput, this technique of handling a number of beds and renewing them in a cycle produces a reduced range over which the feed flow must vary.

Given a particular function for the deactivation, it is possible to calculate the variation in feed rate to maintain a constant conversion (Prenosil, 1979).

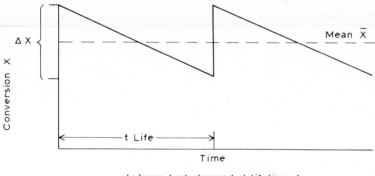

(a) one bed, changed at lifetime t

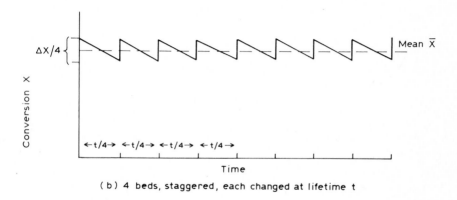

(b) 4 beds, staggered, each changed at lifetime t

Fig. 6.6. Enzymatic hydrolysis of whey by single and multiple beds.

For an exponential deactivation with half-life $t_{\frac{1}{2}}$ in a continuous reactor, the kmol s^{-1} reacted at time t $(F_t X_t)$ is related to the reaction rate initially $(F_0 X_0)$ by:

$$F_t X_t = F_0 X_0 \, e^{-t \, \ln 2 / t_{\frac{1}{2}}} \tag{6.30}$$

If we want to operate at a fixed conversion then $X_0 = X_t$, and we must vary the flow rate with time:

$$F_t = F_0 \, e^{-t \, \ln 2 / t_{\frac{1}{2}}} \tag{6.31}$$

If we define the catalyst life in terms of $L_{\frac{1}{2}}$, the number of catalyst half-lives,

$$L_{\frac{1}{2}} = t/t_{\frac{1}{2}} \tag{6.32}$$

then for a reactor of N beds, which are changed sequentially,

$$F_t = F_0 \, e^{-\left(\frac{L_{\frac{1}{2}}}{N} \ln 2 \right)} \tag{6.33}$$

Equation 6.33 relates the variation in feed to be expected to maintain a constant conversion when a distinct catalyst life is chosen, and the catalyst is to be installed in a number of beds.

This analysis is based upon a known deactivation function and a deactivation which occurs to all the catalyst in the bed equally, and not a deactivation of the type shown in Fig. 6.4.

6.5.c Length profiles for continuous plug flow reactors

Continuous plug flow reactors can benefit from temperature profiles down the length of the reactor if there are side reactions with higher activation energies than the main reaction, or if exothermic reversible reactions are involved. The theory for such profiles is exactly the same as for the batch reactor temperature profiles (section 6.5.a) since the equations governing the two systems are similar.

Contrary to the situation with batch reactors, however, it is comparatively difficult to have a reactor with different temperature levels along its length. This usually has to be done in steps, and each step requires a separately controlled heating circuit. It may also be that each temperature level requires a separate reactor shell. Location of the optimum profile would be made identically to that of determining the profiles for batch reactors.

6.6 ALLOWING FOR UNCERTAINTY

Having carefully optimized our reactor operation and process we must remain aware that the information we have used has not been exact, and uncertainties are present in:

(1) The model inaccuracies through inexact mechanistic descriptions and inaccurate parameter estimates. These are typically ±5-10%.
(2) Market uncertainties resulting in different product demands with time. These in particular affect the size of processes decided upon (Rose, 1976). When a plant is involved with two products, uncertainties also exist concerning the ratio of the two products that will be sold, and this ratio can be a major factor in determining operating conditions.
(3) Raw material quality and composition can vary. Composition of natural raw materials such as crude oil or clay for the cement manufacture is very dependent on their place of origin. A change in political or commercial thinking can resul in a raw material for a plant coming from a new source, with a completely

different composition from that assumed in the original design.

Plant design fashions change; they used to be designed to keep the Works happy (with capacities 50% greater than rated); then to keep the management happy and to prove engineering competence (i.e. capacities 5% greater than rated). The present trend is towards "flexible" plants so that the designer looks into the future and designs a plant that will reasonably handle possible future variation. This means designing the plant a number of times with each of the possible futures. Clearly, reactor and process models are a great help in such an exercise to remove the labour involved in repeated calculations.

It is possible to carry out probabilistic optimizations, attributing probabilities to each possible feed composition, market forecasts, and product ratio, and carrying out a monster optimization to determine the optimum design with a Monte-Carlo simulation as an inner loop, but such methods are of theoretical, rather than practical, use.

In practice, one uses a probabilistic study to determine a suitable plant size, and then a deterministic study to optimize the operating conditions of that plant.

6.6.a   Reactor design safety factors

Reactors are the heart of processes, thus to have a reactor that is too small limits the whole process. Against this, reactors are often very expensive equipment items, and so to install one that would be too big is a considerable waste of money.

How should one handle the problem of the reactor design safety factor to cover the fact that parameters, model, and scale-up contain uncertainties which produce a probability distribution for the output of the reactor and not a single point ?

To determine the safety factor based purely on finding the economic optimum usually shows that safety factors should be small (Rose, 1976). For expensive equipment and not over-profitable projects or projects with a slow market build-up, negative safety factors are usual (i.e. it is not worth building a reactor that might be too big). Only cheap equipment and profitable projects call for safety factors guaranteeing 70-90% confidence of achieving the required output, i.e. substantial positive safety factors. When one takes the analysis further to include the possibility of later plant modifications or earlier plant expansion, then the case for safety factors in reactors is still further reduced.

Methods exist for calculating the optimum safety factor based on economics (Rose, 1976) and these can be used in an open situation, where inter-departmental politics is not a problem.

Usually, however, there is sufficient competition with the different departments in a firm to make it essential that the plant designed by the engineering department achieves its rated capacity, simply to prevent inter-department bickering. In this case, the reactor must have a +95% confidence level of achieving its rated output -

194

no argument! In some cases contracts are signed guaranteeing certain capacities.
In such cases, again, a 95% confidence limit would be a sensible criterion for design

This could be achieved by adding a cool 10-20% safety factor on to the design,
or by predicting the standard deviation of the expected full-scale performance,
based on the modelling work and a Monte-Carlo Simulation, and then design with a
safety factor equal to twice the standard deviation.

6.7 WHEN TO OPTIMIZE

Processes begin in research departments, move through to development departments,
from there to pre-capital sanction engineering, and finally to post-sanction
engineering.

Modelling can begin in research departments and follow through into the
development department, resulting in a process model developing in parallel to
process development. Pre-sanction engineering is normally rather superficial,
certainly no question of process optimization, because the project may not be
sanctioned. Post-sanction engineering is usually done in a great hurry - the project
has been shown to be profitable, and the sooner it operates, the more profitable
it will be. Reactors and process model development for optimization in this phase
is usually considered to be too late.

In this total procedure it is clear that the models must be prepared during the
development stage, ready for immediate use in the post-sanction engineering stage.

If there has been no development stage (repeated engineering of an existing
process, for example) or if no model was produced in the development stage, then
in the post-sanction stage a generalized flowsheet model with a standard reactor
model and a case study optimization is the only possibility open. This is because
setting the process operating conditions is the very first step in a process design,
and it is very rare to find an engineering department willing to devote the first
three months of its post-sanction engineering to their optimization. Three months
is a reasonable estimate of the time it would take an experienced team to produce
a useful specific reactor and process model followed by an optimization study with
that model.

FURTHER READING

Crowe, C.M., Hamielec, A.E., Hoffman, T.W., Johnson, A.I., Woods, D.R. and Shannon,P.T
    1971. Chemical Plant Simulation. Prentice Hall.
Gill, P.E. and Murray, W., 1974. Numerical Methods for Constrained Optimization.
    Academic Press.
Hoffmann, U. and Hofmann, H., 1971. Einführung in die Optimierung. Verlag Chemie.
Murray, W., 1972. Numerical Methods for Unconstrained Optimization. Academic Press.
Rose, L.M., 1976. Engineering Investment Decisions - Planning Under Uncertainty.
    Elsevier.

REFERENCES

Gaddy, J.L., Heuckroth, M.W. and Gaines, L.D., 1976. AIChE J., 22:744.
Prenosil, G., Peter, J. and Bourne, J.B., 1979. Enzyme Engineering. Weetall, H.H.
    (Editor). Plenum Press.
Rippin, D.W.T., 1978. Improvements obtained in the Performance of Batch Reactors
    by the Adjustment of Operating Conditions during the course of the Reaction.
    SIA/FVC Tagung, Basle, Switzerland, October. (SEG/V/13/78, TCL - ETH, Zürich).

QUESTIONS TO CHAPTER 6

(1) Before any type of optimization can be attempted, what must be available ?
(2) What forms of objective function are there ?
    What is necessary before an economic criterion can be used, and what is the
    effect of including taxation in the criteria ?
(3) What methods are available for "optimization", interpreting the word broadly ?
(4) For the gas phase system

$A + B \rightarrow P$ product $\qquad E_1 = 60,000$ kJ kmol^{-1}

$P + B \rightarrow R$
 $\}$by-product $\qquad E_2 = 85,000$ kJ kmol^{-1}
$A + A \rightarrow S$ $\qquad E_3 = 85,000$ kJ kmol^{-1}

list the variables that are free for optimization at the process development
stage.
(5) Describe the various types of model suitable for a process optimization study.
(6) When will a "total process optimization" result in different optimum reaction
    conditions from a "reactor alone" optimization ?
(7) What methods are available for determining optimum profiles for operating
    variables ?
    Name the analytical method and describe some alternative methods.
(8) What do you understand by a "negative safety factor" and when would it be
    optimal ?

Chapter 7

THE DESIGN OF STIRRED TANK REACTORS

For a simple single reaction with zero heat of reaction, no phase changes and no side products, the only requirement for the full-scale tank reactor is that it supplies the residence time demanded by the optimization study which balances conversion against capital cost. There must be adequate mixing to disperse the feeds and this can be achieved by a stirrer designed to provide an adequate pumping rate. Hence the design of such a tank reactor be it for batch, semibatch, or CSTR, is relatively straightforward.

Unfortunately, most systems are not so simple. Heat is usually generated which must be removed, and the relation between heat generation and heat removal is usually different on the pilot and full scale, unless very special measures are taken to prevent this. A change in the heat balance causes changes in reaction temperature which can change the conversion and cause the reaction to work at sub-optimal yields.

When the chemistry is complex, mixing is also more critical:
- inadequate mixing of feeds can cause undesirable side reactions due to local high concentrations before complete dispersion is achieved;
- the mixing rate affects the heat-transfer coefficient and so reflects on reactor temperature;
- the mixing can affect wall temperatures which, during the heating phase, may result in overheating and selectivity loss in the vicinity of the heating surface.
When more than one phase is involved the mixing affects the mass-transfer rate between the phases. This affects stationary concentrations in the liquid and so affects reaction rates in complex reaction situations. Hence sub-optimal yields may be obtained if the full-scale mixing results in a different mass-transfer coefficient to the pilot or optimal case.

Hence, for almost all industrially important reactions care must be taken in the design of the full-scale tank to see that the heat can be removed and the degree of agitation is adequate.

Some mixing theory will now be discussed as a precursor to the design of a full-scale tank reactor.

7.1 MIXING IN STIRRED TANKS

Mixing is carried out in stirred tanks in order to meet a number of criteria:
(1) The incoming feeds must be well and quickly distributed in the bulk liquid (macromixing).

(2) The whole reactor contents must be adequately mixed so that bypassing does
    not occur, and liquid in all parts of the reactor has the same residence
    time distribution and the same concentration.

(3) An acceptably high heat-transfer coefficient for the heat-transfer surfaces
    within the reactor must be achieved.

(4) When mass transfer is involved an acceptably high mass-transfer rate between
    phases must be achieved.

(5) For certain high speed complex reactions particularly thorough mixing must
    be achieved (micromixing) to prevent local high concentrations and low
    selectivities, whilst the mixed pockets of raw material are not yet mixed
    on the molecular scale. This situation is rarely important but it does some-
    times occur and therefore should always be borne in mind.

Stirrer design is aimed at achieving these requirements. This demands a mixture of
qualitative and quantitative thinking. The stirrer type selection, its dimensions,
and arrangement in the tank are qualitative; the calculation of stirrer speed to
achieve required pumping rates and heat- and mass-transfer rates, and the prediction
of power requirements is quantitative and based upon the use of published empirical
correlations.

7.1.a   Stirrer type selection

   Impeller. There are a bewildering number of designs of impeller for tank reactors,
brought about to a large extent by the empirical nature of the subject. Each stirrer
manufacturer has his own design which he has developed from trials and has found to
be satisfactory. There are a number of different principle types, each of which is
suitable for a particular duty and these are summarized in Table 7.1.

TABLE 7.1
Stirrer types.

Marine type impeller
Design similar to normal marine impeller. Flow is axial. Suitable for
achieving high circulation in low viscosity systems.

Turbine - pitched blade
Flat blades angled to provide some axial flow. Suitable for achieving
high circulation rates. Suitable for normal mixing duties in low to
medium viscosity liquids.

Turbine - flat blade
Simple stirrer with radial flow.

Vane-disc
Essentially a flat-bladed turbine with a disc covering the upper side. Operated at high speeds this has high shear on the under side and so is used for dispersion of introduced second phases, the inlet being positioned below the centre of the stirrer. The high shear breaks the introduced phase into small bubbles or droplets which is an essential first step to good mass transfer. Flow is radial.

Paddle stirrer
Essentially slow moving stirrer with large diameter to mix tank contents comparatively slowly, when there is no need to achieve quick mixing. This might find application in reactors for very slow reaction but generally is not used in reactor design.

Anchor stirrer
For use in vessels where heat transfer to the walls is required but the material is viscous and so turbulence at the walls can only be achieved by the stirrer blade being at the walls.

Double motion stirrer
An anchor stirrer with counter revolving blades connected to the shaft. For use in very high viscosity systems over 10000 cp (e.g. grease manufacture).

Helix
Rotating helix for use in extremely viscous systems over 1000000 cp.

Baffles. In most cases vessels should be supplied with four baffles around the edge to prevent the rotating motion of the fluid (Fig. 7.1). Baffle widths are normally 1/10 of the tank diameter (see Fig. 7.8). By means of the baffles, vortex formation is minimized which otherwise would produce problems:
 - in deciding on the operating capacity of the vessel
 - of air being drawn into the impeller
 - of a smooth rotating motion being created instead of a turbulent axial and radial
   mixing motion.
Baffles therefore should be installed with all stirrer types except the anchor, double motion, and helix stirrers.

Number of impellers. With deep tanks it is advisable to provide multiple impellers to ensure that agitation occurs over the whole tank volume (see Fig. 7.2). Multiple impellers should be installed at distances between 1.0 and 1.5 d_I. Tanks should be fitted with multiple impellers when L/D is greater than 2.0-2.5.

Draft tubes. Good circulation within a stirred tank can be achieved by means of a draft tube which induces a full bottom to top flow pattern (Fig. 7.3).

The draft tube can be formed by a coil if this has to be supplied in the vessel.

The draft tube has the disadvantage that it is extra internal surface in the reactor which can cause difficulties in cleaning.

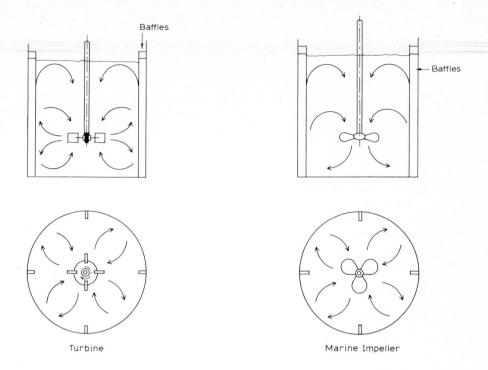

Fig. 7.1. Baffle arrangements and mixing patterns for a stirred tank.

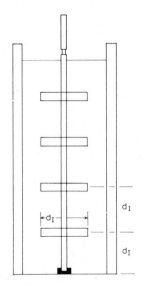

Fig. 7.2. The use of multiple impellers in deep tanks.

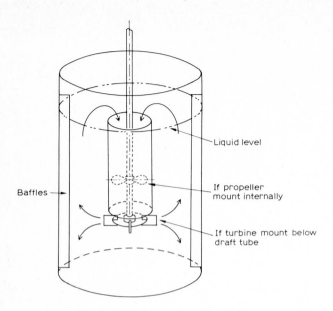

Fig. 7.3. The use of a draft tube to ensure good tank circulation.

7.1.b   Determination of stirrer speed

The degree of agitation, and hence the stirrer speed, depends upon the reason for the stirring. It may be simply to ensure that the materials are well mixed, as in a batch reactor operating adiabatically. It may be to ensure that the inlet is effectively mixed, as is the case with the CSTR reactor. It could be to ensure that a reasonable heat-transfer coefficient is achieved at heat-transfer surfaces in the reactor, or it may be to disperse a second phase with sufficient violence to assure a good surface area for mass transfer. These different criteria are discussed in the following paragraphs.

Impeller pumping rates for CSTRs. A general rule of thumb with both experimental and theoretical backing is that circulation rates should be four to ten times the throughput rate in a CSTR for an adequate approach to perfect mixing (Rase, 1977). The pumping rates for impeller type pumps are given by the relationship

$$Q/N d_I^3 = K'$$

(7.1)

where K' is the discharge coefficient and is dependent on impeller design, d_I is the impeller diameter (m), N is the revolution s^{-1} of the stirrer, and Q is the pumping rate in $m^3 s^{-1}$.

For marine-type impellers, a value of 0.5 is recommended for K'; for turbines the value of $0.93 D/d_I$ has been suggested for a particular type at high Reynolds numbers.

For accurate work this constant has to be determined experimentally or given by the agitator suppliers.

Batch mixing time. For batch reactors an important criterion is that the time for "complete" mixing of the various reactants should be negligible compared to the reaction time.

Various workers have investigated the time required for complete mixing to occur, and correlations in terms of number of circulations of the fluid (Q/V_R) and number of stirrer revolutions (N) have been presented for various stirrer designs.

In general it appears that between 50 and 200 stirrer revolutions are required for complete mixing. This normally yields a time insignificant with the residence time required for reaction. However, if a fast reaction, sensitive to concentration, was under investigation, then this criterion should be reconsidered, and more accurate predictions of mixing time obtained from a text on mixing (Uhl, 1966).

Mixing criteria to achieve adequate heat and mass transfer. These criteria are treated in detail in later sections. Section 7.2 discusses heat transfer in some detail and the following chapter is concerned with mass transfer in tank reactors.

Other mixing criteria. Besides pumping rate/feed rate, batch mixing time and adequate heat and mass transfer, two other useful criteria to characterize the degree of agitation are the blade tip speed and the power input/unit volume of fluid.

The tip speed relates to the shear put on to the liquid, and is particularly relevant when one phase is being distributed into a second. Table 7.2 relates tip speed to the type of stirrer duty required.

TABLE 7.2
Tip speed for various mixing duties.

Duty	Tip speed (m s^{-1})
normal mixing	2.5 - 3.3
heat-transfer duty	3.5 - 5.0
high shear - 2 phase dispersion duty	5.0 - 6.0

It is fairly logical to assume that the degree of mixing is characterized by the power/unit volume put into the liquid. This can be used both to define the degree of mixing and to a limited extent to scale up. Table 7.3 shows the power requirement for various degrees of agitation.

7.1.c   Power requirements for agitators
The power requirement for an agitator is correlated by means of the dimensionless power number (N_p). The power number is defined as:

TABLE 7.3

Degree of mixing and power input/unit volume

Power input/ unit volume kW m^{-3}	Degree of agitation achieved	Application area
0.005	gentle	blending
0.1	mild	homogeneous reaction
0.3	moderate	reaction with a heat-transfer requirement
1.0	high	
2.0	intense	two-phase system, gas/liquid, liquid/liquid

$$N_p = P'/\rho \ N^3 \ d_I^5 \qquad\qquad\qquad\qquad\qquad\qquad (7.2)$$

where P' is the power required by the agitator and N is the stirrer speed (rev s^{-1}).

N_p can be determined from the Reynolds number and impeller design and size, using Fig. 7.4. At high impeller Reynolds numbers, i.e.

$$(\rho \ N \ d_I^2/\mu > 100)$$

the power number is independent of stirrer speed, but it is dependent on stirrer design, as indicated by the hatched area on Fig. 7.4. This Figure can be used to provide rough estimates of stirrer power by using the median curve. For more specific cases the reader should refer to the detailed surveys given in texts on mixing (Uhl, 1966; Nagata, 1975).

Having obtained the appropriate power number for the stirrer design and the degree of turbulence in question, the stirrer power requirement is calculated as

$$P' = N_p \ (\rho \ N^3 \ d_I^5) \quad kW \qquad\qquad\qquad\qquad (7.2a)$$

Agitators can be of such varying designs, viz.
- type
- number of blades
- diameter of agitator
- width of blades
- pitch of blades

that power correlations are available for only a small number of types of stirrer. The texts on mixing discuss the matter in more detail, but it is usual to make contact with mixer manufacturers who have empirical correlations for their own designs. It is usual for instance to have the mixer supplied by the tank manufacturer

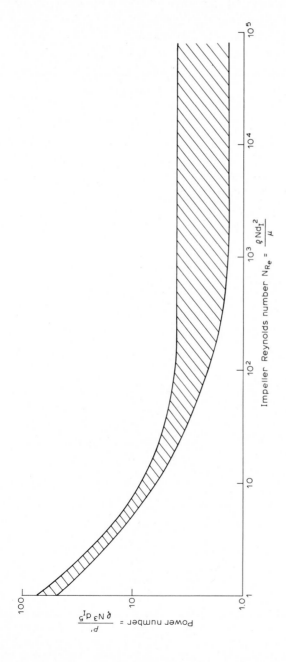

Fig. 7.4. Power number vs. Reynolds number for agitator power prediction.

Hatched region covers the results for various common geometries of agitator

and he can determine the power requirements for the degree of mixing specified by the reactor engineer.

7.1.d Checking the effectiveness of mixing

It is sometimes necessary to check on the effectiveness of mixing in existing equipment. This may be a pilot plant reactor, and its mixing needs to be character- ized so that the reactor behaviour is well understood, or it may be a plant reactor that is not performing well and the mixing is suspected. The techniques described in this section are also particularly useful when novel reactor arrangements are developed, in order to determine their effective mixing patterns to allow modelling of the reactor to proceed.

Problems that can occur due to poor mixing are
- bypassing of less well-mixed zones
- parallel fluid paths
- differing degrees between plug flow and complete mixing behaviour.

These problems can be identified by carrying out tracer tests on the system by injecting an impulse of tracer (dye or sodium chloride solution) into the feed, and observing the distribution of the tracer in the outlet liquid. Figure 7.5 shows the results of such tracer experiments on various malfunctioning reactors. These curves are generally referred to as the "C-curves" and they represent the age distribution of the fluid in the system (Levenspiel, 1972).

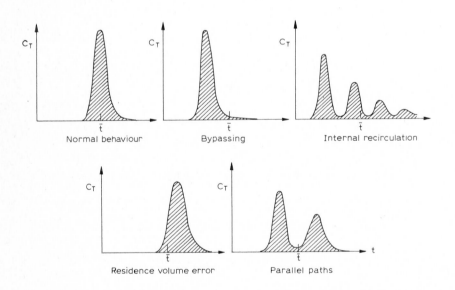

Fig. 7.5. Exit tracer patterns from malfunctioning reactors.

To normalize the results of non-ideal mixing, time is generally quoted as reduced time θ, as a proportion of the mean residence time $(\bar{t})$

$$\theta = t/\bar{t} \tag{7.3}$$

The same tracer techniques can be used to characterize complex reactor systems such as partitioned stirred vessels and wide tubular reactors by comparing the tracer results with the mathematical treatment of a series of stirred tanks.

For one perfectly mixed tank, after the addition of an impulse of tracer T which results in a mean tank concentration C_{T_0}, an instantaneous mass balance gives

$$-V_R \frac{dC_T}{dt} = Q\ C_T \tag{7.4}$$

Therefore

$$t\frac{Q}{V_R} = -\int_{C_{T_0}}^{C_T} \frac{dC_T}{C_T} \tag{7.5}$$

Therefore

$$C_T = C_{T_0}\ e^{-t/\bar{t}} = C_{T_0}\ e^{-\theta} \tag{7.6}$$

since $\bar{t} = V_R/Q$

For N CSTR tanks in series it can be shown that

$$C_T = C_{T_0}\ \frac{N(N\ \theta)^{N-1}}{(N-1)!}\ e^{-N\ \theta} \tag{7.7}$$

The C/θ curves for various values of N are shown in Fig. 7.6.

By comparing the result of a tracer experiment with this Figure it is possible to say how many stirred tanks in series the equipment is equivalent to. This is of considerable use in a qualitative understanding of the equipment under study, and it can also be used with care for quantitative modelling work.

7.2   HEAT TRANSFER IN STIRRED TANK REACTORS

Stirred tank reactors often have to remove substantial quantities of heat from the reacting mixture. This can be done in a number of ways. Refluxing of a boiling solvent is a very common procedure whereby the heat of vaporization of the solvent is removed from the system by the reflux condenser, and the condensed solvent is returned to the reactor.

Cooling the walls of the reactor by means of a jacket with cooling medium is a second method, and insertion of cooling heat-transfer area, usually in the form of coils, in the reactor is a third.

If still further heat must be removed beyond that possible from heat exchange

206

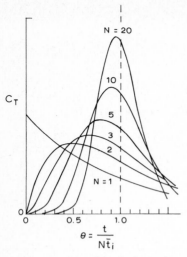

N = number of tanks in series

Fig. 7.6. Tracer results from a series of mixed tanks.

area within the reactor, then external heat exchange area has to be supplied and a pump-around system installed (Fig. 7.7).

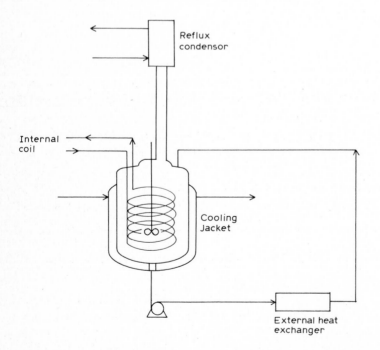

Fig. 7.7. Heat removal possibilities on a stirred tank reactor.

The design of the boiling solvent heat-removal process presents no particular design difficulties. The quantity of solvent vaporized can be calculated from latent heats and care must be taken to check that the vapour disengaging velocity is kept below that which produces excessive entrainment; for example it is advisable not to exceed superficial vapour velocities of 1 m s^{-1}. The only unpredictable problem lies in the tendency of the reacting mixture to foam; stable hexagonal foam can be created at very low gas velocities and the factors causing its appearance are difficult to quantify. If the problem does arise, anti-foam additives can usually be used to overcome it.

The design of heat-transfer area within the reactor requires a knowledge of the various heat-transfer coefficients involved. These are discussed in the next section.

7.2.a Reactor inside film coefficient

The most important resistance to heat transfer is usually the "reactor contents to wall" coefficient. This coefficient is usually low because the fluid velocities are not high, and sometimes there are viscous fluids involved.

The magnitude of this heat-transfer coefficient is basically governed by the intensity of stirring within the reactor. It is correlated by the dimensionless equation (Chapman, 1964)

$$\frac{hD}{\lambda_L} = K' \left(\frac{\rho N d_I^2}{\mu_b} \right)^{0.65} \left(\frac{c_p \mu_b}{\lambda_L} \right)^{0.33} \left(\frac{\mu_b}{\mu_w} \right)^{0.24} \tag{7.8}$$

where λ_L is the conductivity of the bulk liquid, μ_b and μ_w are viscosities of the fluid at bulk and wall temperatures, D and d_I are reactor and impeller diameters respectively, h is the required inside film heat-transfer coefficient, and K' is a constant depending on the stirrer design. For the reactor geometry shown in Fig. 7.8, K' = 0.73. For other geometries this value changes, and for other stirrer types there can also be changes in the powers of some of the dimensionless groups. For a survey of the modified forms of equation 7.8 for different types of stirrer for both reactor walls and coils see Rase (1977) or Nagata (1975). Literature data in this area are incomplete because of the number of systems and geometries possible and so an important source of data for the reactor designer is his pilot plant.

7.2.b Reactor outside (jacket) coefficient

The jacket coefficient is dependent on the structure of and flows in the jacket itself. Common are fully jacketed vessels, and inside such a jacket are usually baffles or a helix to ensure proper coolant distribution and no bypassing. A second type of cooling system is to have a spiral of half-round pipe welded on to the outside of the reactor wall. These two possibilities are shown in Fig. 7.9.

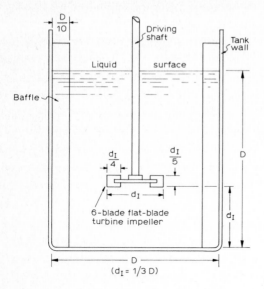

Fig. 7.8. Reactor configuration of Chapman, Dallenbach, and Holland for heat transfe and stirrer power correlations.

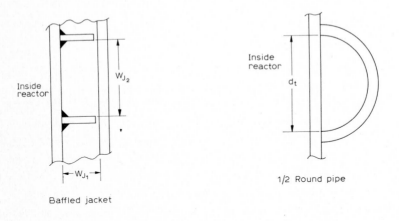

Fig. 7.9. Types of reactor jacket cooling arrangement.

The calculation of the heat-transfer coefficient in the inside of these jackets can be done using normal heat-transfer coefficients for inside pipes using an equivalent diameter (d_e) in place of the pipe diameter.

For the baffled jacket (Fig. 7.9a)

$$d_e = 4 w_{J_1} \qquad (7.9)$$

For the half-round pipe (Fig. 7.9b)

$$d_e = \frac{\pi}{2} d_t \qquad (7.10)$$

For calculation of heat transfer on the inside of coils, the normal inside pipe heat-transfer coefficient can be used with an enhancement factor of $[1 + 3.5\ r]$ where r is the ratio of pipe internal diameter to coil diameter (McAdams, 1954).

Pressure drops in the coolant flow through these various types of jacket can also be calculated using normal pipe pressure drop correlations using an equivalent diameter. That is, for baffled jackets:

$$d_e = \frac{2\ w_{J_1}\ w_{J_2}}{(w_{J_1} + w_{J_2})} \qquad (7.11)$$

and for half-round pipes:

$$d_e = \frac{\pi}{2} d_t / (1 + \frac{\pi}{2}) \qquad (7.12)$$

The fully jacketed, internally baffled arrangement is used for reactors that are comparatively small and do not have high heat-transfer fluid pressures. Note that the internal baffle is essential to prevent bypassing and dead spots in the coolant circuit.

For heat-transfer fluid with pressures over 8 bar the welded half-round pipe is the preferred construction. This can be used up to pressures of 50 bar. This is also the preferred form for large diameter reactors.

The type of reactor jacket is, however, governed by mechanical and manufacturing considerations, and need not be specified by the reactor designer, since it will be determined by the manufacturer constructing the vessel.

7.2.c Some typical values of overall heat-transfer coefficient

For initial studies before the detailed design stage, it is inconvenient to calculate heat-transfer coefficients from detailed correlations when some typically achieved data are available and can be used in the study.

Table 7.4 gives a number of achieved coefficients that can be used for studies in the early stages.

7.3 THE REACTOR ITSELF

Since the stirred tank reactor is a very common piece of equipment, and also quite complicated to fabricate, manufacturers provide standard designs, which designers should make every attempt to adhere to.

Each manufacturer can provide details of his range and these should be obtained during the design. The manufacturer is also capable of providing data on stirrer

TABLE 7.4

Some typical achieved overall heat-transfer coefficients in stirred tanks.

Fluids			
Jacket	Tank	Wall material	OHTC U(kW m^{-2} K^{-1})
water	water	steel	0.15 - 0.3
steam	water	steel	0.5 - 1.5
steam	boiling water	steel	0.7 - 1.7
steam	water	loose lead lined	0.05
water	water	tile-lined	0.04
Half-round pipe			
water	water	steel	0.3 - 0.9
Internal coil			
water	water	steel	0.5 - 1.2

Source: VDI Waermeatlas (1974), VDI Verlag, Dusseldorf.

power, overall heat-transfer coefficient and mass-transfer rates for his equipment.

Standardization is so advantageous for this equipment that there is available a DIN (DIN 28130) standard for a stirred tank with stirrer drive. Figure 7.10 shows the standard form and Fig. 7.11 shows the DIN standard for the enamelled steel impeller.

As can be seen, the jacket, top, bottom outlet, body flange and stirrer drive all demand relatively complex workshop practices, and this has led to the use of standard equipment.

Rather upsetting for the engineer, having carried out his optimization study, is the list of standard reactor volumes (see Table 7.5). These cover from 63 to 40,000 litres, but in steps of a factor of 1.6. If one size is just too small, a 60% overcapacity has to be installed to use standard equipment. This unfortunately is a fact of life, that a 60% overcapacity standard piece of equipment is cheaper than a one-off manufacture of the correct size. The standards should be adhered to simply to minimize investment costs and for this reason they are widely used by industry.

The maximum reactor size in the standard range is 40 m^3. It is possible to manufacture tanks of greater capacities as special designs. For example, wood digesters in the pulp industry are a form of batch reactor which work at 6 bars pressure, and digesters with volumes up to 350 m^3 have been built. These vessels are vertical cylinders with an L/D ratio of about 2:1, both ends are hemispherical, and the vessel has no stirrer; mixing is achieved by circulation of the liquid by

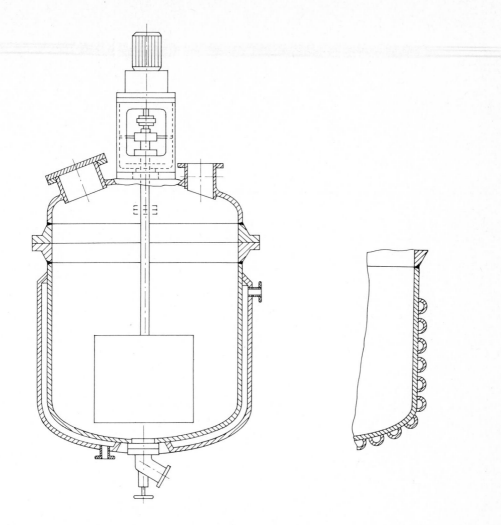

Fig. 10. DIN standard stirred tank with stirrer drive.

means of an external pump around the system.

Because batch reactors so often have to deal with corrosive or very high purity substances, these reactors are available in a variety of materials. Normal steel forms are the cheapest, and can be enamel coated to prevent contamination and reduce corrosion when there are mild corrosive tendencies. Full corrosion resistance is offered by glass-lined reactors, which can be used in strong acid media. Very common are stainless steel reactors because of their reduced contamination of product, improved corrosion resistance compared to mild steel, robustness with respect to the enamel and glass-lined vessels, and ease of cleaning - many batch processes are multi-product, requiring product switching and cleaning between batches. Materials of construction are discussed in more detail in Chapter 11.

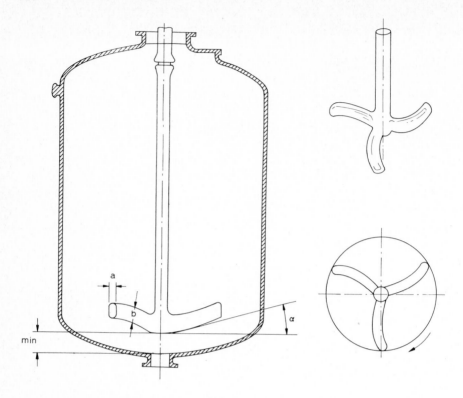

Fig. 7.11. DIN standard enamelled impeller.

Heat-transfer coils inside reactors are not always welcome because they present cleaning problems and modify mixing patterns. However, when extra heat-transfer area must be supplied they are preferred to the pump-around system. The coils can take the form of a helix of approximately 75% of the reactor diameter, with a pitch equal to twice the tube diameter. If this still does not supply enough area, two or more helices can be installed concentrically.

The capital cost of a stirred tank reactor can be approximately represented by the following expression:

$$\text{capital cost} = 8630 \ V_R^{0.47} \ \$ \ (1971)$$

This expression gives the approximate uninstalled equipment cost of a stainless steel reactor in the standard range, including its agitator and agitator drive. It cannot be recommended to be used for obtaining reliable costs, but it gives the correct order of magnitude and correct trend. It can be used, for instance, in optimization studies.

TABLE 7.5

List of DIN standard stirred tanks and some dimensions

Volume m^3	Internal diameter m	Heated surface area m^2
0.063	0.580	0.60
0.100	0.580	0.80
0.160	0.600	1.16
0.250	0.700	1.48
0.400	0.800	2.32
0.630	1.00	2.87
1.000	1.20	3.87
1.600	1.40	5.60
2.500	1.60	7.90
4.000	1.80	9.10
6.300	2.00	13.10
8.000	2.20	
10.000	2.40	18.7
12.500	2.40	
16.000	2.60	25.0
20.000	2.80	
25.000	3.00	34.6
32.000	3.40	
40.000	3.60	46.0

7.4 DIMENSIONING OF STIRRED TANK REACTORS

The decision to use a stirred tank reactor and to use it as a batch, semi-continuous, or CSTR, will have already been made during the optimization studies associated with the process development work. Usually a pilot plant reactor of the required type will have been operated. The last step is therefore to scale up this pilot operation to full scale so as to be able to write a reactor detailed specification. Feed flows, reactor dimensions, stirrer type and speed, heat-transfer area, and coolant flows have to be specified by the designer.

Two approaches are available: either one can arrive at this information by using a quantitative model of the process occurring in the reactor, or by simple direct scale-up methods based on the performance of the pilot plant. Sometimes there is no choice. If a good description of the reactor has not been developed during the process development, only the scale-up method can be applied; if there are no pilot plant results, then the modelling method must be applied. It is preferable, of course, to have a model available and pilot plant results at hand. Then both methods can be used and the results compared.

A particular advantage of the modelling method is that one is not bound to the configuration used in the pilot plant. Relative changes in stirring, heat transfer, and conversion can be calculated and the full-scale plant designed to meet a revised "design basis" which has been modified to include new information which has become available since the design of the pilot plant.

214

The modelling method is more scientific, and more is known about the reactor. Should difficulties arise during start-up or later operation, then one is in a better position to be able to overcome them if the modelling study has been carried out throughout the project.

7.4.a   Design by scale-up

The scale-up method strives to attain the identical conditions on the full-scale reactor that were found to be optimum on the pilot plant.

This is generally not possible because areas/unit volume must change with scale and mixing conditions also alter. It is therefore necessary to decide which criteria are most significant for the reaction under consideration and then to choose a scale-up method which holds these critical conditions consistant on the two scales, and lets the inevitable differences occur in the less significant conditions.

For example, if heat removal was a problem, then one would ensure that the agitation on the two scales produced identical heat-transfer coefficients; if the gas dispersion was important, scale-up based on equal stirrer shear rates would be more appropriate.

The problems in scaling-up complex single phase reactions are that
- changes in temperature can affect selectivity and yield;
- changes in heat-transfer coefficient can affect film temperatures and degradation when heat sensitive materials are involved;
- changes in heat-transfer area/unit volume alters the heat removal potential and this can lead to hazardous situations.

It is therefore understandable that when the scale-up approach is being used and the reactor mechanism is not particularly well understood, a very conservative attitude must be taken.

The various scale-up criteria are discussed below.

(a)   Constant stirrer power/unit volume. This is a suitable criterion when the main duty of the agitator is liquid mixing. The reactor volume is scaled up proportionally to the larger output required and the power is scaled by the same factor. The stirrer speed required to input this power can then be calculated as follows:

The relation between power and dimensions for a fixed physical system can be defined from equation 7.2 as

$$P' = K' \, d_I^5 \, N^3 \tag{7.13}$$

Since we are scaling up at constant power/unit volume

$$P'_1/V_1 = P'_2/V_2$$

and since the scale-up is based on geometrical similarity:

$$\frac{D_2}{D_1} = \frac{d_{I_2}}{d_{I_1}} = \left(\frac{V_2}{V_1}\right)^{1/3} \tag{7.14}$$

hence, for constant power/unit volume

$$\frac{d_{I_1}^5 N_1^3}{V_1} = \frac{d_{I_2}^5 N_2^3}{V_2} = d_{I_1}^5 \left(\frac{V_2}{V_1}\right)^{5/3} \frac{N_2^3}{V_2} \tag{7.15}$$

hence

$$N_2 = N_1 \left(\frac{V_2}{V_1}\right)^{-2/9} \tag{7.16}$$

Notice that the stirrer speed (expressed rev s^{-1}) must reduce on the larger scale.

(b) <u>Constant heat-transfer coefficient.</u> This criterion would be suitable when the main problem of the reactor was with heat removal. From equation 7.8:

$$h = K'' \, D^{-1} \, d_I^{1.3} \, N^{0.65} \tag{7.17}$$

Hence to scale up a reactor from V_1 to V_2, maintaining geometric similarity, to give the same heat-transfer coefficient we can derive a new stirrer speed as follows:

$$D_1^{-1} \, d_{I_1}^{1.3} \, N_1^{0.65} = D_2^{-1} \, d_{I_2}^{1.3} \, N_2^{0.65} = D_1^{-1} \left(\frac{V_2}{V_1}\right)^{-1/3} d_{I_1}^{1.3} \left(\frac{V_2}{V_1}\right)^{1.3/3} N_2^{0.65} \tag{7.18}$$

hence

$$N_2 = N_1 \left(\frac{V_2}{V_1}\right)^{-0.15} \tag{7.19}$$

Having achieved the same heat-transfer coefficient on the larger scale, the heat removal facilities must be increased because the heat generation is proportional to V_2/V_1 but the surface area of the vessel has increased by only $(V_2/V_1)^{2/3}$.

This can usually be done by additional area in the form of coils in the reactor itself. In extreme cases, larger areas can be added by using external heat exchangers and a pump-around system (until the volume of the exchanger is significant compared to the volume of the reactor).

In some cases it may be possible to lower the coolant temperature and so increase the heat flow through the existing surface, but this is usually fixed by stability considerations which normally require the coolant temperature to be within a few degrees of the reaction temperature.

(c) <u>Constant tip speed</u>. This criterion maintains constant shear in the liquid and this would be expected to maintain the same gas distribution qualities in the pilot and full scale.

To do this, and maintain geometric similarity of the vessels, the equation 7.14 holds, and, defining the tip speed (s') by

$$s' = N \pi d_I \quad m \ s^{-1} \tag{7.20}$$

then

$$N_1 \pi d_{I_1} = N_2 \pi d_{I_2} = N_2 \pi \left(\frac{V_2}{V_1}\right)^{1/3} d_{I_1} \tag{7.21}$$

hence

$$N_2 = N_1 \left(\frac{V_2}{V_1}\right)^{-1/3} \tag{7.22}$$

(d) <u>Constant pumping rate/unit volume</u>. This can be taken as a criterion when the mixing time is thought to be a significant factor in the system. From equation 7.1, since

$$Q = K' N d_I^3 \tag{7.23}$$

$$Q_2 = \frac{V_2}{V_1} Q_1 = K' \frac{V_2}{V_1} N_1 d_{I_1}^3 = K' N_2 d_{I_2}^3 \tag{7.24}$$

combining 7.14 with 7.24 we get

$$N_2 = N_1 \tag{7.25}$$

<u>Notes on various scale-up criteria</u>. These four scale-up criteria produce different relationships between the full-scale plant stirrer speed and the pilot stirrer speed. They all have the general form

$$N_2 = N_1 \left(\frac{V_2}{V_1}\right)^n \tag{7.26}$$

but n varies between 0 and -1/3 depending on the criterion.

Although this range suggests that the different criteria are in fairly good agreement, because scale-up ratios are necessarily high, full-scale stirrer speed predictions can be significcnatly different. For example, take a 40 litre pilot plant operating satisfactorily with a stirrer speed of 2 rev s^{-1}. A 40 m^3 full-scale reactor (scale-up 1000) would have a stirrer speed of either 2 rev s^{-1} or 0.2 rev s^- depending on the criteria chosen!

The engineer scaling up must therefore use his judgement to decide which criteria

to employ and must then check that the resulting power consumption at tip speed
is reasonable. When this is not the case, various manipulations in dimensions
must be made to produce a viable design that reproduces as well as possible the
conditions used in the pilot experiments.

As a general rule, it is possible to put more agitation per unit volume into
a smaller reactor than is physically or economically possible on the larger scale.
Hence these scale-up criteria can result in impossible full-scale reactor designs
if the pilot plant reactor has been designed to be "well stirred" without reference
to how it would look on the full scale.

For example, it is not unreasonable to use a 250 W stirrer on a one litre vessel
in the laboratory. Scaling up to a 40 m^3 vessel using the same power per unit
volume would require 10,000 kW, requiring a motor more suitable for a ship than a
reactor.

On both feasibility and cost grounds one would hope for a 200 kW duty on such
a reactor, equivalent to consuming only 5 W in the laboratory reactor.

7.4.b   Design by modelling

If the process development has resulted in models of the reactor and pilot scale
reactor, these can be used for the design of the full-scale stirred tank.

By including in the simulation model the correlations for the prediction of
heat-transfer coefficient and stirrer power requirement, in fitting this to the
pilot scale reactor one then has a useful tool for design.

The model can be used to simulate large scale reactors, starting off by using
geometric similarity and some of the scale-up rules mentioned in the previous
paragraphs to provide initial data for the simulation. The result of this simulation
produces a first design and this can be used as a basis for discussion, various
factors changed, and new results obtained. Wherever compromises have to be made
they can be done quantitatively. For example, the heat may have to be removed at
a slightly higher temperature, with an addition of some extra area, and a lower
coolant temperature. The model will adequately make all the quantitative corrections
for the changes (as long as they are not major).

Once the best conditions have been decided upon, and this may also include a
simplified economic optimization, the model can be used to test the design for
stability by making perturbations in the control variables. This is discussed
further in Chapter 10.

When large scale-up ratios are involved, it is unlikely that correlations on the
small scale still hold for the large scale. It is difficult to believe that the
measurements of stirrer power and heat-transfer coefficient in a 1/2 litre glass
pilot reactor can be scaled up reliably using proportional formulae as described in
the last section. Hence, when large scale-up ratios are involved, the scale-up
method begins to lose its conviction and the modelling approach begins to show

advantages.

Small scale correlation for glass equipment can be included in the pilot model and then replaced by correlations developed for industrial scale equipment for the full scale model to be used for the design. The inevitable changes that then occur are predicted from an understanding of the process and the reaction mechanism occurring in the reactor. Hence this gives greater confidence when large scale-up ratios are used than is possible with the scale-up method alone. This in turn means that the use of modelling enables much smaller pilot plants to be built.

The main advantage of the scale-up method which must not be overlooked is that, given results from a reasonably-sized pilot, a full-scale reactor can be designed with minimum demand for information on physical properties and reaction mechanisms, and minimum expenditure on theoretical work and computer programming.

7.5   CASCADES OF STIRRED TANK REACTORS

As explained in section 1.4.c, it is possible and usual to have CSTR reactors in a cascade to overcome the severe disadvantages of the CSTR of the low reactant concentrations and low reaction rates. The only time when a cascade would not be advisable is when there is a need to maintain low reactant concentrations for reasons of selectivity.

The cascade enables a fraction of the reaction to proceed in each reactor, and so each reactor handles a fraction of the heat liberated. Cascades are normally arranged to flow automatically under gravity by having the reactors installed at progressively lower levels (see Fig. 7.12).

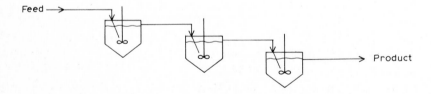

Fig. 7.12. Gravity flow through a cascade of CSTRs.

Such a cascade introduces a number of degrees of freedom that the engineer must satisfactorily handle:
- how many reactors should be in the cascade ?
- what should be the volume of each ?
- should reactants be fed to the first only or distributed over more than one reactor ?

The number of reactors used should be kept to a minimum because this keeps the capital cost to a minimum. Generally the improvement obtained in reduction in total volume in going from three to four reactors does not make the installation of the fourth reactor economical. Also the difference between a three reactor system and a plug flow one in terms of product distribution from consecutive reactions is small. Therefore, one rarely needs to go above three reactors in a cascade.

Appendix 3 contains a model of a CSTR reactor that can be run in cascade to investigate the relative importance of the various degrees of freedom available to the designer.

This point is demonstrated by Table 7.6 which gives the relative selectivity to the monochloride product for chlorination of an organic which can form mono-, di-, and trichlorinated products. The table shows that the relative improvement in selectivity from three to four reactors in the cascade is 0.8%, with a loss in selectivity compared with a plug-flow reactor of 3.5%. However, this 3.5% is reduced by such a small increment by each extra reactor installed that it does not usually justify the extra investment necessary to reduce it.

TABLE 7.6
Chlorination of dodecane by a reactor cascade
organics:chlorine ratio 3:1 (Simulation Results, Horak (1978))

Number of stages	Selectivity:	$\dfrac{\text{Monochloro}}{\text{Mono-+ di-+ trichloro}}$
1		78.98
2		82.88
3		84.40
4		85.11
5		85.62
plug flow		87.51

The volume of each reactor must be determined, and it is usual to assume the reactors are going to be of equal volumes and then with two, three, or four reactors the volume to obtain the desired conversion is calculated. Having all the reactors the same size reduces design costs and affords advantages in holding spares.

There may be situations when different sized reactors could be worthwhile, particularly if heat dissipation is a problem. Changing the size of the first reactor would change the heat generation in it and this might be used to overcome heat-transfer problems in the initial stages. The heat generation could then be increased in the later reactors where heat removal is less of a problem. In such cases it may be advisable to size reactor cascades on the fraction of heat removed rather than the fraction of the total volume supplied.

Excess heat liberation problems can also be overcome by adding only part of one of the reactants to the first reactor, and the remainder to the next and subsequent reactors. This would again provide the facility to even out heat removal across all the reactors in the cascade. When one of the reactants is a gas, there is no choice but to add part of the gas to each reactor, since otherwise the flow between reactors will be disturbed and the gas, having disengaged after the first reactor, will not react in subsequent reactors.

A thorough investigation of reactor cascades and their ensuing design is ideally suited to modelling. The one reactor model is simply re-used to represent all the reactors in the cascade. The engineer tries various combinations until he is satisfied that he has obtained a good compromise for the final design. The model as described in section 2.4.a is a good basis for such a design tool.

Predictions of overall heat-transfer coefficient by given stirrer speeds, stirrer powers, installations of additional coil or extended heat-transfer area to achieve the heat removal can be made. The model should not determine the temperature of operation; this can be given by the designer. The model effectively includes the temperature controller when it is programmed to operate at a given temperature, and it can then print out whether the heat-transfer area is capable of removing the heat or not. This approach also simplifies the mathematics compared with asking the model what equilibrium temperature will be achieved when full cooling is applied.

7.6   MODIFIED FORMS OF CSTR CASCADE

As reactor volumes increase, and as the number of stages increases owing to gas addition requirements or particularly sensitive kinetics, a cascade of CSTRs looks complicated and becomes expensive. This can lead to the development of equipment somewhat cheaper and different-looking, which is really a CSTR cascade to which a little design effort has been given.

7.6.a   Compartment tank reactors

In various large-scale hydrocarbon reactions and alkylations, very large reactor volumes are required (e.g. 300 m^3). This has led to the design of the compartment reactor (Fig. 7.13). This reactor design is particularly suitable when one of the

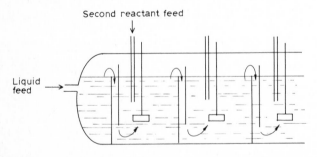

Fig. 7.13. A compartment reactor cascade.

reactants is a gas. Here, in one horizontal cylindrical shell fitted with baffles, it is possible to obtain the same flow pattern as a cascade of independent CSTRs, with almost no backmixing.

7.6.b   Cascade towers

On a smaller scale, a set of CSTRs can be replaced by a baffled, stirred column (see Fig. 7.14).

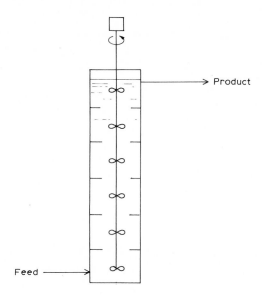

Fig. 7.14. A cascade tower reactor.

The operation of such a column is identical to a cascade of CSTR reactors but the staggered horizontal lay-out is spared, it being replaced by a very compact, trouble-free piece of equipment. This equipment does not have facilities for removing heat of reaction, although it is conceivable that a jacketed form could be developed which would extend applications somewhat.

7.6.c   Determination of the mixing characteristics in modified forms of cascade reactor

Once the concept of individual reactor stages connected by piping is abandoned, the number of stages represented by the equipment is less certain. An undefined mixing pattern results, and it is important to know to what extent the mixing is not ideal.

Investigation can be simply carried out with the pulse tracer technique as described in section 7.1.d. A trace is recorded at the end of the equipment and

222

this can be compared with Fig. 7.6 to determine the number of equivalent, perfectly-mixed stages the equipment represents.

Such experiments on the cascade tower (Fig. 7.14) show it to be effectively plug flow.

FURTHER READING

Denbigh, K.G. and Turner, J.C.R., 1971. Chemical Reactor Theory. Cambridge University Press. - Good section on mixing.
Levenspiel, O., 1972. Chemical Reactor Engineering. Wiley. - Good section on residence time distributions.
Nagata, S., 1975. Mixing - Principles and Applications. Wiley. - Thorough survey of published work on agitation.
Rase, H.F., 1977. Chemical Reactor Design for Process Plants. Wiley, New York. - Review of mixing correlations and stirred tank reactor design.
Uhl, V.W. and Gray, J.B., 1966. Mixing - Theory and Practice, Vols. 1 and 2. Academic Press. - General survey of agitation.

REFERENCES

Chapman. F.S., Dallenbach, H. and Holland, F.A., 1964. Trans. Inst. Chem. Eng., 42: T398.
Deutsche Industrienormen (D.I.N.) 28130, Beuth Verlag, Berlin 30.
Horak, J. and Pasek, J., 1978. Design of Industrial Chemical Reactors from Laboratory Data. Heyden, London.
McAdams, W.H., 1951. Heat Transmission. McGraw-Hill.
Perry, R.H., 1973. Chemical Engineer's Handbook (5th edition). McGraw-Hill.

QUESTIONS TO CHAPTER 7

(1) What factors enter into the choice of impeller type ?
(2) What are the various mixing criteria that can be used to determine the level of agitation required in a reactor ?
(3) How are the stirrer power and stirrer speed, as well as impeller diameter, decided for a specific reactor application ?
(4) How can the overall heat-transfer coefficient for a jacket round a stirred tank be predicted ?
(5) What are the various criteria that can be used to "scale-up" pilot stirred tank reactors to full scale ? What is the difficulty in the "scale-up approach"
(6) What are the differences in the modelling and "scale-up" approaches to sizing industrial-scale equipment from pilot plant data ?
(7) For the system:

$$A + B \xrightarrow[30\% \ H_2SO_4]{90°C} C \qquad H = -50,000 \ kJ \ kmol^{-1}$$

$$C + C \longrightarrow P \ product$$

$$C + A \longrightarrow R \ by\text{-}product$$

intermediate C never attains a high concentration.
What reactor types come into question for this reaction ?
(8) How can a partially mixed reactor be adequately modelled without resorting to the use of back-mixing diffusion models ?

Chapter 8

THE DESIGN OF GAS-LIQUID REACTORS

8.1   MODELS FOR REACTION WITH MASS TRANSFER AND HEAT GENERATION

When a gas reacts with a liquid the reaction takes place in the liquid phase. This means that the gas molecules must transfer from the gas phase into the liquid phase where they remain in solution until they collide with molecules of the liquid reactant and thereby react.

The whole reaction mechanism is complicated by the inclusion of a mass-transfer step, and this gives the possibility for the reaction to be mass-transfer-controlled when the mass-transfer rate is slow relative to the reaction rate; kinetically controlled when the reaction rate is much slower than mass transfer; and controlled by a balance of the two rates when they are of the same order.

The gas has to set up a concentration in the liquid and it is this concentration that enters into the kinetic equation:

$$A + B \xrightarrow{k_1} P \quad \Delta H_1$$

With C_B as the concentration of B in the liquid phase, the kinetics for a batch reaction become

$$\frac{dn_A}{dt} = - V_R (1 - \varepsilon_g) k_1 C_A C_B \quad \text{kmol s}^{-1} \tag{8.1}$$

where V_R is the total volume of fluids and ε_g is the fraction of gas in the mixture, leaving volume $V_R(1 - \varepsilon_g)$ of liquid for reaction. This concentration of B results from mass transfer into the liquid and reaction in the liquid. The mass-transfer rate is given by

$$\frac{dn_B}{dt} = K_g a V_R (p_B - C_B H_B') \quad \text{kmol s}^{-1} \tag{8.2}$$

where K_g is the overall mass-transfer coefficient (kmol m^{-2}s^{-1}bar); a is the interfacial area per unit volume of fluid (m^{2}m^{-3}); and p_B is the partial pressure of the reactant B in the gas bubbles (bar). It is assumed that the gas concentration can be represented by a single value of p_B. This will be so if the gas is well mixed in the system or if no inerts are present. H_B' is the Henry's law solubility coefficient for B where

$$H_B' = \frac{p_B^*}{C_B} \quad \text{m}^3\text{bar kmol}^{-1} \tag{8.3}$$

which relates liquid concentration C_B to its equilibrium partial pressure p_B^*. Hence the accumulation of B in the liquid is given by

$$\frac{dn_B}{dt} = K_g \, a \, V_R \, (p_B - C_B H_B') - k_1 \, V_R \, (1 - \epsilon_g) \, C_A C_B \qquad (8.4)$$

the heat evolved by the reaction is given by

$$\frac{dH}{dt} = - \Delta H_1 \, k_1 \, V_R \, (1 - \epsilon_g) \, C_A C_B \qquad (8.5)$$

and the resulting temperature change is

$$\frac{dT}{dt} = \frac{1}{\rho V_R \, (1 - \epsilon_g) \, c_p} \left(\frac{dH}{dt} - UA' \, (T - T_J) \right) \qquad (8.6)$$

where $UA' \, (T - T_J)$ is the heat removed from heat transfer area A' (neglecting any heat capacity of the gas). The change in temperature will change H_B' and k_1, since

$$k_1 = A_1 \, e^{-E_1/RT} \qquad (8.7)$$

and, from the van't Hoff equation which relates H' and T:

$$\ln H_B' = \frac{k'}{T} + k'' \qquad (8.8)$$

where k' and k'' are constants. Hence the effect of temperature can be fully included in the model.

The model of a batch reactor now corresponds to equations 8.1 - 8.8. For the CSTR reactor, the liquid and gas reach standing concentrations and the system is represented by a set of algebraic equations - from a balance on component A:

$$(C_{A_0} - C_A)Q = V_R \, (1 - \epsilon_g) \, k_1 \, C_A C_B \qquad (8.9)$$

from a component B balance over the liquid:

$$C_B Q = K_g \, a \, V_R \, (p_B - C_B H_B') - V_R \, (1 - \epsilon_g) \, k_1 \, C_A C_B \qquad (8.10)$$

and a heat balance (neglecting thermal capacity of the gas):

$$(T - T_0) \, \rho \, Q \, c_p = - \Delta H_1 \, V_R \, (1 - \epsilon_g) \, C_A C_B - UA' \, (T - T_J) \qquad (8.11)$$

Together with equations 8.7 and 8.8, equations 8.9 - 8.11 describe the CSTR with mass and heat transfer.

For the design of such reactors, besides having the kinetic parameters A_1 and

E_1, one must have physical data on gas solubilities (k' and k''), and data based on experimental measurement or empirical correlations for ($K_g a$), U and ε_g.

Methods of obtaining the chemical kinetic data have occupied chapter 3. It is worth noting here that if the chemical mechanism is complex, there is no problem in replacing the simple second order kinetics used here for demonstration purposes with complex kinetics.

Solubility data can often be obtained from the literature. Table 8.1 gives some typical values for common gases in water.

TABLE 8.1

Examples of some Henry's law solubility coefficients in aqueous solution, H' (bar m^3 kmol^{-1})

Gas	273 K	293 K	353 K
Hydrogen	1056	1247	1378
Nitrogen	966	1467	2303
Oxygen	463	729	1254
Methane	407	685	1243
Ethane	220	481	1207
Carbon dioxide	13.3	25.8	-
Hydrogen sulphide	4.88	8.80	24.7

Values of ($K_g a$), ε_g and U are available in the literature as correlations based on experimental measurement. There is often no opportunity for these to be checked out before they are used in design, because the checking needs a large experimental rig which is never available. If the process is being piloted on a fairly large scale (say 50 litres reaction vessel), then there is an opportunity during the pilot work to verify the available correlations with the reacting system itself. However, this is often not possible and one has to take the literature values without experimental checking, even though there is always the feeling that they were obtained with the wrong system, in the wrong equipment, and on the wrong scale ! Typical correlations will be given in later sections of this chapter where the various reactor types are described in detail.

The discussion so far has been concerned with straightforward mass transfer where the gas and liquid phases have no concentration profile. With each type of reactor and system one should carefully consider whether these conditions hold, and if necessary different models should be derived to represent the particular reactor in question.

8.2   GAS-LIQUID REACTOR TYPES

The most common type of gas-liquid reactor is the stirred tank reactor, with gas introduced below a vane-disc impeller. Such reactors can be batch, semi-batch or continuous. Bubble column reactors are probably the next most common, where the gas is bubbled into the bottom of a tower containing the liquid. There is no mechanical agitation involved because there is adequate mixing created by the bubbles as they rise through the liquid.

Trickle bed reactors are particularly common in the oil industry where the liquid is passed down a tower packed with inert packing (or catalyst particles). For example, such reactions use partially vaporized feed, mixed with hydrogen to carry out desulphurization reactions.

Also in this group comes the "absorption with chemical reaction"-type of equipment. This takes the form of conventional packed or tray-type columns. They are used for fast reactions (the liquid residence time is very low) and the process has more the characteristics of enhanced absorption than reaction. They are discussed separately in section 8.4.

Figure 8.1 shows diagrammatically the various types of gas-liquid reactor that are available.

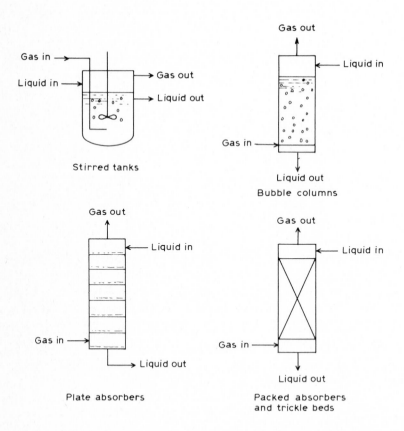

Fig. 8.1. Types of gas-liquid reactor.

8.3 RELATIVE MERITS OF THE VARIOUS REACTOR TYPES

The two main categories of reactor are those with a large liquid hold-up - stirred tanks and bubble columns - and those with very small hold-up - trickle beds and packed or plate absorption towers. For reactions which are very fast, with the whole reaction being in the liquid film, the second type of reactor is to be preferred. These are basically mass-transfer devices and the reaction is of secondary importance. If the reaction is not extremely fast then the plate column is useful, as it does provide considerably more hold-up than the trickle bed or packed column. Trickle beds are required for fast reactions when the solid in the bed is also the catalyst of the reaction.

Fast reactions suitable for absorber-type equipment can have selectivity problems should there be a consecutive reaction capable of consuming the product. This is because in the film the concentrations of gaseous reactant and the product are higher than in the bulk of the liquid. This can result in a faster consecutive reaction than would be expected from kinetics based on bulk concentrations for all components. In this case it is advisable to make the reaction more kinetically controlled (e.g. by lowering the temperature) and then use reactors suitable for lower reaction rates - i.e. a plate column or stirred tank.

The stirred tank and bubble column reactors are best for kinetically controlled reaction schemes since they provide considerable liquid residence time for the reaction. The choice between these reactor types is often difficult because each is equally suitable for many systems.

There are however some specific situations where the two types of reactor do not overlap in suitability. Bubble columns are unsuitable for viscous liquids but the extra degree of freedom given by the agitator allows these systems to be handled by stirred tanks. Secondly, when high gas volumes are involved, dispersion by an agitator is unsuitable because of impeller flooding, and a bubble column then gives a better overall design. Bubble columns are also preferred for liquid volumes above 40 m^3, which is the largest available size of standard stirred tank reactor. The stirred tank has an agitator which consumes power, but so does the pressurisation of the gas to be fed to the taller bubble tower. Capital costs seem to be quite similar, and although the stirrer enables higher mass- and heat-transfer rates to be achieved, the bubble column can be taller and so can overcome this deficiency quite cheaply. The bubble column has no stirrer, and therefore no glanding problems (which may be critical at high pressures) and less maintenance and running cost if the gas is already available at a sufficiently high pressure not to require its own compressor.

8.4 ABSORBER REACTORS FOR VERY FAST REACTIONS

As the reaction rate becomes fast with respect to the rate of absorption, the dissolved gas concentration in the bulk liquid drops as progressively more of the

228

reaction is completed in the liquid film before the gas can diffuse into the bulk
liquid.

When the rate has increased such that the gaseous component bulk concentration
is negligible, there is no point in supplying bulk liquid volume, and thin film
mass-transfer equipment is the preferred choice.

As the rate of reaction increases still further, the reaction occurs at the
interface and not in the film and so liquid mass transfer resistance also becomes
negligible. In this limiting case the rate is determined by the gas film mass-
transfer coefficient.

Of these various stages, it is of particular interest to the reactor designer
to know when a reaction rate is so fast that there is no point in choosing equipment with high liquid hold-ups, i.e. when bulk concentration of the gaseous
component will be very low. Having identified this situation, he knows whether he
should design the reactor to be purely an enhanced mass-transfer device or as a
conventional reactor.

8.4.a Identifying fast reactions

Thus the designer must recognize when he has a very fast reaction in order to
make a choice of equipment type. This can be done either experimentally, or by
means of a mixture of theory and experiment.

The purely experimental approach would be to carry out the reaction in the
laboratory in suitable equipment to show that virtually the total resistance is
in mass transfer. Characteristics of mass-transfer-controlled reactions are that
the rate is directly proportional to the mass-transfer area and independent of
liquid volume (as long as the same surface area is presented), and also that the
reaction is not highly temperature-dependent - its rate being no more than directly
proportional to temperature. If the results of such experiments show that the
chemical kinetics are not important, it would be fairly reasonable to choose an
absorber-type reactor for the proposed duty.

Looked at more theoretically, if the reaction is complete within the time
taken for reactants to diffuse through the film to the bulk liquid, then the
reaction will be absorption controlled. The correlations for k_g and k_L for equipment such as plated columns, tray columns and trickle beds exist, and so the time
taken for the diffusion across the liquid film can be determined from such
literature data.

Assuming that the reaction is mass-transfer controlled, the gas concentration
in the liquid bulk will be zero, and the mass-transfer rate will be

$$\frac{dn_B}{dt} \approx K_g \, a \, V_R \, C_B^* \, H_B^! \quad kmol \ m^{-3}bar^{-1}$$

(8.12)

where C_B^* is the liquid concentration of B in equilibrium with the gas. According to the two film theory of mass transfer the film thickness (Z) is given by

$$Z = \frac{D_B}{k_L} \qquad (8.13)$$

where D_B is the diffusivity of B, and k_L is related to K_g by

$$\frac{1}{K_g} = \frac{1}{k_g} + \frac{H_B'}{k_L} \qquad (8.14)$$

Assuming that the film contains a mean concentration of $\frac{C_B^*}{2}$, the mean residence time t_F of B for it to pass through the film must be, from equation 8.12:

$$t_F = \frac{kmol\ B\ in\ film}{throughput\ of\ B} \approx \frac{V_R\ a\ Z\ C_B^*/2}{K_g\ a\ V_R\ C_B^*\ H_B'} \qquad (8.15)$$

and including equation 8.13:

$$t_F \approx \frac{D_B}{2\ k_L\ K_g\ H_B'} \qquad (8.16)$$

If the gas film has negligible resistance, equation 8.16 simplifies further to

$$t_F \approx \frac{D_B}{2 k_L^2}\ s \qquad (8.17)$$

t_F is a very rough approximation to the time that B will remain in the surface film before it reaches the bulk of the liquid. If the reaction time for 90% conversion of B (t_R) is shorter than this, it can be assumed that the system will be effectively mass-transfer controlled. From plug flow second order kinetics and assuming that A has a mean concentration of $\frac{C_A}{2}$ in the film,

$$t_R \approx \int_{C_B^*}^{0.1C_B^*} -\frac{d\ C_B}{k_1\ C_B\ C_A/2} = \frac{4.6}{k_1\ C_A}\ s \qquad (8.18)$$

Hence, if $t_F \gg t_R$, it can be assumed that the system will be mass-transfer controlled.

When mass transfer is controlling, trickle beds and packed beds can be used as reactors. When there is some reaction occurring in the bulk or when the dominance of mass transfer is not completely certain, tray columns can be used, since these present a greater liquid residence time to allow reactions to be completed. When there is known to be reaction occurring in the liquid bulk, it is better to design a stirred tank or bubble reactor than to run the risk of having incomplete reaction owing to inadequate liquid residence time.

8.4.b Design of absorbers with reaction

The effect of a fast reaction on an absorption system is to reduce the concentration of the gas (B) in the liquid film by reaction (Fig. 8.2). Hence the concentration gradient will be larger than without reaction. This will enhance the mass transfer, relative to the case of mass transfer without reaction. It has been shown, both for first-and second-order reactions, that it is theoretically possible to account for the reaction occurring by using an enhancement factor E' so that the apparent mass-transfer coefficients with reaction, k_L', can be related

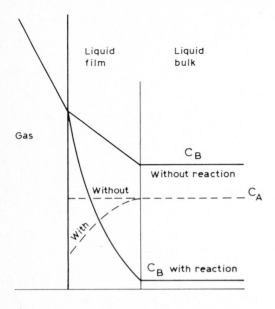

Fig. 8.2. Enhancement of k_L by fast reactions.

to the mass-transfer coefficients without reaction, k_L, by

$$k_L' = k_L E' \tag{8.19}$$

where E' is greater than 1. For first order reactions, E', sometimes called the Hatta Modulus, is given by:

$$E' = t_F \left(\frac{k_L}{D_B} \right)^{\frac{1}{2}} \Bigg/ \tanh \left\{ t_F \left(\frac{k_L}{D_B} \right)^{\frac{1}{2}} \right\} \tag{8.20}$$

For second order reactions it has been shown that

$$E' = \frac{\left(M \dfrac{E'' - E'}{E'' - 1}\right)^{\frac{1}{2}}}{\tanh \left(M \dfrac{E'' - E'}{E'' - 1}\right)^{\frac{1}{2}}} \qquad (8.21)$$

where

$$M = \frac{D_B \, k_1 \, C_{A_0}}{k_L{}^2} \qquad (8.22)$$

and

$$E'' = 1 + \frac{D_A \, C_{A_0}}{D_B \, C_{B_i}} \qquad (8.23)$$

For further details of theoretical estimation of the enhancement factor see Horak (1978).

Although equations 8.20 and 8.21 give the theoretical relationships required to determine the enhancement factor, to enable standard design methods to be applied, they are difficult to use in practice because they need to have the kinetic rate constants for the fast reaction. These generally have to be determined experimentally and it is easier to determine the effective mass-transfer coefficient k_L' (i.e. $E'k_L$) from absorption measurement than to calculate E' from the separate measurements of k_1, D and k_L.

It is, however, comforting to know that there is theoretical backing for using the mass-transfer approach when fast reactions are involved. Laboratory work can be carried out on a variety of types of equipment to measure $E' k_L a$, and this can be compared with a standard system on the same apparatus. When the performance of the standard system is known in the full scale equipment, the performance of the new fast reaction system can be determined by direct proportionation.

Some experimentally measured values of mass-transfer coefficients for reaction systems are published by Rase (1977). If it is assumed that the gas film resistance is negligible, it is possible to derive the enhancement factor E' for these examples. These are displayed in Table 8.2. The figures are only approximate and are included only to demonstrate the order of magnitude of the enhancement due to reaction.

Having obtained the value of k_L' from experimental work, the absorber design itself is often complicated by the absorption being non-isothermal - usually exothermic. This can be extremely important since many systems, e.g. HCl and NH_3, produce so much heat on absorption in water that temperatures, and hence partial pressures, increase such that the required concentrations cannot be obtained without intermediate cooling. This cooling can be done either by using a pump-around system with an external circulation around the bottom half of the tower, or by liquid removal, cooling, and return to some point within the tower; liquid

TABLE 8.2

Some approximate enhancement factors for k_L in reacting absorption systems

Gas	System Solvent	Reactant	Enhancement Factor E'
Cl_2	H_2O	NaOH	50
H_2S	H_2O	MEA	20
H_2S	H_2O	NaOH	20
SO_2	H_2O	NaOH	10
SO_2	H_2O	Na_2SO_3	5
CO_2	H_2O	MEA	25
CO_2	H_2O	NaOH	15

removal part way down the absorber is facilitated by using a tray-type column. Alternatively, cooler-absorber equipment could be used. This has been specifically designed for HCl absorption, and takes the form of a graphite block with vertical holes down which the gas and liquid flow, and separate horizontal holes through which cooling water is passed. The resulting cooled graphite block is ideally suited to absorption of gases involving high heats of solution, particularly when corrosive liquids are produced.

8.5 STIRRED TANK GAS-LIQUID REACTORS

The design of stirred tanks is discussed in detail in chapter 7. Their use in gas-liquid reacting duties involves the addition of a gas stream beneath the impeller and the use of appropriately modified correlations for power, $K_g a$, ε_g and the heat transfer coefficient U.

When a gas is directed to the eye of an impeller it is broken into small bubbles by the shear in the system at this point. These bubbles then circulate in the liquid following the normal liquid flow patterns. Hence the liquid contains a homogeneous mixture of gas bubbles with a random age distribution, and so it is adequate to assume that the gas, as well as the liquid, is perfectly mixed and at a constant composition.

8.5.a Power consumption and impeller flood point

As the gas displaces some of the liquid at the impeller there is less effective area for stirrer contact with the liquid and hence there is a drop in power for a given stirrer speed. Over the normal working range this power drop is of the order of 20 - 50%. As the gas rate increases, progressively more liquid is displaced from the impeller by the gas until, fairly sharply, the point is reached where the agitator is spinning in a bubble of gas. The power requirement drops sharply and the agitator is no longer capable of stirring the liquid. This point

is called the flood point of the impeller and it is the limit to the gas feed
rate to the stirred tank. It occurs at superficial gas velocities around $0.1\,\mathrm{m\,s^{-1}}$.

In the normal operating region the power consumed by an agitator can be
calculated from the power in the ungassed condition by the following empirical
correlation (Nagata, 1975):

$$\log_{10}\left(\frac{P'_g}{P'}\right) = -192 \left(\frac{d_I}{D}\right)^{4.38} \left(\frac{d_I^2 N\rho}{\mu}\right)^{0.115} \left(\frac{d_I N^2}{g}\right)^{1.96\,d_I/D} \left(\frac{Q}{Nd_I^3}\right) \qquad (8.24)$$

where P'_g is the power required by the agitator with gas (kW); P' is the power
required by the agitator without gas (kW), and can be determined from equation 7.2;
N is the stirrer speed (s^{-1}); d_I is the impeller diameter (m); D is the reactor
diameter (m); Q is the gas flow to the reactor ($m^3 s^{-1}$); μ is the liquid viscosity
($kg\,m^{-1}s^{-1}$); ρ is the density ($kg\,m^{-3}$); and g is the gravitational constant
($9.81\,m\,s^{-2}$).

The available correlations for power input are scarce and flooding capacity
correlations appear to be non-existent. So many variables are involved that it is
always a worry as to whether the power correlation used is applicable in a specific
instance. The advantage of working with a stirrer manufacturer from an early stage
in a project is that he has specific correlations for his own stirrer; he has
measured the effect of physical properties and float point and can supply correlations
which inspire much more confidence than those available in the literature. Furthermore, he probably has pilot mixing test facilities and will be willing to run a
number of tests to provide a potential customer with data.

8.5.b Mass-transfer predictions

The rate of transfer across an interface is dependent on the mass-transfer
coefficient which is composed of two terms: the gas film and liquid film resistances.

However, the gas film resistance in most stirred tank reactors can be neglected
because the gas is usually pure and this eliminates the gas film resistance. In
addition, measurement of the individual film resistances in stirred tanks has shown
that the gas film has much less resistance than the liquid (Calderbank, 1958).
Hence it is reasonable to simplify most cases by assuming that the overall resistance
is in the liquid film, and use liquid film correlations alone for the mass transfer.
It should be borne in mind, however, that should a system be encountered where a
high gas film resistance is to be expected (e.g. a dilute, soluble gas) then
correlations including the gas film resistance must be included to obtain a proper
estimation of the mass transfer.

In practice, a certain degree of agitation produces a certain mass-transfer
coefficient and a certain interfacial area. For practical purposes there is no

234

advantage in separating K_g and a (the interfacial area per unit volume of mixture), hence the correlation of K_g a is perfectly adequate for design work. These data can be determined from mass-transfer experiments on the pilot plant, they can be obtained from stirrer manufacturers' correlations, or they can be obtained from correlations available in the literature.

If the data are to be experimentally determined, K_g a can be evaluated as a single variable. If data come from literature correlations, then K_g and a must be separately determined because literature correlations handle these two variables separately.

In common with other literature correlations of stirrer characteristics, data are sparse and never completely relevant to the situation under consideration, but they are a good, and often the only available, guide for the designer.

The work of Miller (1974) details a method of calculating K_L, ε_g and a based on his work on different scales (maximum vessel size 0.25 m³!). The method is described in the following paragraphs.

Consider a bubble rising in a liquid; the terminal velocity of the bubble can be predicted from a wave equation as:

$$u_t = \left(\frac{2\sigma}{\rho_L d_B} + \frac{g d_B}{2} \right)^{0.5} \qquad (8.25)$$

The contact time of the liquid film around the bubble can be calculated, assuming the bubble "slips" through the stationary liquid, as

$$t = \frac{d_B}{u_t} \qquad (8.26)$$

and the mass transfer for this idealized situation is given by:

$$K_L = \left(\frac{4 D_A}{\pi t} \right)^{\frac{1}{2}} = \left(\frac{4 D_A u_t}{\pi d_B} \right)^{\frac{1}{2}} \qquad (8.27)$$

Let the gas hold-up be ε_g; assuming the gas is present in the form of spherical gas bubbles of diameter d_B, then by the geometry of the situation the surface area per unit volume is given by

$$a = \frac{6 \varepsilon_g}{d_B} \quad m^2 \ m^{-3} \qquad (8.28)$$

Now considering a gas with superficial velocity u_s, defined by

$$u_s = \frac{4Q}{\pi D^2} \quad m \ s^{-1} \qquad (8.29)$$

flowing through a column of liquid with a terminal bubble velocity u_t, the gas hold-up ε_g for this idealized model can be shown to be

$$\varepsilon_g = \frac{u_s}{u_s + u_t} \qquad (8.30)$$

These equations so far are theoretically derived for a bubble swarm rising freely in a liquid. We have yet to predict d_B, and to relate this theoretical model to experimental measurements in mechanically agitated vessels.

The diameter d_B has been investigated by Calderbank (1958) and he presented the following correlation for d_B for mechanically agitated bubble swarms:

$$d_B = 4.15 \ N_{Ca} \ \varepsilon_g^{0.5} + 0.0009 \qquad (8.31)$$

where

$$N_{Ca} = \left(\frac{1}{\left(\frac{P'_g}{V_R(1-\varepsilon_g)} \right)^{0.4}} \cdot \frac{\sigma^{0.6}}{\rho_L^{0.2}} \right) \qquad \text{(dimensionless)}$$

P'_g is the effective power input to the system.

This is basically the power required for the agitation for the gassed system. Miller (1974) details some minor corrections for the power input due to the sparged gas, but these are of minor importance for normal mechanically agitated systems.

The gas hold-up in a mechanically agitated vessel is understandably going to be somewhat different to the idealized equation (8.30). Calderbank (1958) investigated the prediction of ε_g and found the following modified equation necessary to predict gas hold-up in a mechanically agitated vessel.

$$\varepsilon_g = \left(\frac{u_s}{u_s + u_t} \right)^{0.5} \left(\varepsilon_g + \frac{0.000216}{N_{Ca}} \right) \qquad (8.32)$$

Calderbank also derived an expression for a, which is based on d_B and ε_g and equation 8.28, but which has been simplified somewhat in view of the empirical nature of the correlation for d_B and ε_g:

$$a = \frac{1.44}{N_{Ca}} \left(\frac{u_s}{u_s + u_t} \right)^{0.5} \ m^{-1} \qquad (8.33)$$

Miller (1974) found that to get an adequate fit to his experimental results using these equations he had to introduce an empirical correlation to the mass-transfer coefficient which was a function of d_B. Hence, substitution of

236

$$k_L = 683 \; d_B^{1.376} \left(\frac{4 \; D_A \; u_t}{\pi d_B} \right)^{\frac{1}{2}} \tag{8.34}$$

in place of 8.27 enabled his results to be adequately fitted.

Hence, k_L, a, and ε_g can be predicted from the set of equations 8.25, 8.29, and 8.31-8.34.

Finally, to use these correlations with the model equations derived at the beginning of this chapter, there has to be a relationship between K_g and k_L. Since in most cases the gas film resistance is negligible in industrial reactors, the normal equation 8.14 reduces to

$$K_g = \frac{k_L}{H'} \tag{8.35}$$

where H' is the Henry's law constant.

Although Miller's is one of the more recent and complete studies reported in the literature, maximum vessel sizes were of the order 0.25 to 1.0 m³ and equipment dimensions were far from "standard". It must be emphasized that such correlations cannot be expected to be very accurate, but they are useful in giving first estimate and in defining suitable model forms for modelling of pilot reactors.

For very rough calculations mean values of k_L and a can be simply used without resort to the full correlations. In general, k_L is between 1×10^{-5} and 5×10^{-5} m s⁻¹ and a is of the order of 500-2000 m² m⁻³.

8.5.c   Heat-transfer coefficient

The presence of gas reduces the heat-transfer coefficient (h_g) compared with that of gas-free stirred tanks (h) by the following relationship:

$$h_g = h \left[\frac{P_g' + P''}{P'} \right]^{0.25} \tag{8.36}$$

where P_g' and P' are the agitator power requirements with and without gas respectively, P" is the power input into the system from the gas when it is expanded adiabatically from the input pressure (P_{in}) to the pressure above the liquid (P_{out}), and h is the heat-transfer coefficient for gas-free reaction operating at the same conditions (Rase, 1977). The correlation for h is given in section 7.2.a.

From normal theory for work done in adiabatic gas compression, the power released on adiabatic expansion of Q m³ s⁻¹ of gas from a pressure of P_{in} bars to P_{out} is given by

$$P'' = 1 \times 10^5 \frac{Q \; P_{in} \; \gamma \; (RT_{in})}{(\gamma - 1) \; M} \left[\left(\frac{P_{out}}{P_{in}} \right)^{1-\frac{1}{\gamma}} - 1 \right] \quad \text{kW} \tag{8.37}$$

where γ is the ratio of specific heats (for diatomic gases such as air γ = 1.31).

8.5.d The design of gas/liquid stirred reactors

The correlations described in the previous section allow the power, $K_g a$, ε_g, and h for a proposed stirrer design and speed to be determined. These correlations can be included in a model of reaction kinetics in order to calculate standing concentrations of dissolved gas in the liquid phase, the resulting reaction rates, and hence the reactor performance. The overall heat-transfer coefficient can be used either to predict resulting reaction temperatures or to check that operation at any particular temperature is within the cooling capacity of the reactor. Such a complete model allows many alternatives to be evaluated and the sensitivity of the reactor performance to various factors to be investigated before the final design is decided upon.

Care must be taken not to lay too much weight on the published correlations because they may well have been determined on conditions and scales very different from the proposed equipment. A detailed literature research may provide more suitable data for a particular duty, or at least further data with which to make comparisons. Manufacturers may have more realistic data available. Failing this, one can at least determine the sensitivity of the design to possible errors in the correlations and decide whether these errors could have serious consequences.

The pilot results can be used to provide information by determining the power, ε_g, and h_g by direct measurement, and $K_g a$ by fitting to experimental results. These results should be compared with the published correlations so that any discrepancies can be detected and corrected for in some way before being used for the full-scale design.

When no model is available, and yet the reactions are complex and selectivity is sensitive to agitation, the "scale-up" method must be used where one of the scale-up criteria given in Chapter 7 is employed. The criterion chosen may be constant tip speed to produce a similar shear for the gas distribution, or constant heat-transfer coefficient, depending on the relative importance of these two processes for the reactor in question.

Example. Consider that we want to extend the CSTR model, begun in Chapter 2 and developed further at the beginning of this chapter, resulting in equations 8.7, 8.8, 8.9-8.11, so that this model also includes agitation and calculates the power required by the agitator, $K_g a$, and U for use in equations 8.10 and 8.11.

The type of model most convenient, both from a mathematical point of view and from an engineering standpoint, would be a simulation of the reaction and its temperature controller: i.e. the independent variables would be feeds, temperature, pressure, equipment dimensions, and stirrer speed, $(C_{A_0}, Q, T, p_B, V_R, d_I, D, N)$, and dependent variables would be stirrer power, coolant temperature, and reactor exit concentrations (P_g', T_J, C_A, C_P)

238

The model steps can be divided into two stages:

(a) Calculate stirrer performance to get $K_g a$ and U.

(b) Use the kinetic equations to determine the reactor performance.

Since the set of equations is beginning to look a little complicated, the individual steps are listed below:

(1) Evaluate ungassed power requirement P' (equation 7.2a).

(2) Evaluate gassed power requirement P_g' (equation 8.24).

(3) Assume a bubble diameter (e.g. $d_B = 0.004$ m).

(4) Calculate its terminal velocity u_t (equation 8.25).

(5) Calculate fraction gas hold-up ε_g from equation 8.32. This will require iteration because the equation cannot be solved analytically. A reasonable starting value would be $\varepsilon_g = 0.15$. Either repeated substitution or the secant method could be tried to find the root.

(6) Calculate the bubble diameter (d_B) from equation 8.31. If this does not agree with the initial estimate (step 3) make a new estimate and iterate to convergence. Repeated substitution or the secant method would be suitable algorithms.

(7) After convergence of d_B, calculate the interfacial area/unit volume (a) from equation 8.33.

(8) Calculate the liquid film mass-transfer coefficient with equation 8.34.

(9) Calculate the overall gas mass-transfer coefficient from equation 8.35.

Now $K_g a$ and ε_g are known.

(10) Calculate the ungassed film heat-transfer coefficient (h) from equation 7.8.

(11) Calculate the gassed film heat-transfer coefficient (h_g) from equation 8.36, which requires P" from equation 8.37.

(12) Calculate the outside film coefficient as described in section 7.2.b, and hence the overall heat-transfer coefficient (U) from addition of the individual film resistances.

Now U is also known.

(13) Evaluate k_1 and H_B' for the given operating temperature from equations 8.7 and

(14) Solve equations 8.9 and 8.10 simultaneously to determine the reactor concentration C_A and C_B. For simple kinetics this could be done analytically, for complex kinetics a multi-dimensional non-linear equation solving routine would be necessary (e.g. Secant or Newton-Raphson algorithms).

(15) Determine the required jacket temperature T_J from equation 8.11.

A model so constructed is of considerable value in relating the interaction betw the stirring and the kinetics. The effect of stirrer speed on the concentration lev of the gas in the liquid and its consequential effect on reactor conversion and selectivity can be admirably demonstrated.

8.6   BUBBLE COLUMN REACTORS

In these reactors, in place of the stirrer, agitation is provided by the swarm
of bubbles rising through the reactor. By having a comparatively tall reactor,
adequate mass transfer can be achieved without mechanical agitation. This type of
reactor is very well established in industry and very large reactors are in operation.
Its total volume is not limited as is that of the stirred tank reactor, and the
absence of intense agitation makes it suitable for biological systems such as single
cell protein production and water purification. Reactors of $3000\,m^3$ volume are used
in protein production from methanol, and $20000\ m^3$ vessels in waste water treatment
(Gerstenberg, 1979).

8.6.a   Correlating equations

The design of such reactors again depends upon the prediction of $K_g a$ and ε_g
for a given superficial gas velocity in the column (u_s). The calculation of the
compression required for the gas feed gives the energy demand, and hence operating
cost, of such equipment.

Superficial gas velocities used in industrial equipment are usually around $0.05\,ms^{-1}$
but velocities up to $1\ m\ s^{-1}$ have been reported (Gerstenberg, 1979). Superficial
liquid velocities of the order of $0.1\ m\ s^{-1}$ are possible.

There are numerous empirical correlations for $k_L a$ and ε_g for bubble columns
(Calderbank, 1959, Towell, 1965, Akita, 1974, and Miller, 1974). We will describe that
of Miller because it indicates the close relationship between the sparged tower and
the mechanically agitated tank.

The theoretical equations for bubbles rising freely in a liquid produce equations
8.25 to 8.30, from which a and ε_g can be determined once the bubble diameter is
known.

Lehrer (1971) presented a correlation for bubble size in the bubbling region in
which industrial contactors operate (with bubble sizes of the order of 0.001 and
0.006 m) as being a geometric mean of the bubble diameter at the sparger orifice
(d_{bo}) and of the stable diameter d_{bs}. d_{bo} is given by

$$d_{bo} = 1.49 \ (\frac{Q}{\pi N})^{0.4} \tag{8.38}$$

where N is the number of holes in the sparger, Q is the flow rate in m s^{-1}, and
d_{bs} is given by

$$d_{bs} = 1.93 \ \frac{\sigma^{0.6}}{\left(\frac{P''}{V_R(1-\varepsilon_g)}\right)^{0.4} \rho_L^{0.2}} = 1.93 \ N_{Ca} \tag{8.39}$$

where

$$\frac{P''}{V_R(1-\varepsilon_g)}$$

is the energy given to the system by the sparger per unit volume of unaerated liquid and

$$\left(\left(\frac{1}{\frac{P''}{V_R(1-\varepsilon_g)}}\right)^{0.4} \cdot \frac{\sigma^{0.6}}{\rho_L^{0.2}}\right) = N_{Ca}$$

the dimensionless group introduced in equation 8.31. Hence,

$$d_b = (d_{bo} \ d_{bs})^{\frac{1}{2}}$$

$$= 1.70(\frac{Q}{\pi N})^{0.2} \ N_{Ca} \qquad\qquad (8.40)$$

These equations are modified somewhat when the initial bubble diameter is more than 3/4 the distance between the holes in the sparger. For accurate work the original literature should be consulted.

The power input given by the gas to a sparged system (P'') is the sum of the fraction of jet energy developed at the sparger holes, which is given to the bulk liquid and the pressure difference between the pressure at the sparger (P_0) and that above the liquid, P.

For a gas volumetric flow of Q m s^{-1} the power is given by Lehrer (1968) as:

$$P'' = Q \ \rho_g (0.03u^2 + \frac{RT}{M} \ln(\frac{P_0}{P})) \text{ kW} \qquad\qquad (8.41)$$

where u is the gas velocity through the sparger holes and M is the gas molecular weight.

Miller found that, using equation 8.40 to predict bubble size, equations 8.26 and 8.28-8.30 to predict a, and ε_g, the same empirically corrected equation could be used to predict k_L (equation 8.34), as was found for agitated tank systems.

Heat can be removed from the bubble column by building coils or tube bundles in the tower, or by cooling the reactor walls. The heat-transfer coefficient is dependent on the gas velocity and reaches a peak around 0.05-0.1 m s^{-1}. Hence, to maximize heat-transfer rates, a design gas velocity of 0.1 m s^{-1} is recommended.

At this velocity the heat-transfer coefficient $(h)_{0.1}$ is given by

$$(h)_{0.1} = 0.12(\frac{g^2 \ \rho_L}{\mu_L})^{1/6} \ (\lambda_L \ \rho_L \ c_p)^{1/2} \text{ kW m}^{-2} \text{ K}^{-1} \qquad\qquad (8.42)$$

where g is the gravitational constant and λ_L is the heat conductivity of the liquid

(Mersmann, 1977).

At gas velocities below 0.1 m s^{-1}, the heat-transfer coefficient is given by

$$h = (h)_{0.1} \left(\frac{u_s}{0.1}\right)^{0.3} \quad kW \ m^{-2} \ K^{-1} \tag{8.43}$$

(Burkel, 1972).

8.6.b   Design of bubble column reactors

·The liquid hold-up and agitation on industrial-scale reactors is such that complete liquid mixing can be assumed. The gas flow can be considered to be plug flow, and so a plug flow gas model should be employed in any modelling work. The use of the correlations for a, ε_g, and k_L (equations 8.25, 8.28-8.30, 8.34, 8.40 and 8.41) enable a model of the reaction to be developed and used in a similar way to a stirred tank reactor model, For example, various reactor diameters, heights, and gas rates can be investigated by the model until an optimal design is found.

Pilot plant work is particularly important in the case of bubble columns because the physical properties of the liquid and small concentration of surface active components can alter bubbling characteristics and foaming tendencies in a way that cannot be predicted. In particular, maximum gas superficial velocities can only be determined by pilot trials.

As with the gas/liquid stirred tank reactor, once pilot plant results are available they can be fitted by a model to give information on $K_g a$ which can later be used to substantiate the correlations to be used for the full-scale design.

The power requirement for a bubble column reactor is the work done in compressing the gas to the inlet pressure of the reactor. The reactor pressure drop involves the pressure drop across the distributor plate plus the hydrostatic head of the liquid/ gas mixture above the distributor with a density $\rho_L(1-\varepsilon_g)$.

When it is necessary to obtain a high conversion on the component in the liquid phase, or when the reactor product remains in the liquid phase and there is a possibility of it undergoing further reactions to produce unwanted by-product, then the liquid flow should be in the form of a cascade to improve conversion and selectivity for exactly the same reason that CSTR reactors are used in cascade. This can be done either by having a series of independent columns, or by having one column divided off by sieve plates into discrete sections which reduce the back mixing and result in cascade flow for the liquid. An alternative method for very large reactors is a compartmental type reactor (Fig. 8.3).

8.6.d   An example - design details of bubble column reactors

The opportunity is taken here to discuss the considerations that are made when an engineer comes to the detailed design of a reactor. We have discussed how the volume, diameter, and height of bubble column reactors can be determined from

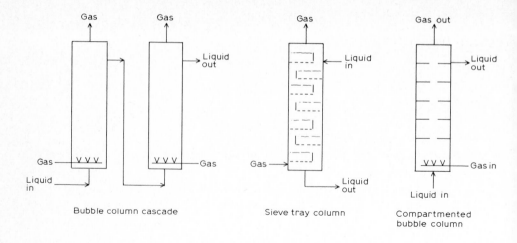

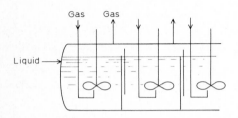

Horizontal compartmented reactor

Fig. 8.3. Various forms of cascade gas/liquid reactor.

superficial velocities, mass-transfer coefficients, and reactor residence times, but having reached this stage there are still many engineering design decisions to be made before a practical design can be considered to have been made.

A particularly good practical survey was made by Gerstenberg (1979) and this is used here to demonstrate the detailed engineering that goes into good reactor design.

Figure 8.4 shows the simple conventional bubble column.

The total reactor must perform the following functions:

- gas distribution
- mass transfer
- reaction
- heat removal
- liquid circulation
- gas disengagement.

The gas distribution can be achieved with various forms of gas distributor, sintered plate, or perforated plate, as shown in Fig. 8.5.

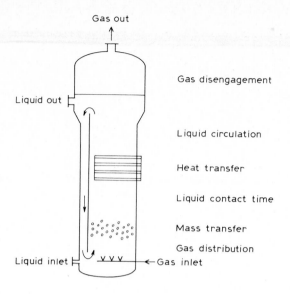

Fig. 8.4. The simple bubble column and its duties.

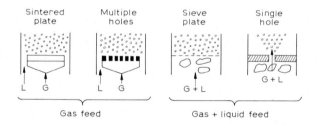

Fig. 8.5. Possible forms of gas distributor.

Alternatively, the gas can be introduced by jets, leading to another set of possible designs (Fig. 8.6). The use of jets is particularly advantageous to improve liquid recirculation, because the jets can be used to induce a liquid flow.

A separate path for liquid recirculation will enable a higher gas loading to be achieved before the column becomes overloaded. Various ways of inducing the circulation are shown in Fig. 8.7. This can be done by a form of draft tube around which the liquid can flow, or an external pump can be used, or incoming gas can simply be fed through jets to induce the circulation motion. The mass-transfer performance of the column can be improved by using internals, and various forms from simple baffles, pulsed sieve plates to packing have been employed (Fig. 8.8).

Heat transfer is achieved by installing heat-transfer area and various ways of doing this are shown in Fig. 8.9.

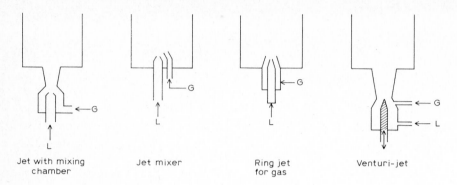

Fig. 8.6. Bubble formation by means of jets.

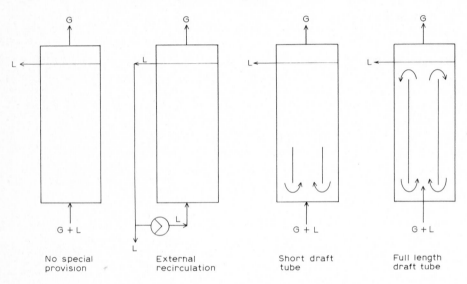

Fig. 8.7. Liquid recirculation in bubble columns.

Finally, the gas must disengage from the liquid, and this leads to a variety of designs for the head of the reactor, each design having specific advantages for particular gas rates or varying needs to reduce entrainment of liquid (Fig. 8.10). The basic principle common to all is to increase cross-section area for the gas to reduce velocities and hence entrainment.

This example with the bubble column reactor serves to show that the engineering work is far from complete when the basic dimensions have been determined, and good engineering judgement should be exercised well into the detail design of the equipment to ensure a trouble-free design is achieved.

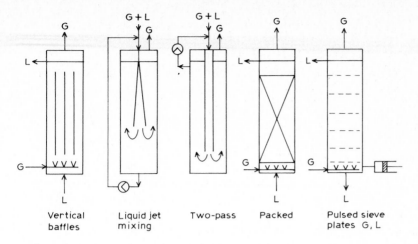

Fig. 8.8. Examples of bubble column internals.

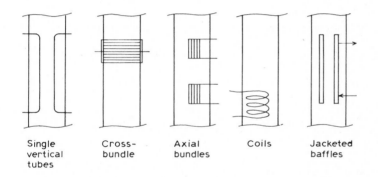

Fig. 8.9. Different forms of internal heat exchanger arrangement.

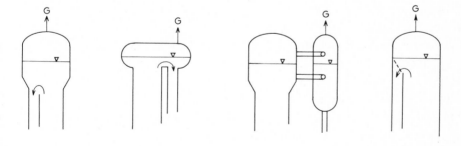

Fig. 8.10. Methods of disengagement at the head of bubble columns.

8.7 FERMENTOR DESIGN

Fermentors represent a special form of gas/liquid reactor. There are multitudes of micro-organisms (single cell cultures such as yeasts or various fungi) that grow very quickly given a suitable nutrient (substrate) and the correct balance of trace elements. The majority of these living organisms (the aerobic species) need oxygen to grow and use the substrate. Hence, the majority of fermentors have the problem of gas/liquid mass transfer, and the limitation to the growth and production rate is often the rate at which oxygen can be transferred into the liquid.

Fermentors are used for a wide range of products; antibiotics such as penicillin and streptomycin production, ethyl alcohol production, single cell protein from hydrocarbon fractions or methanol, animal protein feedstuff from sugar wastes, are to cite but a few examples.

They represent a special form of gas/liquid reactor because:

(1) They require high gas dissolution rates, leading to particularly high agitation rates.

(2) There are often large residence times involved, resulting in very large volumes. Hence, conventional stirred tanks are often too small, and large bubble columns are used, but impellers of special design are built into the bubble column to improve mass transfer. Because of vessel sizes, special designs are developed, e.g. bottom driven impellers to prevent the need for long impeller shafts that would be necessary with top mounted impellers. Large vessels (e.g. 300 m^3) with small impellers require careful design to ensure that bypassing and incomplete use of the total volume supplied does not occur.

(3) The liquid phase has a high concentration of biological matter in it. This changes its physical properties considerably, depending on the structure of the organism growing. A fibrous organism such as a filimentous fungus will provide a liquid with very different physical properties from a yeast culture.

(4) Conditions should be aseptic. This means that all equipment must be sterilized before use, and all feeds must contain no living organisms. This can be achieved by temperature or pH control for liquids, and by efficient filtering for the air feeds.

(5) Heat removal is necessary as the growth of these organisms on the large scale produces heat and the fermentor must be kept within a closely defined operating temperature range. The heat transfer in the fermentor and the tendency to fouling by deposition of the organisms on the tubes are further difficulties that may be encountered.

These conditions have led to the development of special designs for fermentors which have been developed by companies specializing in this area.

In principle, fermentor design does not differ from normal gas/liquid reactor design; $k_L a$ is calculated from correlations described in the earlier sections of the chapter, and organic growth kinetics can be determined in the laboratory, following some form of Michaelis-Menten kinetics:

$$\frac{dC_P}{dt} = \frac{k_1 \, C_S}{K_M + C_S} \tag{8.44}$$

where S is the substrate, P is the product, and K_M is a constant.

Difficulties arise when fermentor volumes become greater than the standard stirred tank, and the liquid properties fall outside the range of $k_L a$ and bubbling correlations. When this occurs, the specialized firms can be of much assistance.

Fermentors are preferably run in a continuous rather than a batch mode so that the culture achieves a steady state. A stream is continuously removed from the fermentor and the biological growth is harvested by filtration.

8.8   SOME NOTES ON LIQUID/LIQUID REACTORS

Reactors in which two immiscible liquid phases are mixed together, and the reaction occurs after one component has transferred into the second phase are called liquid/liquid reactors.

Such reactions are common in the organic chemical industry where sulphonations, nitrations, hydrolyses, and alkylations are often done with two liquid phases present.

In principle, liquid/liquid reactions have similar requirements to gas/liquid reactions in that one phase must disperse in the second to provide mass-transfer area so that the reactants can transfer between phases.

The production of a dispersed phase is more difficult than the dispersion of a gas and so a mechanically agitated reactor becomes essential. This can take the form of a stirred tank or an agitated liquid/liquid extraction tower.

Assuming that the reaction occurs in the continuous phase (1) and one of the reactants (B) is dispersed in phase (2), it is possible to derive mass-transfer and kinetic equations analogous to equations 8.1-8.11 at the beginning of the chapter.

For example, equations 8.1 and 8.4 can describe the processes occurring in an isothermal liquid/liquid batch reactor. In place of H_B' (Henry's constant) being defined by equation 8.3 it must now represent the distribution coefficient between the two phases: H_B''

$$(C_B^\star)_{phase\ 2} = H_B'' \, (C_B^\star)_{phase\ 1} \tag{8.45}$$

where * denotes equilibrium concentrations.

As the reaction proceeds, the dispersed phase is reduced as it transfers and reacts in the continuous phase. If the dispersed phase is pure component B then the quantity available to be the dispersed phase is reduced proportionally to the conversion of the reaction, and if this reduces the drop size and not the number of drops then the surface area during the reaction can be related to the initial area a by:

$$a(C_A/C_{A_0})^{2/3}$$

when the reactor is fed with a stoichiometric quantity of B (Nagata, 1975).

Because we have reactions in the second phase, this can enhance mass transfer if the reaction is fast and occurs within the liquid film. Hence the use of the enhanced mass-transfer coefficient should be included in the analysis. This can be done by insertion of the enhancement factor E'.

Taking these points into consideration, equation 8.4 becomes

$$\frac{dn_B}{dt} = E' \, K_L \, a \left(\frac{C_A}{C_{A_0}} \right)^{2/3} V_R((C_B)_{phase \ 2} - (C_B)_{phase \ 1} \, H_B'') - k_1 V_R(1-\varepsilon_g) \, C_A(C_B)_{phase \ 1}$$

(8.46)

Equation 8.1 remains unaltered.

The form of this equation can be used for modelling liquid/liquid reactions. The two extreme cases, when the system is mass-transfer-controlled or reaction-kinetic-controlled , are easily derived and the reactor can be described as follows:

mass-transfer-controlled:

$$\frac{dn_A}{dt} = -E' \, K_L \, a \left(\frac{C_A}{C_{A_0}} \right)^{2/3} V_R(C_B)_{phase \ 2}$$

(8.47)

kinetically controlled:

$$\frac{dn_A}{dt} = -V_R(1-\varepsilon_g) \, k_1 \, C_A \, H_B'' \, (C_B)_{phase \ 1}$$

(8.48)

The prediction of K_L, a, and ε_g of the dispersion achieved by agitation is more difficult than the prediction of bubble size and area in a gas system, because of the different tendencies for liquids either to form stable emulsions or to coalesce into two phases. Also the presence of very small quantities of surface active agent affect the stability of liquid/liquid dispersions.

The maintenance of a good dispersion is very dependent upon good agitation, and it is likely that agitation of small reactors is more uniform than agitation of large ones when equal power/unit volume is installed. This means that reactor perform ance is different on different scales because of better dispersion on the small scal The impeller dimensions also become more critical, since a too small impeller leads to coalescence near the reactor walls, and a too large one results in phase separati near the agitator shaft (Nagata, 1975).

All this makes the design of liquid/liquid reactions from published correlations unreliable and pilot plant experiments are essential. Furthermore, difficulties can be expected in scaling-up pilot plant results, and scale-up must be done conservatively, in comparitively small steps.

In such a situation modelling can be of assistance in correlating the results of various scales together and in the prediction of the full-scale plant.

The model used can include the chemical reaction together with empirical mass-transfer relationships that can be fitted to the pilot scale results. The equations upon which the experimental model is based should bear some relation to the mechanisms occurring in the liquid/liquid emulsions, and these are given in more detail in the literature. The reader is referred to the book of Rase (1977) which gives a detailed survey of the fundamental models published for liquid/liquid reactions, and Nagata (1975) who surveys the experimental work published on estimation of interfacial areas. There seems to be very little work reported on measurements of mass-transfer coefficients, but one can get an order of magnitude prediction from the k_L correlation for gas/liquid dispersed systems.

FURTHER READING

Astarita, G., 1967. Mass Transfer with Chemical Reaction. Elsevier.
Danckwerts, P.V., 1970. Gas-Liquid Reactors. McGraw-Hill. - Surveys industrially important gas-liquid reactions.
Horak, J. and Pasek, J., 1978. Design of Industrial Chemical Reactors from Laboratory Data. Heyden. - Theory of gas-liquid reactions treated in detail.
Nagata, S., 1975. Mixing Principles and Applications. Wiley. - Detailed agitation survey.
Rase, H.F., 1977. Chemical Reactor Design for Process Plants. Wiley. - Concise survey of available correlations for gas-liquid systems.
Vusse, J.G. and Wesselingh, J., 1976. Multiphase Reactors. 4th ISCRE, Heidelberg. DECHEMA, Frankfurt. - State of the art review from a practical standpoint.

REFERENCES

Akita, K. and Yoshida, F., 1974. Ind. Eng. Chem. Proc. Des. Dev., 13: 84.
Burkel, W., 1972. Chem. Ing. Tech., 44(5): 265.
Calderbank, P.H., 1958. Trans. Inst. Chem. Eng., 36: 443.
Calderbank, P.H., 1959. Trans. Inst. Chem. Eng., 37: 173.
Gerstenberg, H., 1979. Chem. Ing. Tech., 51(3): 208.
Lehrer, I.H., 1968. Ind. Eng. Chem. Proc. Des. Dev., 7: 226.
Lehrer, I.H., 1971. Ind. Eng. Chem. Proc. Des. Dev., 10: 37.
Mersmann, A., 1975. Chem. Ing. Tech., 47(21): 869.
Miller, D.N., 1974. AIChE J., 20: 445.
Towell, G.D, Strand, C.P. and Ackerman, G.H., 1965. AIChE-Inst. Chem. Eng. Symp. Series, 10: 97.

QUESTIONS TO CHAPTER 8

(1) Does mass transfer enhance reaction, or reaction enhance mass transfer ? Describe the mechanisms occurring in absorption plus chemical reaction.
(2) When would an absorption column be a suitable reactor design ?
(3) How does the presence of gas in a stirred tank affect the power consumption and heat-transfer coefficient ?
(4) What are the differences in area/unit volume and voidage developed in a bubble column and in an agitated tank ?

250

(5) Describe the different gas/liquid reactor types, and the different situations
    in which each is preferred.
(6) What are the various functions that a piece of equipment designed to be a
    gas/liquid reactor must perform ?
(7) Compare and contrast liquid/liquid reactor systems and gas/liquid systems.

Chapter 9

THE DESIGN OF PLUG-FLOW-TYPE REACTORS

Treated in this chapter are those reactors where the reactants are fed continuously into one end of a tube and the reactor products flow out of the other end, with the diameter of the tube such that there is effectively no back mixing along it, and so plug-flow conditions are achieved. Hence, these reactors adhere to the PFR treatment given in Chapters 1 and 2 and can be modelled by sets of differential equations.

As already discussed in Chapter 1, the advantages of PFR reactors are that their particular concentration profile results in a high selectivity when there are consecutive reactions involved in which the product formed can react further with the reactants to form side products.

The main disadvantage of the PFR is that heat generation along the tube is uneven, and at the beginning of the reaction it can be so great that uncontrollable temperature rises occur and the reactor is unworkable. Because of this heat removal problem, the tubes are usually very narrow to ensure a high surface/volume ratio for heat transfer. Thus, in order to get adequate reactor capacity at an acceptable pressure drop, the industrial reactor must contain several thousands of such tubes, built together in a single shell in which coolant is flowing.

With reactions involving no heat of reaction, there is no critical limit on the tube diameter, and this can be selected to satisfy the space time required of the reaction. Such reactors take the form of vertical cylindrical vessels, and are tubular only in the broadest sense of the word. However, given a good gas inlet distribution these reactors are also effectively plug-flow and so can be treated by a plug-flow model for selectivity predictions.

For endothermic reactions, there is no need to have low diameters because the cooling of the reactor which occurs during the reaction reduces reaction rate, and thus is a stabilizing rather than a destabilizing influence. Hence, expensive, narrow, multi-tubular reactors can be replaced by a series of short reactor beds. Between each reactor the fluid passes through a heat exchanger which raises the temperature back to the inlet temperature for the next bed.

9.1   GASEOUS REACTIONS ON SOLID CATALYSTS

Since gas/solid catalyzed reactions are, industrially, a very important, if not the most important, class of reactions, and since they are almost always carried out in plug-flow reactors, we should now look in more detail into the theory of them.

Chapter 1, section 1.2.d, describes the chemical mechanisms that occur in this form of catalysis: active sites are present on the catalyst surface, and after one or all of the components have adsorbed on these sites, they are held there in a strained form which renders them more susceptible to reaction. After reaction the new species detach themselves from the surface.

When this mechanism is translated into the behaviour in a reactor, where we must present the catalyst surface by having a catalyst pellet or granule in the flowing gas stream, a number of mass- and heat-transfer steps must occur consecutively at steady state:

(1) Bulk (film) diffusion. Components must diffuse out of the bulk gas through the the stagnant film (according to the film theory) around the catalyst pellet, before they reach the surface of the pellet.

(2) Pore diffusion. The components must then diffuse from the pellet surface onto the active sites on the catalyst, as must the product diffuse back from the site to the pellet surface. Most active sites are not on the pellet surface, since the "surface area" of the catalyst by absorption measurements is generally many hundreds of times the external geometric surface area of the pellet. Hence, the active sites are situated throughout the catalyst pellet. Some are near the surface, and some near the centre. Therefore, the pore diffusional resistance is not the same for all sites. Furthermore, the internal structure of the cataly pellet is not a series of equal resistance channels. Some surface is easily available, others have a large diffusional resistance, resulting in concepts of macro and micro pores and tortuosities to define different hindrances to diffusi (Fig. 9.1). Such theories can be verified by pore volume measurements.

(3) Heat transfer. Heat is generated where the reaction occurs and hence some heat is generated in the middle of the catalyst and some generated on the outside. The heat generated in the middle must transfer by conduction to the outer shell on the pellet, and all the heat generated on the pellet must transfer across the gas film into the bulk gas. This results in a temperature profile in the system, giving the catalyst pellet surface temperatures above the gas (for exothermic reactions) and the temperature in the centre of the pellet being greater than on its outside surface.

These three effects mean that the kinetic reaction rate, as defined by the rate of reaction on the catalyst site, no longer represents the rate that will be observe when the pellet is in a reactor. Bulk and pore diffusion will be resistances which reduce the rate from this intrinsic kinetic rate, and the resistance to heat transfe will result in higher temperatures than the bulk gas (for exothermic reactions) whi will result in a higher reaction rate than expected, if the bulk gas temperature is taken as the reaction temperature.

This means that a reactor will display a retarded reaction rate, enhanced rate or a rate equal to the rate determined by reaction kinetics, depending on whether bulk diffusion, heat conduction, or kinetic rate is the dominant (slowest) step in the overall process.

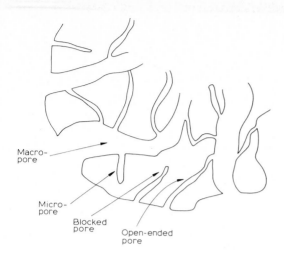

Fig. 9.1. Pore structures in a catalyst pellet.

If the intrinsic kinetic rate is known, or can be measured in the laboratory (for example an SGS reactor at high stirrer speed, crushed catalyst particles, low concentration, and low temperatures enables the kinetic rate to be measured), the rate used for design must be modified to take the various resistances into account.

This is done by including an effectiveness factor η which is the ratio of the reaction rate of the catalyst in the reactor to the reaction rate of the catalyst in the absence of diffusion and heat effects. This is included in the kinetic equations by using ηk_1 in place of k_1. Whether η is greater or less than 1 depends on the relative magnitude of the contributing diffusions and thermal conductivity effects.

It is possible to theoretically calculate the effectiveness factor, and relate this to the diffusion coefficient, via the Thiele modulus ϕ, and the pellet thermal conductivity, via the factor β where:

$$\phi = \left(\frac{1}{a}\frac{k_1}{D_A}\right)^{\frac{1}{2}} \quad \text{for first-order reaction}$$

$$\text{or} \quad \frac{1}{a}\left(\frac{k_1\ C_A}{D_A}\right)^{\frac{1}{2}} \quad \text{for second-order reaction}$$

and

$$\beta = \frac{\text{adiabatic } \Delta T}{T \text{ inside pellet}} = \frac{-\Delta H_1\ D_A\ C_{A_0}}{\lambda T} \quad \text{(a is the catalyst surface area/unit volume).}$$

Theoretical predictions of effectiveness factor for various values of ϕ and β are given in Fig. 9.2.

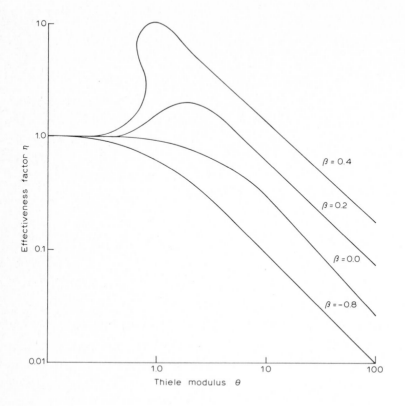

Fig. 9.2. Theoretical predictions of effectiveness factor.

In addition to producing faster or slower reaction rates than expected, the mass- and heat-transfer effects can cause unexpected selectivity problems in complex reactions.

In consecutive reactions, where the product can undergo reactions to give unwanted by-products, diffusional resistance is detrimental because this resistance means that the product must have a higher concentration at the catalyst surface to provide the driving force to overcome the mass-transfer resistance. The higher concentration increases the following reaction rate, and lower selectivity results.

In systems where side reactions are present with higher activation energies than the main reaction, the higher catalyst particle temperatures due to the heat-transfer resistance mean that the side reactions are enhanced, and again lower selectivities result than one would predict from reactor bulk temperature measurements.

It is also possible that the higher catalyst temperature causes reactor instability. The high sensitivity of catalytic reactions to temperature is described in Chapter 1.

and hot catalyst particles will augment this sensitivity.

When catalytic reactors are to be designed from data obtained with the same particle size as that to be used on the final plant, with velocities comparable to full-scale (as is the intention with most reactor development work), it is not necessary to know the effectiveness factor, since the overall performance (i.e. $n \ k_1$) is being measured directly.

However, if catalyst development work is being carried out on the manufacture of the catalyst or its support, then the effectiveness factor is a very useful piece of information in pointing out whether diffusional or other resistances are present, and what potential there is for improving productivity and selectivity by modifying the support pore structure and catalyst particle size.

9.2   TYPES OF PLUG-FLOW REACTOR

9.2.a   Tubular reactors for liquids

As long as the initial heat generation is not excessive, tubular reactors can be used for liquid reactions. Because of the high density of liquids and the relatively low reaction rates as compared to gases, liquid tubular reactors have low velocities, and hence low pressure drops/unit length. This means that the reactors designed are long, and often single tubes, since any multi-tubular design requires a considerable pressure drop across the tubes to ensure equal distribution of flow between the tubes. If the tube diameter is not limited by heat-transfer considerations, the liquid reactor tube diameter is high to obtain the required residence time in a reasonably practical tube length.

Since few tubes are involved, the reactor takes the form of a jacketed pipe, often U-shaped, with the coolant flowing in the outside jacket.

Tubular reactors for liquids are fairly rare because the usually high residence times associated with liquid reactors, and initial heat removal problems, favour the use of a CSTR cascade. When tubular reactors can be used, however, they are more convenient and cheaper. This can be decided by calculating the reactor volume and heat liberation rate from the kinetics of the reactions occurring.

9.2.b   Reactors for endothermic gaseous reactions (catalytic and non-catalytic)

In this case, since there are no heat removal problems, the total flow is put through one tube, and the tube diameter and length are arranged so that the required residence time is provided, the L/D ratio is such that an adequate approximation to plug-flow is achieved, and the pressure drop is acceptable.

As the reaction proceeds, the temperature falls, and this may necessitate re-heating the gas mixture part way through the reaction to obtain reasonable residence times. This is usually done by dividing the reactor into a number of shells and passing the gas through a heat exchanger between the shells (Fig. 9.3).

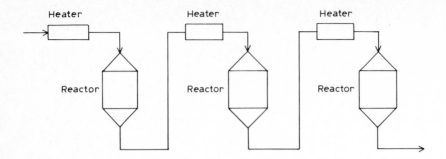

Fig. 9.3. Multibed reactor with multi-stage heating.

When the reaction is an equilibrium reaction, since it is endothermic, the formation of product will be favoured by high temperature (see section 1.2.a). Hence, to achieve a good final conversion the end temperature must be high. This is also achieved by the series of reactors and heat exchangers as shown in Fig. 9.3. In some cases it is economic to use up to six such reactors in series, but each case should be considered on its own merits.

If the kinetics are available from the development work, the optimal number and volume of reactors can be determined quantitatively. Figure 9.4.a is the standard diagram showing how each bed lifts the conversion, and the conversion in each bed is an asymptotic approach to the equilibrium line.

9.2.c   Adiabatic reactions for mildly exothermic gas reactions (catalytic and non-catalytic)

When the adiabatic temperature rise of a gas is not detrimental to reactor selectivity, it is possible to carry out the reaction in a single bed, as with the endothermic reaction. The reactor volume is determined by the required contact time, and the diameter/length again chosen to achieve plug-flow.

When the reaction is an equilibrium reaction, then in this case low temperatures are required to favour the formation of the product. Hence, as the reaction proceeds the heat generated is preventing a high conversion being obtained. When this is the case, the reactor is again divided into a number of beds, but in place of providing heating between the beds, coolers are provided. Figure 9.4.b shows the course of such a reaction diagrammatically. The lowest temperature to which the reactor can be cooled is determined by the rate of reaction at the low temperature. There is a critical temperature where the reaction rate becomes so low that it effectively does not start reacting when entering the next reaction vessel (the initiation temperature). This temperature can be determined from the kinetics of the reaction, and the coolant temperature must always be kept above this.

It is generally not possible to put an exothermic reaction in a series of

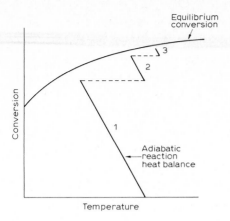

Fig. 9.4.a. Reaction conversion/temperature plots - endothermic reaction.

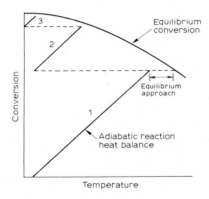

Fig. 9.4.b. Reaction conversion/temperature plots - exothermic reaction.

uncooled vessels with interstage cooling if the adiabatic temperature is detrimental
to the selectivity. Theoretically, a reactor could be designed to only partly react
the gas before it leaves the reactor and is cooled, so that it never reaches too
high a temperature and affects selectivity. This would indeed work for a fixed
reactor throughput and very steady conditions, but if the throughput needed to be
reduced, or if inlet temperature varied a few degrees, then there would immediately
be a serious reactor temperature problem because of the exponential nature of the
temperature rise in an adiabatic reaction.

Once the adiabatic temperature rise is not acceptable, then in practice, cooling
during the reaction becomes essential, and this usually requires the use of a
multi-tubular reactor to enable a reasonable heat removal capacity to be available.

9.2.d   Multi-tubular gas phase reactions (non-catalytic)

Strongly exothermic gas phase reactions can be carried out in multi-tubular heat exchangers to provide the heat-transfer surface to remove the heat of reaction. They are not very common, and the reason lies in their very limited range of applicability. The provision of more than a few seconds' residence time by means of a shell and tube exchanger is very expensive, and so this can only be considered for short contact time reactions. However, short contact time plug-flow reactions have very high heat fluxes at the beginning of the reaction - probably greater than can be handled by a heat exchanger. Hence, these reactions come into consideration in those few instances where a short contact time is required, but not so short that initial heat generation rates are excessive.

The reactor mostly used for long contact time, thermally sensitive, exothermic consecutive gas reactions has a mixture of CSTR, plug-flow, cascade, and split feed characteristics, as the next paragraphs describe.

9.2.e   Empty volume gas phase reactors for exothermic reactions

By designing a reactor as a fairly wide tube with an L/D ratio of, say, 10:1, and by arranging the gas inlet to be in a form of jet, a degree of mixing is induced in which the hot reacting gases mix with the cold inlet gas. The cold inlet gas is therefore providing a heat sink in place of having to provide heat-transfer equipment internally. Such a design has some CSTR characteristics but will also have a section with plug-flow characteristics at the end of the reactor. Should the heat developed from complete reaction be greater than the heat capacity of the cold feed, then part of one reactant can be added in one reactor, the outlet gas cooled in an external heat exchanger, and fed to a second reactor where more of the second reactant can be added. If necessary this can be done a third, or more, times. Two forms of such a reactor are shown in Fig. 9.5.

Such a reactor satifies the requirement stated at the end of the last section. Being an empty volume it is comparatively cheap to have long residence times. Temperature control is possible by controlling the fraction of the second reactant put into each reactor shell. Hot spot and heat-removal problems are overcome by having the mixed zone at the beginning of the reactor, and experimental work has shown that selectivities more nearly approach that expected from a PFR than a CSTR.

The reactor has a major disadvantage in that start-up is difficult, unless extra equipment is installed for use only at start-up. The reactor or the feed must be heated up by some external means during the start-up phase in order to get the reaction underway.

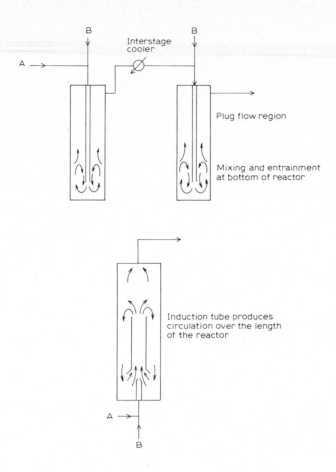

Fig. 9.5. Forms of adiabatic gas phase reactor for exothermic reactions.

9.2.f   Fluidized bed reactors with inert solids

The empty volume reactor relies upon creating a suitable gas mixing pattern to warm the cold feed and reduce the temperature of the hot spot. Alternatively, this evening-out of temperature can be achieved by carrying out the reaction in a bed of fluidized solid particles. These particles freely move around the reactor, and, because of their high surface area, have good heat transfer with the surrounding gas. Hence, the completely mixed solid particles have the effect of evening-out the temperature very effectively, in a more positive way than inducing a particular gas mixing pattern and hoping that it will be a stable pattern over a wide range of feed rates.

However, this inert fluid bed has problems. The inert solid must be strong enough to withstand the fluidization, and the equipment must be built to withstand the considerable erosion that occurs in fluid beds. Fluid beds are discussed in more depth in section 9.7.

9.2.g Fixed bed multi-tubular gas phase reactors

When a gas phase system reacts over a solid catalyst and heat is evolved, it is usually necessary to remove that heat from the system. If this is not done the high temperatures usually cause catalyst damage and often induce side reactions, which reduce yield and selectivity. The heat is generally removed by means of heat-transfer area within the reactor, and this takes the form of a bundle of tubes with the catalyst being packed inside the tubes, and the coolant flowing outside.

With gaseous systems, and typical residence times of 1-10 seconds, very large reactors are necessary for any respectable plant output, as already described in section 9.2.d. However, in the case of catalytic reactions there are no alternative design options open except in exceptional cases where a fluidized bed can be used, and so exothermic catalytic gas phase reactions are generally carried out in large expensive multi-tubular reactors, sometimes containing up to 20000 tubes in each shell (Fig. 9.6).

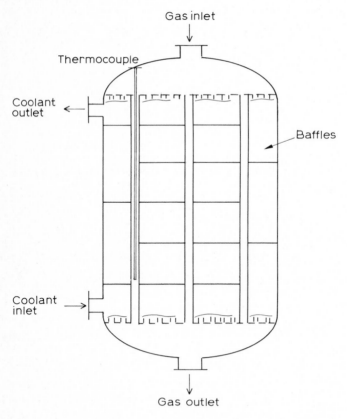

Fig. 9.6. Multi-tubular catalytic reactor.

The coolant flows through the shell side, which must be adequately baffled to provide a reasonably high velocity across the tube bank and, more importantly, to ensure that there are no "dead spots" in the coolant flow. Wherever there are "dead spots" the tubes will behave as adiabatic reactors, since there will be effectively no cooling for them.

The reactor design must also have provision for removing the top and bottom reactor covers so that the catalyst can be charged and discharged. There must be provision for the bottom catalyst retaining plate to be withdrawn carefully for the emptying of the catalyst, and it is sometimes necessary, particularly if the flow direction is chosen to be upwards, that there should be some form of retaining plate at the top of the reactor.

Tube lengths up to about 3 metres for 25 mm tubes and 5-10 metres for 50 mm tubes are practical. Longer tubes can present difficulties in emptying should there be any tendency for the catalyst to fuse together. Tube diameters are usually 25 mm, with possibly 50 mm for duties where the heat removal requirement is not severe. 25 mm is generally considered the minimum practical operating diameter under industrial conditions. There is not a great deal of incentive for going to larger tube sizes because the heat-removal problems increase more than directly proportionally to the tube diameter and the seriousness of this is magnified by the exponential nature of the effect of temperature.

Catalyst pellet diameters are often about 1/5th the internal tube diameter. Larger pellets cause bridging problems during filling, and smaller pellets create pressure drop problems.

Having accepted that there is a heat-removal problem, and piloted a 25 mm tube reactor, changing the tube diameter to 50 mm would require further piloting with a pilot reactor four times as big. This would result in a less controllable reactor with greater hot spot temperature which may cause selectivity and catalyst life problems even if it proved to be operable, in order to save the difference in cost between a lot of small tubes and fewer large tubes. Once a process has been established, and when the second generation of reactors is being designed is the right time to be considering whether investment cost can be reduced by using larger diameter tubes.

The capital cost of multi-tubular reactors has been satisfactorily modelled to manufacturers' data by assuming that the cost is a function of total weight (W), the number of tubes (N), and the total length of circumferential welding round each tube in the tube plate, by a model which reduces to

Capital cost = 690 W + 900 N + 39000 $ (1965)   (9.1)

The reference year is too old for the equation to be useful for costing purposes, but it is interesting to note that equipment cost models can be derived from modelling the work content of their manufacture.

9.3   DESIGN PRINCIPLES FOR TUBULAR REACTORS

Tubular reactors are comparatively straightforward to design from pilot plant data because the pilot plant will normally have operated with the same catalyst particle size as is to be used on the full-scale plant, with the same contact time, and if there is a heat-removal problem, with the same tube diameter.

If there is no heat-removal problem and the reaction is to be carried out adiabatically, then the pilot work will have been done on an insulated tube which can be considered to be a representative plug taken from the full-scale reactor.

Only the liquid tubular reactor will have been operated with different diameters even if it has a heat-removal problem, and therefore special attention is needed to predict the effect of the different heat-transfer coefficient that will result in the much larger diameter tubes. This can be done using normal heat-transfer correlations for fluids inside tubes.

The reactor length on the pilot plant (particularly if a mini-pilot plant in a laboratory was used) is, however, often smaller than the length that the full-scale reactor will be, and one needs care to see that the effect of increasing the velocities and feeding more reactant to each tube is properly allowed for on scale-up.

As has been repeatedly described in earlier chapters, this effect can be thorough investigated on a model of the reactor, if it has been produced during the development stage of the process. The model can include the correct form of heat-transfer coefficient as a function of velocity, and hence can indicate the changes that can be expected as a result of tube length changes and enable further modifications to operating conditions to be investigated to bring performance back to optimum.

The models developed in Chapter 2 and coded in Appendix 3 are adequate for this purpose, given an adequate form of correlation for heat-transfer coefficient (see next section). The literature contains a great deal of work concerned with two-dimensional fixed bed catalytic tubular models, where it is not assumed that conditions are constant along the tube radius, but temperature and hence concentrations vary across the diameter because of the heat conduction through the catalyst particles to the reactor wall. These models become quite complex in their descriptio of the processes occurring along the radius even for only very simple chemical reactions.

Comparisons between the two types of model show very similar results, except that noticeable differences occur on the point of incipient instability. Since these comparisons were made on two entirely theoretical models, it can be assumed that a one-dimensional model fitted to experimental results, as is the way in which these models are used, would show even less discrepancy from a two-dimensional model also fitted to the experimental results.

Hence, for modelling for process development design purposes a one-dimensional model would appear to have adequate accuracy, and it has the advantage of being simple enough to be able to include an accurate description of the chemistry involved, which is essential for reactor design purposes. The two-dimensional model is necessary for predictive purposes, particularly the prediction of the effect of tube diameter. However, as yet, no reliable design method for predicting the effect of tube diameter changes on a real reactor have been developed.

Although the pilot reactor will have produced information on heat-transfer coefficient (via the parameter fitting), it is still of interest to know of correlations for these factors; firstly, so that the model can have the correct equation form in it even though the significant parameters will be fitted by regression, and secondly, it is always useful to compare experimental results with literature results, and also to be aware of what to expect before the experiments actually start.

9.4 CORRELATIONS FOR HEAT TRANSFER

Heat-transfer coefficients for the liquid and empty gas tubular reactors can be determined using perfectly standard heat-transfer correlations for fluids inside the tubes (McAdams, 1951).

In general, for the fluid flowing through a tube the following correlation holds:

$$\left(\frac{h\ d_t}{\lambda}\right) = 0.023 \left(\frac{d_t\ \rho\ u}{\mu}\right)^{0.8} \left(\frac{c_p\ \mu}{\lambda}\right)^{0.4} \tag{9.2}$$

where u is the fluid velocity in the pipe (m s^{-1}), of internal diameter d_t (m), for turbulant flow (Reynolds number > 2000) and low viscosities ($\mu < 0.2$ kg m^{-1} s^{-1}).

For high viscosity liquids the modified form

$$\left(\frac{h\ d_t}{\lambda}\right) = 0.027 \left(\frac{\mu_b}{\mu_f}\right)^{0.14} \left(\frac{d_t\ \rho\ u}{\mu}\right)^{0.8} \left(\frac{c_p\ \mu}{\lambda}\right)^{0.33} \tag{9.2a}$$

is more appropriate where μ_b/μ_f is a correction for the viscosity difference between the bulk liquid and the bulk temperature (μ_b) and the liquid at the heat-transfer surface temperature (μ_f).

For gaseous systems equation 9.2 approximates to a more convenient form:

$$h = 8.18 \times 10^{-5} \frac{c_p\ (\rho\ u)^{0.8}}{d_t^{0.2}} \qquad \text{kW m}^{-2}\ \text{K}^{-1} \tag{9.3}$$

to represent the inside film heat-transfer coefficient for a gas flowing in a tube.

When the tubes are packed with catalyst there is a complex picture of heat being generated in the catalyst which is transferred to the walls by conduction through the catalyst particles and by conduction and convection through the gas.

The prediction of the effective inside tube heat-transfer coefficient in terms of the physical properties of the particles and the gas, and the size and shape of the particles, has yet to be satisfactorily achieved.

In the absence of a predictive method, all that can be done in modelling work is to assume a form of relationship similar to that for the empty tube, and fit the constants to pilot plant work for the system in question.

It is worth noting here that pilot plant experiments can be very conveniently done to obtain an overall heat-transfer coefficient by passing a cold feed gas through the heated reactor tube and measuring the resulting temperature profile without reaction. Fitting this profile gives the overall heat-transfer coefficient, uncorrelated with any kinetic parameters. Values of the order of 0.05-0.1 kW m^{-2} K^{-1} are usually obtained.

9.5 CORRELATIONS FOR PRESSURE DROP

9.5.a Pressure drop in an empty tube

Pressure drops in empty pipes can be calculated by the normal pressure drop correlations for friction loss in pipes. This can be obtained by using the Fanning equation

$$\frac{dP}{dL} = \frac{f \, \rho \, u^2}{2 \, d_t} \qquad (Pa \; m^{-1})$$

$$= \frac{f \, \rho \, u^2}{2 \, d_t} \times 10^{-5} \, (bar \, m^{-1}) \qquad (9.4)$$

for a tube L metres long and d_t m diameter with the fluid moving with a superficial velocity of u m s^{-1}.

The value of f is dependent on the pipe roughness and Reynolds number, but for normal 25 mm tubing and Reynolds numbers over 5000 it is adequate to assume f is constant at 0.024 for rough assessments. For accurate work and for low Reynolds numbers where the value of f increases, the reader should consult a standard text.

9.5.b Pressure drop across a packed bed

Pressure drop through packed beds has been studied for a variety of different packing shapes (spheres, cylinders, and irregular granules) and the following correlations derived (Ergun, 1952).

A friction factor can be defined depending only on the Reynolds number and the voidage ε_p

$$f = 1.75 + \frac{150(1 - \varepsilon_p)}{\left(\dfrac{d_e \, \rho \, u}{\mu}\right)p} \qquad (9.5)$$

where u is the superficial velocity and d_e is the equivalent diameter of the particle, defined as

$$d_e = \frac{6}{\text{surface area/unit volume of particle}}$$ (9.6)

e.g. for a cylinder, diameter d, length L,

$$d_e = \frac{3d}{\left(2 + \dfrac{d}{L}\right)}$$ (9.7)

The pressure drop is given by

$$\frac{dP}{dL} \simeq \frac{f \, u^2 \, \rho}{d_e} \left(\frac{1 - \varepsilon_p}{\varepsilon_p^2}\right) \times 10^{-5} \text{ bar m}^{-1}$$ (9.8)

Ergun's correlation is known to be very conservative (i.e. it predicts higher pressure drops than are achieved) but no more suitable correlations are available in the literature.

It is particularly important to be able to predict the pressure drop given by various designs because this is an important criterion in the design of the reactor, and as will be explained in section 9.6, pressure drop measured in pilot plant equipment is often not accurate enough for design purposes.

Too large a pressure drop will cause excessive compression cost and so result in an uneconomic process. Too small a pressure drop will allow mal-distribution in the tubes to occur. This is because some tubes will inevitably be nearer the gas inlet and parallel to the incoming gas flow; others will be far from the inlet with the gas needing to make two direction changes before it can enter the tubes (see Fig. 9.7).

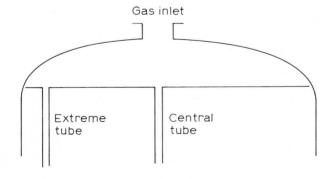

Fig. 9.7. Inlet gas distribution to a multi-tubular reactor.

In order to get the gas flow through the tubes as even as possible, the pressure drop across the header plus a velocity head for the turns should be small compared with the pressure drop through the tubes (say less than 10%). Hence, the convenientl placed tubes will receive a flow of u m s^{-1} and the inconveniently placed tubes u'. Since the overall pressure drop must be equal, the pressure drop in the convenient tubes must be 1.1 times that in the tubes at the edge of the reactor. Since pressure drop is proportional to the square of flow we can say that $u \simeq 1.05 \, u'$. A 5% tolerance on flow rate per tube is usually acceptable.

A point frequently discussed, and never agreed upon, is the advantage in filling each of the catalyst tubes with the same quantity of catalyst, and then checking and adjusting the pressure drop across each tube individually (all 10000 of them). The alternative is for them to be filled by a man with a broom, brushing catalyst across the tube plate until all tubes appear full.

In many cases the tube filling will not be critical. In some cases, where hot-spots develop and the tubes are near temperature runaway conditions, it may be that tubes with less catalyst get more gas, overheat, runaway in temperature and result in catalyst, or even reactor, damage.

There are a number of features that reduce the seriousness of this situation. A 20% imbalance in tube filling will produce only a 10% increase in the flow through it (by the argument already given above) and it would be unwise to run a reactor so near the runaway that a 10% change would result in disasterous consequences.

When a reactor model is available it is possible to investigate the effect of flow rate on the temperature profile and so determine whether a variation in packing density between various tubes would result in serious consequences or not. As a result of the study, the operating manual could be appropriately written.

Catalyst particle diameters have to be decided in some way. Sometimes the reactio is improved by reducing the particle diameter; when it is pore-diffusion-controlled the small particles will at least increase the rate of reaction, if not also improve the selectivity. When this is the case there is a lower limit on the particle size, set on the one hand by the pressure drop and design of the resulting full-scale reactor, and on the other hand by the lowest practical size below which bed blocking occurs because of scale which is always present in mild steel equipment.

In general it is inadvisable to use particles of less than 3 mm mesh in industria equipment. In a specific reactor study it may be found that these particles produce a short, large diameter reactor in order to keep the pressure drop down to manageable proportions. In this case the extra cost of the squat reactor and any extra compression costs, and the likelihood of bed plugging, should be weighed against any quoted conversion and selectivity advantages of the smaller particles before the final decision is made to employ small particles on the full-scale plant.

9.6   AN EXAMPLE: FIXED BED REACTOR DESIGN

As an example of reactor design, let us take our PFR reactor example begun in Chapter 2 to the final design of the industrial-scale reactor.

In producing a design for a fixed bed tubular reactor, the designer must decide on the following:

1. Reactor type
2. Feed ratios
3. Catalyst type, composition, particle size, and shape
4. Reactor conversion
5. Reactor residence time
6. Reactor temperature
7. Reactor pressure
8. Tube diameter
9. Materials of construction
10. Heat-transfer media
11. Heat-transfer medium temperature
12. Heat-transfer medium flow
13. Gas flow direction
14. Reactor length
15. Number of tubes
16. Reactor design: gas distribution, shell side baffles
17. Stand-by spare capacity
18. Reactor control system
19. Reactor safety system.

These decisions can be divided into those made during the reactor development stage, and those finalized at the design stage. During the development stage the decision will have been made which reactor to use. In our example we have an exothermic, temperature sensitive, consecutive reaction, and so a plug-flow, tubular, fixed bed reactor is a sensible choice.

Also during the development stage the decision concerning the optimum operating conditions will have been made. The model developed in Chapter 2 will have been used for fitting to laboratory and pilot plant results and values for kinetic, pressure drop, and heat-transfer parameters will have been fitted (these are A_1, A_2, A_3, E_1, E_2, E_3, K', and K''). This fitted model will then have been used to look for optimal selectivities to the product, and then to look for minimum cost processes where costs other than raw material costs have been brought into the model. At this stage, therefore, decisions concerning operating conditions (points 1 to 8) will have been made. The design engineer is now faced with finalizing the remaining decisions.

This does not mean that these later points would not have been thought about before the design stage. There must not be clear demarcation lines between which

process points are considered at which stage, but the decisions need not be
finalized until the design stage.

From discussions with metallurgists and results of corrosion tests, the first
decision to be made is the materials of construction of the equipment. Following thi
the heating medium must be selected. This could be steam, or heat-transfer oil or
other heat-transfer fluids, or molten salts, depending on temperature ranges
involved. These points are discussed in more detail in Chapter 11.

From a heat balance over the reactor, operated at its proposed conversion, it is
possible to determine a suitable heat-transfer medium coolant flow rate, and this
will be needed for the design of the shell side baffles as well as the design of
the associated fluid heating and circulation systems.

The calculation of this flow rate requires a decision to be made on the temperature difference between inlet and outlet fluid. Must this be very close, say 2°C
difference, or would a 10 or more degree difference be satisfactory? This depends
upon the reactor stability, and also its stability at different throughputs.
Larger temperature differences produce cheaper but less-stable equipment.

The model can be used to investigate such questions. A coolant jacket temperature
varying with length can be approximated by including the following equation in
the model

$$\frac{dT_J}{dt} = - \frac{U A'}{(Q \rho c_p)_J} (T_R - T_J)$$
(9.9)

where T_{J_0}, the jacket temperature at the gas entrance of the reactor, is the initia
condition for the integration. The effect of this temperature profile on the
chemistry and selectivity will then be shown by the model. The comparative advantage
or disadvantage of co- or counter-current coolant flow are shown by changing the
sign in equation 9.9. As displayed, the - sign represents counter-current flow.

Co-current flow has the advantage that the coldest jacket temperature is adjacen
to the highest reaction rate, and the highest jacket temperature is where the
reaction rate could advantageously be increased.

This results in an upward gas feed arrangement because circulating liquids
should always be fed from the bottom to prevent gas locking. A general disadvantage
with upwards flow is that at high bed pressure drops (which may occur during
occasional gas surges) the bed is lifted and then dropped again, or even lifted
out of the tubes. This movement is detrimental to the catalyst, and breakage and
tube plugging can result.

Now we can look at the size of the reactor itself. The tube length should be the
maximum length which practical experience and pressure drop limitations allow becau
this yields the lowest capital cost reactor.

The flow rate through the longest practical tube, that gives the required contac time, should be calculated and the pressure drop then determined.

Say the laboratory reactor is L_{lab} metres and the proposed plant length is L_{pt}. Then for the same contact time the gas velocity in the tube will be increased by

$$u_{pt} = u_{lab} \, L_{pt}/L_{lab} \qquad (9.10)$$

and the plant pressure drop will increase because the gas velocity has increased and the tube is longer

$$\Delta P_{pt} = \Delta P_{lab} \frac{L_{pt}}{L_{lab}} \frac{u_{pt}^2}{u_{lab}^2} = \Delta P_{lab} \left(\frac{L_{pt}}{L_{lab}}\right)^3 \qquad (9.11)$$

If the plant reactor is three times longer than the lab-reactor, then the pressure drop in the lab will have been 1/27 of the industrial pressure drop, probably negligible in comparison to the other pressure drops existing in the laboratory system. For this reason, pressure drop data from the laboratory or pilot plant are often very unreliable, and one must put more reliance on published data, e.g. equation 9.8. This equation can predict the pressure drop for the reactor, and if this pressure drop is less than 0.5 bar, and is acceptable to the process, then the tube length has been determined.

Pressure drops over 0.5 bar are to be treated with caution because, as the reactor becomes dirty, two- or three-fold pressure increases can occur and if one starts with a high pressure drop initially the plant will not be robust enough to handle even a moderate degree of plugging. If the pressure drop from this maximum tube length is too high, a shorter tube must be considered and a new pressure drop worked out even though this will impose a greater capital cost on the process.

If the reactor feeds have to be compressed, and the reactor pressure drop can be directly related to a specification of some compression equipment, there is an optimization to be carried out between the total annual compression cost and the annual reactor cost associated with the shorter reactor. Such an optimization can be made by comparing the two factors directly, unless the process has already been subjected to a total process optimization, in which case these points may have been modelled and decided upon at that stage. This optimization means that as energy costs rise we are going to see many more shorter reactors being constructed.

The effects of the new tube length on the conversion and stability of the reactor should then be checked on the reactor model. The modified temperature patterns will affect conversion, and so tube operating conditions will have to be revised to be once more at the optimum condition for the process.

Having arrived at a reactor tube length, the number of tubes is calculated simply as the number which will produce the total output required by the full-scale

270

plant, either on the total residence time basis or on an output basis.

Shells with 10000 tubes are quite common and 20000 tubes are reported (Rase, 1977). Should your reactor contain more than 10000 tubes, then consideration should be given to having two shells to ease manufacture. This point would have to be decided with the proposed equipment manufacturer.

Consideration of the catalyst replacement procedure and its frequency, and the philosophy toward installed spares can also affect how many shells are used.

If the catalyst has a short lifetime and frequent changes are required, it is economic to have a spare reactor installed that is refilled whilst the second is in operation. When the catalyst in the first reactor is deactivated, the feeds can be switched over to the second reactor and minimum time is lost. When only one reactor is installed, the plant must be shut down frequently for a few days at a time, and so it must handle a higher instantaneous load, resulting in an overall bigger plant to produce the annual product requirement. Cooling, opening-up, cleaning blocked tubes, refilling, closing, and re-heating these huge reactors is no easy operation and takes a number of days to complete.

Alternative philosophies, depending on the catalyst lifetime, might be to have three by 50% capacity (or four by 33% capacity) shells which enables one always to be off-line for refilling whilst the plant is operating at full output. A further possibility might be two at 50% capacity to leave 50% plant capacity at each catalyst change. Such decisions are analogous to the philosophy of installed spares for other important plant items requiring maintenance, such as compressors. Each case must be considered on its own merits.

Once the number of tubes has been set, and the outside diameter and pitch are known (e.g. 1.25 triangular pitch is usual) then the shell diameter can be found which contains this number of tubes. This is given by the following empirical equation for N tubes on a triangular pitch of s metres.

$$D_{shell} = s\left(0.94 + \left(\frac{N - 3.7}{0.907}\right)^{\frac{1}{2}}\right) \qquad (9.12)$$

Having obtained the shell diameter we are now in a position to determine the shell baffle layout. This is done using methods for shell-side heat exchanger design (Kern, 1950).

We have now met all our decisions in the table at the beginning of this section, with the exception of deciding on a control system and defining a safety system. These are discussed specifically in Chapter 10.

Needless to say, the model also features here in being able to identify instabilities and hazards which have to be guarded against by the control scheme developed

9.7   FLUID BED REACTORS

Fluidized beds are beds of small particles of solids (50-300 μm) which become mobile when fed with a gas above a certain minimum fluidizing velocity (u_{min}). The bed has every appearance of a mobile fluid, the solid moves around very freely, and complete solid mixing is easy to obtain. As the gas increases to commercially interesting velocities, the majority of the gas flows through the bed in the form of rising bubbles which contain very little solid (Fig. 9.8).

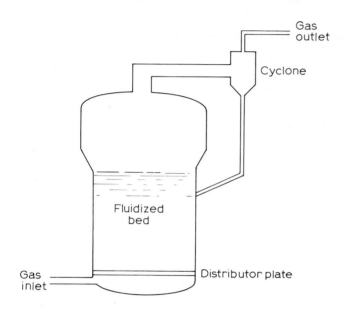

Fig. 9.8. The fluidized bed reactor.

The potential of having a freely circulating solid phase means that temperatures in the whole bed are very even and there is no possibility of a gas hot-spot existing. Thus a reaction can be operated isothermally at the optimum temperature to obtain an optimal yield.

Fluid beds also have the advantage that because the solid phase is very mobile, catalyst removal for regeneration is very simple. This can be done by blowing the catalyst from the reactor to a regenerator where it can undergo regeneration. If the solid is part of the reaction, the feeding of new solid reactant is simple because it can be added to the gas feed and so blown into the reactor.

These advantages of fluid beds make them very attractive for very many reactions and a great deal of theoretical work on the hydrodynamics occurring in the beds has been carried out in research departments, reflecting the general optimism that fluidized beds have great potential. Add to this list of advantages the success

story of the development of the first fluid bed catalytic cracker in 1942 and the
speedy impact that it had on the petroleum processing industry and it is easy to
understand this optimism.

Unfortunately, based on the performance since 1942, and considering the vast
effort that has gone on in the subject, this optimism has not been warranted. Very
few fluid bed projects have ever reached full-scale plant, and of those that have,
only a part is still competitive with newer fixed bed processes (Geldart, 1968).

There are disadvantages associated with fluid bed reactors and these should be
clearly stated to prevent development effort being wasted on fluid beds which later
prove to be unfeasible.

Fluid beds operate normally at 10-100 times the minimum fluidization velocity.
This means that the majority of the gas passing through the bed is in large bubbles.
There is some diffusion of the gas from the bubble into the solid / gas phase around
the bubble (the so-called "cloud") and there is also interchange in the "wake" of
the bubble where there are solids drawn up into the bubble itself (Fig. 9.9).

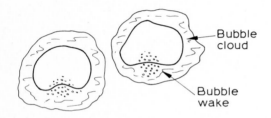

Fig. 9.9. Rising bubbles with cloud and wake.

However, these interchanges are comparatively minor and this means that for the
majority of the gas, contact with the solid is low. Hence, if it is a solid-gas
reaction, very little reaction occurs in the bubbles rising through the bed. It
appears that the majority of the contacting between gas and solid occurs during the
formation of the bubble and also at the bottom of the bed when the bubbles are still
small and before they have coalesced. Therefore, there is no linear relationship
between bed height and extent of reaction. As a general rule, unless reaction is
very fast and can occur at the bottom of the bed, a fluid bed is unsuitable because
increasing bed height will not provide opportunity for further reaction.

Although the fluidized bed keeps the temperature constant, which should improve
yield, the back mixing of the solids can have a negative effect on yield if there
is a possibility of a consecutive reaction consuming product, and if the product
is adsorbed to any extent on the solid. The solid on the top of the bed can adsorb
product, circulate to the bottom of the bed and present product for further reactic
with the feeds, resulting in an overall reduction in yield.

The solid used for a catalyzed fluid bed must be both a good catalyst for the reaction and also one that will fluidize well and is not subject to attrition due to continual movement in the fluid bed. The extra demands on the catalyst which are additional to the requirements of a fixed bed catalyst make the search for catalysts more difficult, and probably result in a compromise where a sub-optimal catalyst (as far as yield is concerned) is chosen because of its good fluidization characteristics.

The fluidization characteristics are not amenable to prediction because they depend on the nature of the solid and its size distribution. Hence pilot plant work is essential. Small pilot plant (e.g. 10 cm tubes) cannot be considered to give representative data because bubble coalescence occurs, and the bubbles move up the reactor in slugs and not as freely moving bubbles in a continuum of fluidized solid. Hence scale-up has to be done carefully on a number of scales, some of which are comparatively large. This inevitably delays the development of a process and this is often reason enough to decide against the use of a fluidized bed.

Finally, the lay-out of the fluid bed, particularly if it has a regenerator attached and there is continuous solid circulation between the two, is difficult because it is in essence a solids handling job, and these are difficult to design to run trouble-free. Solid is circulating:
o What gas velocities are necessary to transport the solids ?
o How steep can the transportation pipe work be ?
o How are bends and pipe junctions to be constructed so as to give minimum problems due to erosion ?
o Under which conditions will the catalyst "bridge" and stop flowing ?
o Is it necessary to strip the leaving catalyst with steam to remove adsorbed product before regeneration ?
o How should the cyclones be designed and installed to reduce loss of fines ?
All these points are difficult to handle for the first time, and this usually results in any firm interested in developing a fluid bed full-scale process calling in an engineering firm with cat-cracker design experience so that these detailed questions can be answered from experience.

Having seen their advantages and disadvantages it is now possible to define those areas where fluid beds may be worth consideration. These are:
(1) when only a gas phase reaction is involved, and a circulation of inert solid is necessary to obtain an even temperature to enable the reaction to take place autothermally; or
(2) when frequent catalyst regeneration is necessary and the regeneration is best done at a controlled temperature (e.g. the burning off of carbon from a catalyst, ensuring that the catalyst does not go over a particular temperature which is known to deactivate the catalyst); or

(3) when the solid is a raw material or product of the reaction.

Fluid beds should not be used for slow reactions, or for complex reactions where the product is strongly adsorbed on the catalyst.

9.7.a   Design of fluidized beds

As already explained, it is difficult to predict the fluidization behaviour of solids. It is also difficult to predict the mass-transfer behaviour when the solid takes part in the reaction, in order to predict reaction rate in the bed. Furthermor it is difficult to predict the concentration pattern of reactant in the bed because of adsorption and solid circulation.

The design of a fluidized bed reactor must rest heavily on pilot plant results, and before any large-scale process is built a reasonably-sized pilot reactor (e.g. 600 mm diameter bed) must be operated. Then the design of the full scale is made from a geometric scale-up of this comparatively large pilot result, and reliance on the empirical correlations is kept to a minimum.

However, it is necessary at the beginning of a research project to be able to approximately size a fluid bed so that its general dimensions and cost can be estimated to enable the decision to be made whether to develop it or not.

To get a rough impression of a full-scale fluid bed reactor the bed size (diamete and height) should be calculated and any internal cooling surface determined. When solid removal and regeneration are involved, this can also be very roughly sized for initial process assessment purposes.

Approximate methods to determine bed sizes.

Bed diameters. Bed diameters are determined on the basis of superficial velocitie The maximum velocity is determined by the loss of fines which are carried away with the exit gas. When the quantity is so great as to result in excessive loadings on the cyclones which are installed in the outlet gas to prevent catalyst loss from the reactor, the gas velocity should be reduced. In practice, it may be found that much lower gas velocities should be used to prevent excessive gas bypassing in the bed. Superficial velocities employed industrially are so much more than the minimum fluidization velocity that there is little use in basing any recommended operating velocity on a function of the minimum velocity.

Taking some examples from operating plant data, for catalytic cracking, a super-ficial gas velocity of 0.8 m s^{-1} is considered high (requiring 10 m of disengagement height above the bed), and 0.5 m s^{-1} is considered low (requires 7 m of disengagemen height) (Saxton and Worley, 1970). A published example for an oxychlorination reaction used a superficial velocity of 0.15 m s^{-1} (Vries et al., 1972). This latter case involved a gas phase reaction in an inert bed, where a high gas residence time is advantageous in allowing the reactor to proceed to completion. Werther (1980) shows that as the gas velocity is reduced from 0.15 m s^{-1} in a 1 m high bed, the conversion rises, suggesting that to get reasonable conversions with catalytic systems, velocities of 0.05 m s^{-1}, or even less, are required. This work proposes a quantitative model of fluidization which appears to successfully relate conversion to gas velocity.

Bed height. As a general rule, on hydrodynamic grounds, it is recommended that the bed height should equal the bed diameter. It is possible, if a large diameter shallow bed is used, for instability to occur and gas to flow up one part of the bed leaving the fluidized solid blown to the other part. Cat-cracker fluid beds have bed height of 3 m even though their diameters are many times this. At these diameters it is likely that the 1:1 ratio can be modified somewhat. Bed heights can be greater if this is necessary to provide a larger residence time for the gas.

Working on small diameter pilot reactors is often done with bed heights much greater than the diameter. In this case the hydrodynamics alter in that a slugging bed is produced in which the bubble diameters equal the tube diameter and there are alternate slugs of gas bubbles and fluidized catalyst passing up, and draining back down, the reactor. As the diameter increases, the bubbles can coalesce further before reaching the slugging condition. It must therefore be appreciated that with small-scale fluid bed work a different hydrodynamic region is operating, and so results from it must be treated with caution.

Bed expansion. A well-fluidized bed performing satisfactorily will have the fraction of gas in the expanded bed of the order of 0.2 - 0.5. This is dependent upon the gas velocity, and also very dependent on the quantity of fines in the catalyst. It is frequently reported that catalyst with a distribution of particle sizes, in particular a proportion of fines (e.g. 10-20% of particles being < 40 μ), has greatly improved fluidization properties.

Pressure drop. The pressure drop across a fluid bed is equal to the weight of the bed once the solids start to fluidize. The pressure drop remains constant whatever the gas velocity, though naturally the pressure drop through the distribution will vary with velocity. By inference, the minimum fluidization velocity can be calculated as the velocity that produces a pressure drop through the catalyst bed equal to the bed weight/unit cross-section area, i.e. by equation 9.8.

Heat exchange surface. When heat of reaction has to be removed from a fluid bed, this is done by installing a heat exchange tube bundle through which coolant is passed. A rough guide for estimating the reactor-to-wall heat-transfer coefficient is given by the work of Gelperin and Einstein (1971). For vertical tube bundles in a fluid bed

$$\left(\frac{h\, d_e}{\lambda_g}\right) = 0.75 \left[\left(\frac{g\, d_e^3\, \rho_g^2}{\mu_g^2}\right)\left(\frac{\rho_s - \rho_g}{\rho_s}\right)\right]^{0.22}\left[1 - \frac{d_t}{s}\right]^{0.14} \tag{9.13}$$

where h is the film heat-transfer coefficient, λ_g, μ_g and ρ_g are the corresponding physical properties of the gas, ρ_s is the density of the solid, d_e is the equivalent particle diameter (see equation 9.6), d_t is the tube (outside) diameter, and s is the pitch of the tube bundle inserted in the bed.

There has been intense academic interest in fluid beds over the last thirty years, and this has produced a vast literature of correlations for gas flow patterns, entrainment velocities, pressure drops, and heat transfer, most of which have been determined on small equipment and contain parameters that must be experimentally determined. These correlations are interesting in that they indicate the factors that are important, but they have not reached the stage where they can be used for design, or even where they can be combined with chemical kinetic models to predict reaction behaviour in fluid beds. Hence, the general recommendation remains: after a preliminary paper study using very approximate methods, data must be obtained experimentally and not from published correlations.

Readers wishing to go deeper into the type of correlations available for fluid bed systems are recommended to refer to Rase (1977).

9.8   FLUIDIZED REACTOR/REGENERATION SYSTEMS

The most important use of fluid beds is the catalytic cracker where heavy oils are broken down into lower molecular weight hydrocarbons by means of a fluidized catalyst. The reaction produces coke by side reactions, which deposits on the catalyst and which must then be burned off. The fluid bed reactor/regeneration system has been developed for this duty (Fig. 9.10) and consists of two fluid beds arranged so that a stream of catalyst flows continuously out of the reactor into the air feed pipe of the regenerator (Fig. 9.11), where it is carried pneumatically to a second fluid bed (the regenerator), where the carbon is burned off the catalyst. From the regenerator a stream of catalyst is returned to the reactor.

As the process operates an equilibrium is set up where the beds contain particles with a range of carbon concentrations, but the mean carbon content in the regenerator is lower than that in the reactor. The catalyst is being regenerated continuously, and cyclic operation and unpleasant solid handling jobs are circumvented.

The total system is amenable to mass- and heat-balance calculation. Once the rate of carbon lay-down by the reactor in question is known, regenerator sizing can be done from coked catalyst oxidation kinetics (which are published for catalytic cracker catalyst regeneration), and mean carbon levels, catalyst recycle rates, and heat removal requirements can be determined by calculation.

Consider the reactions occurring forming x kmol s^{-1} carbon lay-down on the catalyst at the production rate required by the full-scale plant. The heat generated in the regenerator by burning the carbon to CO_2 is therefore $-\Delta H_2.x$ kW, taking $-\Delta H_2$ as the heat of combustion of the laid-down carbon.

If we define the mean carbon content on the catalyst in the reactor and regenerator as y_1 and y_2 mol fraction respectively, then we have fixed the catalyst recycle rate (r kmol s^{-1}) by carbon mass balance (for low carbon loadings):

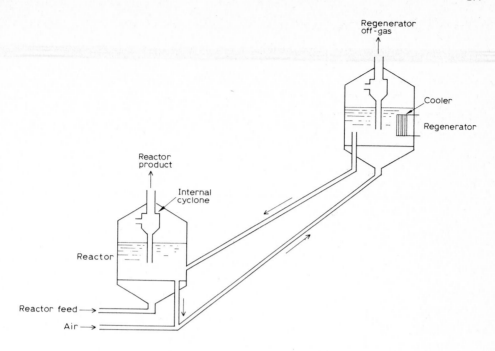

Fig. 9.10. A fluid bed reactor/regenerator system.

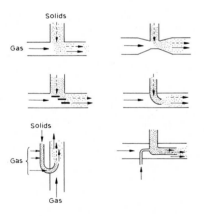

Fig. 9.11. Details of catalyst transport arrangements.

$$y_1 r - y_2 r = x \tag{9.14}$$

hence

$$r = \frac{x}{(y_1 - y_2)} \tag{9.15}$$

If the reactor operates at T_1, and the regenerator at the higher temperature T_2, the recycling catalyst is removing heat from the regenerator at the rate of

$$(T_2 - T_1) \; r \; (c_p M)_{cat} \quad kW \tag{9.16}$$

The heat required by the air fed to the regenerator in heating it up from 293 K to T_2 K is given by:

$$Q \; (c_p M)_{air} \; (T_2 - 293) \quad kW \tag{9.17}$$

where Q is the molar gas feed rate.

For the regenerator system to work in heat balance, these three loads must balance

$$Q \; (c_p M)_{air} \; (T_2 - 293) + (T_2 - T_1) \; r \; (C_p M)_{cat} \; = \; -\Delta H_2 \, x \tag{9.18}$$

It may be possible to obtain such a balance by varying Q, r, or T_2 if
o they do not violate the constraints presented by the kinetic requirements;
o sufficient air is supplied to burn-off x kmol s^{-1};
o the regenerator temperature is such that an acceptable reaction rate is achieved.

Variation of r alters carbon levels on reactor and regenerator which alters performance of both units.

To continue this example further requires numerical values to be supplied and the kinetics of regeneration to be given. These data are supplied in Exercise 9 of Appendix 3, and to go further with this example would spoil the fun of this exercise!

Stripper columns. It is interesting to note that the systems which have been developed for catalytic crackers include a steam stripping section in the downleg of the catalyst exit from the reactor (Fig. 9.12). This is to desorb reactants and products from the catalyst surface and return them to the reactor. Without this they would burn in the regenerator with a resulting loss in yield.

Concluding this section on the design of fluid bed reactors, it must again be emphasized that pilot results must be obtained. After large-scale pilot operation, the scale-up is done by maintaining similar conditions on the full-scale plant. However, there also remains much detailed design work which will not have been clarified by the pilot plant, and it is advisable to consult a petroleum engineering contractor who already has experience in design and start-up of fluid beds to reduce the risk of constructing a system that is difficult to operate or even simply will not work.

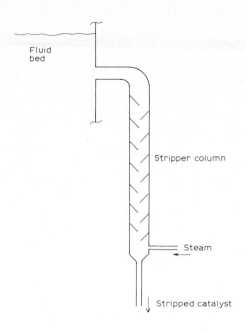

Fig. 9.12. Steam stripping of catalyst leaving reactor.

FURTHER READING

Botterill, J.S.M., 1975. Fluid Bed Heat Transfer. Academic Press. - Summarizes
    research findings in this field.
Carberry, J.J., 1976. Chemical and Catalytic Reactor Engineering. McGraw-Hill.
    - Thorough treatment of theories on catalysis.
Franks, R.G.E., 1967. Mathematical Modelling in Chemical Engineering. Wiley. - Useful
    section on handling partial differential equations.
Holland, C.D. and Anthony, R.G., 1979. Fundamentals of Chemical Reaction Engineering.
    Prentice-Hall. - Detailed treatment of heterogeneous catalysis.
Horak, J. and Pasek, J., 1978. Design of Industrial Chemical Reactors from Laboratory
    Data. Heyden. - Useful discussion on catalytic reactions.
Rase, H.F., 1977. Chemical Reactor Design for Process Plants. Wiley. - Details of
    industrial designs of reactor.

REFERENCES

Ergun, S., 1952. Chem. Eng. Prog., 48: 89.
Geldart, D., 1968. Chem. Ind., Jan.: 47.
Gelperin, M. and Einstein, V.G., 1971. In Davidson and Harrison (editors),
    Fluidization. Academic Press.
Kern, D.G., 1950. Process Heat Transfer. McGraw-Hill.
McAdam, W.H., 1951. Heat Transmission. McGraw-Hill.
Saxton, A.L. and Worley, A.C., 1970. American Petroleum Institute Proceedings,
    Refining Division (Spring Meeting): 74.
DeVries, R.J., Van Swaaij, W.P.M., Mantonvani, C. and Heijkoop, A., 1972.
    Proceedings of 2nd International Symposium on Reaction Engineering. Elsevier.
Werther, J., 1980. Chem. Eng. Sci., 35: 372.

QUESTIONS TO CHAPTER 9

(1) Why is the reaction rate over bed of catalyst pellets different from the
    kinetic reaction rate ?
    Under what circumstances is it greater, and less, than the kinetic rate ?
(2) What equipment can be used for homogeneous gas phase reactions, and under what
    conditions would each type be preferred ?
(3) What equipment is available for solid-catalyzed gas phase reactions ?
    Give the advantages and disadvantages of each type.
(4) What criteria determine the tube diameter, tube length, and number of tubes
    of a catalyst-packed gas phase tubular reactor ?
(5) What are the problems of too-low and too-high reactor pressure drops ?
    How can pressure drops be determined for the full scale ?
(6) How is the number of reactor shells determined for a tubular reactor installation
    Give arguments for and against spare reactors and describe methods of reducing
    the capital investment in providing spare reactors.
(7) What are the details concerned in the design of the shell side of a tubular
    reactor that deserve careful attention ?
(8) What are the advantages and disadvantages of the fluid bed reactor ?
    How can the bed diameter and bed height for a fluid bed reactor be determined ?

Chapter 10

CONTROL AND SAFETY

10.1   THE OBJECTIVES OF A REACTOR CONTROL SYSTEM

A control system is required on a reactor to ensure that it operates under
optimum economic conditions, and that it operates safely at all times.

Having located optimum process operating conditions in the design stage, it
is necessary to ensure that these are adhered to in operation, and this is achieved
by the control system that is specified by the reactor designer. In particular,
feed flows of each component and reactor temperature must be controlled. The primary
objective is the minimization of the operating costs of the plant but this can only
be instantaneously measured by a very extensively instrumented plant, together
with a dedicated computer to evaluate all the functions contributing to the economics
of the operation. This is generally not technically or economically feasible and so
it is usual to use a secondary objective, which is to hold the feed flows and reactor
temperature at those values which the design calculation showed to represent
optimum operation. This approach is widely used and is generally satisfactory, but
when uncontrollable, or difficult-to-measure, variations occur this method of
controlling the reactor at fixed operating conditions (which is basically a form of
feed forward control) is not entirely satisfactory; this provides incentive for the
use of process computers to evaluate more sophisticated objective functions for
feed-back control.

For example, when catalyst decay occurs, resulting in continuously changing
operating conditions, a control system, including an on-line optimization or on-line
parameter fitting technique, may be justified to follow the optimum economic
operation for each stage of the catalyst life.

Further examples are changes in feed composition and flow, either due to new
batches of material, or new feed stocks, or due to changes in integrated upstream
processes, all of which can call for complex control schemes to locate and keep
operation at the optimum economic operating point.

Safe operation of the reactor means that the reactor simply must not burst or
leak. These latter occur either as the result of an explosion when the composition
in the reactor has passed into the explosive region and a source of ignition is
present, or because the vessel is too weak to withstand the pressure put on it,
even though no explosion is involved. The vessel may be too weak because the pressure
in the vessel has risen accidentally above the operating pressure, or because the
reactor has become weakened by a temperature rise over the design temperature.

Hence, for safe operation the control system should ensure that compositions cannot drift into explosive regions, and temperatures and pressures must not become out of range. In addition, safety control systems should alarm and come into operation when any safety limit is overstepped. This requires the installation of pressure relief valves, high temperature alarms, analytical instrumentation such as explosion meters, and dump tamks.

10.2   CONTROL OF FEED FLOWS

10.2.a   Feed forward methods

Feeds must be controlled to ensure that the correct ratios of the various components are fed to the reactor. This usually has a very significant effect on the process economics because, unless there is a recycle of unreacted material in the process, material that is fed in in the incorrect ratio is lost. For example, take a simple high-conversion reaction involving two feeds fed in their stoichiometr ratio; any deviation from this ratio represents a simple loss of the excess material Hence a 10% error in the ratio is as serious as a 5% reduction in the reactor efficiency!

Since the average plant flow meter has an accuracy of only ±5%, and worse when it becomes dirty, it is clear that it is often worth having a feed control system somewhat more sophisticated than a simple flow control on each feed ratioed to each other (see Fig. 10.1), although this is probably the most common form of feed control.

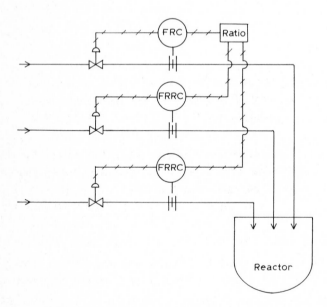

Fig. 10.1. Flow ratio feed control.

To overcome the poor accuracy of flow controllers operating under industrial conditions, feed tanks are often employed, which enable the quantity of feed over a period to be accurately measured. Level indication in the tank provides a form of measurement, but weighed feed tanks are to be preferred because of the much better accuracy obtained with industrial weighing machines. Since these machines give only cumulative flows over discrete time intervals, continuously operating reactors still require instantaneous rate indication which involves a flow control system (see Fig. 10.2).

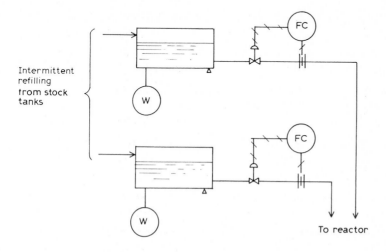

Fig. 10.2. Feed control with feed weigh tanks.

If the process recycles unreacted raw materials as liquids, the feed weigh tank system requires duplicate tanks with one tank being filled whilst the second is collecting the recycled material. When raw materials are recycled as gases, the weighed feed tank method cannot be employed.

Feeding batch reactors presents less of a problem, since cumulative volumetric meters are usually accurate to ±1%, or alternatively the reactor can be mounted on a weighing machine, or the feed tanks can be mounted on weighing machines.

The weighed feed tank method has the disadvantage that it is labour intensive and does not deliver rate information instantaneously.

10.2.b   Feed back method

These disadvantages can be overcome using a feed back system where the reactor outlet is analyzed in some way so that the imbalance in reactor feed can be recognized and some trim control reaction taken (see Fig. 10.3). This method is very common off-line, i.e. reactor outlet samples are taken, analyzed in the laboratory, and the feed trimmed manually by the operator.

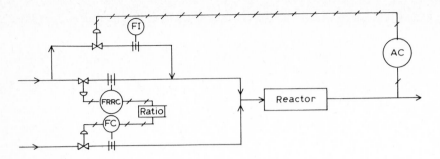

Fig. 10.3. Reactor feed control by feed-back trim control.

The automation of this method is entirely dependent on being able to find suitabl[e]
automatic, robust analytical methods to detect the concentration of each raw
material. This will be the theme of section 10.6.

When the reactor effluent is gaseous, and single phase, this analysis can usually
be made by gas chromatography. Liquid phase presents more difficulty unless some
liquid physical property can be measured which can be related to composition. Two-
phase mixtures can present almost insurmountable sampling problems, even if the
liquid phase is present as no more than a slight "dew" in the system. It is these
problems that prevent feed back methods having a wider application.

When the raw materials are separated in the process down-stream from the reactor,
for example when they are to be recycled, it is possible to devise a flow control
scheme which measures the flow of the recovered stream and uses this rate informatio[n]
for the trim controller. This overcomes both the problems of analysis and sampling
of the reactor outlet (see Fig. 10.4).

This method in practice is not simple, as long dead times, variable process
hold-ups, and variable stream compositions can cause wrong control action to be
initiated, resulting in a completely unstable process.

The problem of choosing a suitable reactor feed control system is not an easy
one to solve satisfactorily, and in some installations no satisfactory solution
exists as yet - and it is then left to the operator to devise his own "touch"!

10.3 CONTROL OF REACTOR TEMPERATURE

10.3.a Temperature stability of the CSTR

Consider a CSTR with volume V_R, operating at temperature T, with feed rate
Q m^3 s^{-1}, feed concentration C_{A_0}, and conversion X_A, carrying out the simple first-
order exothermic reaction

$$A \xrightarrow{k_1} P$$

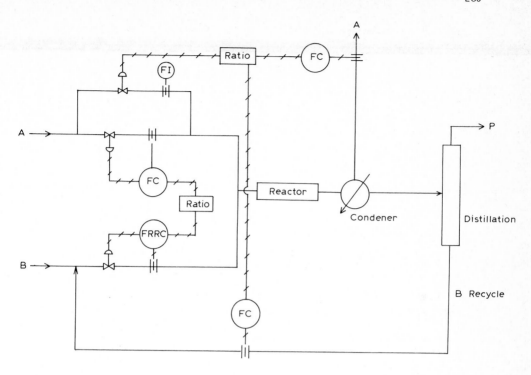

Fig. 10.4. Reactor feed control with feed back time control based on isolated reacted flows.

The rate of production of P is

$$V_R k_1 (1 - X_A) C_{A_0} = C_{A_0} X_A Q \quad \text{mol s}^{-1} \tag{10.1}$$

from which we can define the reactor conversion

$$X_A = \frac{V_R k_1}{Q + V_R k_1} = 1 - \frac{Q}{Q + V_R k_1} \tag{10.2}$$

The heat generated by this reaction, q_g, is:

$$q_g = - V_R \Delta H k_1 (1 - X_A) C_{A_0} \tag{10.3}$$

Replacing k_1 by the Arrhenius expression to include the effect of reactor temperature, and substituting equation 10.2 in 10.3 gives:

$$q_g = \frac{-V_R \Delta H Q C_{A_0} A_1 e^{-E_1/RT}}{Q + V_R A_1 e^{-E_1/RT}} \tag{10.4}$$

286

The heat removed from the system, q_r, is achieved through the difference in temperature between inlet and outlet streams

$$Q\ c_p\ \rho\ (T - T_{in})$$

and the heat removed by the external cooling

$$U\ A'\ (T - T_J)$$

hence

$$q_r = Q\ c_p\ \rho\ (T - T_{in}) + U\ A'\ (T - T_J) \tag{10.5}$$

When the CSTR operates under equilibrium, the heat gain must equal the heat removed from the system:

$$q_r = q_g \tag{10.6}$$

Expressions for q_r and q_g are very different with respect to temperature; equation 10.4 is a complex exponential function, and equation 10.5 is linear.

Figure 10.5 shows these two expressions plotted against reactor temperature.

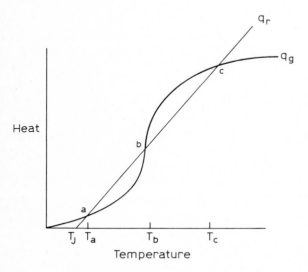

Fig. 10.5. Heat generation and heat removed from a CSTR as a function of the reactor temperature.

Wherever these two curves cross, condition 10.6 holds, and this represents a potential operating point for the reactor. In this case we have three points: (a), (b) and (c). Inspection of these points reveals that some points are "stable" operating points and others are unstable. Consider the reactor operating at point (b). A small rise in temperature would produce a greater generation of heat than removal, hence the temperature would tend to increase further. Similarly, a drop in temperature would induce a greater drop in temperature. Hence, this point would be unstable, and would probably never be observed in practice.

This does not apply to points (a) and (c), and so in practice this reactor could operate at either point quite stably, for exactly the same operating conditions of T_{in}, V, T_J, and C_{A_0}.

Which of these two points the reactor operates at depends upon its earlier temperature history. If the reactor started cold, then it would stay at point (a). If it were pre-heated above temperature T_b, it would operate at point (c).

The reactor is exhibiting "multiple steady states". This subject occupies much of the academic literature as it is, mathematically speaking, a fascinating problem. In practice it is less important, but it is as well to understand what is likely to occur before the reactor is built.

Only exothermic reactors exhibit this phenomenon; endothermic reactions have q_g and q_r curves of opposite slope and so only one intersection, and hence one stable state, is possible.

In the case under discussion, point (a) represents virtually no reaction - the reaction has not initiated - whereas point (c) is a normal operating condition with the reaction occurring. This behaviour is fairly common in practice, the simplest example being flame reactions. High temperature chlorinations also exhibit this effect. T_b is the ignition temperature that must be achieved in the reactor before the reaction will start.

In practice a much more serious problem is introduced by the simultaneous nature of industrial reactions. Often there exist side-reactions - parallel or consecutive - that become significant at higher temperatures (e.g. pyrolysis and burning reactions). These reactions produce considerable heat, and so once they start, the reaction temperature "runs away" and the reactor operates at a stable point involving complete burning as a result of temperature disturbance at the required operating point. Figure 10.6 shows the q/Q diagram for such a situation.

Changing process conditions (V_R, C_{A_0}, T_{in}, T_J) changes the relationship of q_r and q_g to each other, and this can result in more or less stable states, and in moving the position of the stable states. Potentially this gives some latitude for the designer of the reactor to find a set of conditions which gives a state stable over a wide range of operating conditions.

Autothermal reactions. Reactions which provide enough heat to sustain themselves at the reaction temperature through the heat generated by the reaction itself are

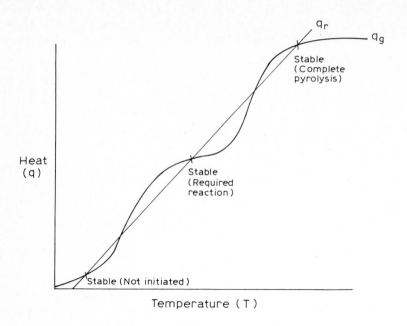

Fig. 10.6. q/Q curves for a set of simultaneous reactions.

called autothermal reactions. When autothermal reactions occur under conditions of complete back mixing the heat balance equations are the same as those developed in the previous section, but since there is no heat removed, the $U A' (T - T_J)$ term of equation 10.5 is zero. There still remain the two factors q_r and q_g, and q_r is still linear.

Autothermal reactors are very cheap to construct because they have no heat-transfer area, but they can be inconvenient to operate because there is no temperature control by means of a cooling medium. The only control is by feeding excess of one of the components or the use of a diluent.

Any disturbance that causes the temperature to fall (a feed ratio imbalance or stoppage) results in the reactor "going out", and the reactor must be reheated by some inconvenient means such as an open flame torch before the reaction can be re-started. Though burning reactions are not too difficult in this respect, chlorination reactions which have an upper temperature limit to prevent pyrolysis (see Fig. 10.6) are difficult to operate because operation must always be a balance between the reaction "going out" and a temperature "runaway".

10.3.b   Temperature stability of plug flow reactors

An ideal plug flow reactor has no back mixing and this means no back transfer of heat of reaction into the unreacted mixture. Hence, autothermal reaction is impossible and for the reaction to occur the feed mixture must be at its initiation temperature.

As the inlet temperature rises beyond the initiation temperature, the reaction takes place and if it is an exothermic reaction, the temperature rises sharply as the reaction rate increases until the reaction is reaching completion. If the reaction is a complex set of simultaneous reactions involving the appearance of unwanted reaction at higher temperatures, then the reactor can progress through to the final decomposition products unless the temperature is adequately controlled (see Fig. 10.7).

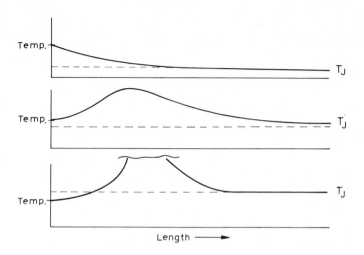

Fig. 10.7. Temperature profiles along a plug flow tubular reactor.

The intiation temperature in such a system is the point where the rate of heat gain is equal to the rate of heat removal. For our first-order reaction this would be:

$$-\Delta H_1 \; \frac{\pi}{4} \; d_t^2 \; C_{A_0} \; A_1 \; e^{-E_1/RT} \; \delta x \;\; = \;\; U \; \pi \; d_t \; (T - T_J) \; \delta x \qquad (10.7)$$

or

$$-\Delta H_1 \; d_t \; C_{A_0} \; A_1 \; e^{-E_1/RT} = 4 \; U \; (T - T_J) \qquad (10.8)$$

where d_t is the reactor tube diameter, and T in this case is the initiation temperature.

The hot-spot temperature is the point where the rate of heat gain is again equal to the heat removed, as the heat gain is falling. This is dependent upon the inlet temperature, prior reaction rate, and heat removal prior to the hot-spot, and cannot be determined without a model integration.

Catalytic plug-flow reactor instability phenomena. The presence of a catalyst causes a number of extraordinary effects and one of these is the inverted temperature effect noted in some catalytic gas phase tubular reactors. If a reactor is operating stably and the feed temperature is dropped as a step function to a lower value, the temperature of the hot-spot within the reactor bed rises. When the reaction involves competing side reactions which release more heat, this could lead to instability and a reactor runaway and very high bed temperatures.

This pathological situation is brought about by the reaction not occurring at the same rate at the beginning of the bed, so presenting higher concentration in the hotter part of the bed. This hot bed plus high concentration produces the higher temperature rise (see Fig. 10.8)

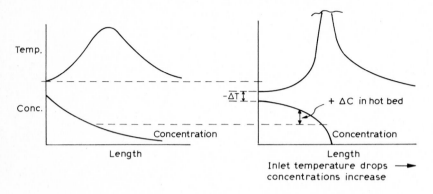

Fig. 10.8. Temperature inversion effects in catalytic tubular reactors.

Although such effects can be modelled for simple reactions using partial derivative models with heat capacity effects, this would not be suggested as a standard feature of a design procedure. In practice, all changes in reactor conditions should be carried out slowly, and consideration given to reducing gas compositions before other changes are made.

A second pathological phenomenon of catalyst reactions is their ability to produce higher conversion in pulsating conditions rather than stable operation. Reasons for this are concerned with the complex nature of catalysts and adsorption of reaction species on their surfaces. In some cases the pulsations are a continuous form of catalyst regeneration, e.g. as with the case of SO_2 catalytic oxidation to SO_3 where pulsations with a cycle time of 40 minutes produce noticeable improvements in conversion (Abdul-Kareem, 1980).

Pulsed catalytic reaction still remains a promising field for research, and cannot yet be considered to be a recommended method of operation.

Temperature measurement within reactor tubes. Tubular reactors must have a reactor bed temperature measurement so that they can be successfully operated.

Basically two possibilities exist:
- a moveable thermocouple in a central pocket
- a fixed thermocouple in a central tube,
(see Fig. 10.9). For operational reasons the fixed method is preferred for operating plants, although this always'poses the question of whether a hot-spot is lying between two readings.

In a large tubular reactor three or more tubes may be fitted with temperature measurement. Usually no provision is made for correcting for the reduced volume as a result of the thermocouple, although the matter is often a point of discussion.

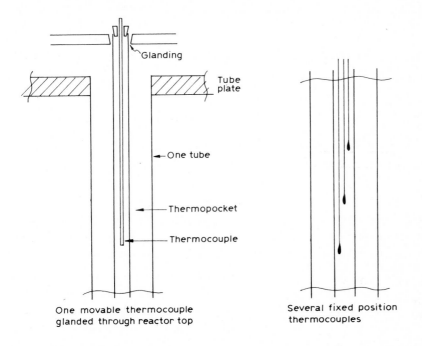

One movable thermocouple
glanded through reactor top

Several fixed position
thermocouples

Fig. 10.9. Methods of tube temperature profile measurement.

10.3.c   Temperature stability of batch reactors

Mathematically, batch reactors are similar to plug flow reactors, and so similar stability characteristics of ignition temperature and hot-spot temperature exist. In practice there is the convenient possibility of heating the reaction mixture up to the initiation temperature whilst it is in the reactor.

A further convenience in the batch reactor that is difficult to provide in a tubular reactor is the provision of a coolant temperature that is a function of conversion. It is fairly convenient to firstly heat the reactor, then later cool it to avoid hot-spot difficulties, and at the end of the reaction, heat again to ensure complete conversion.

10.4 THE USE OF MODELS IN CONTROL STUDIES

Although section 10.3 presents some convincing mathematics and figures
describing the thermal stability points of the reactors, it is not suggested that
these be used as design methods. They have been presented as an aid to understanding
the thermal stability phenomena exhibited by reactors.

For actual design and the testing of proposed designs for thermal stability we
should return to our reactor simulation models developed in Chapter 2 and carry out
sets of simulation runs to determine operability and stability conditions.

The first lesson taught by running such models is that the coolant temperatures
must always be reasonably close to the reactor temperature to achieve a stable
design: 5 - 10°C temperature difference would be a reasonable estimate. This rule
is valid for both CSTR and plug flow reactors. In the plug flow reactor, when
coolant temperatures are lower than this, it is easy for the feed to become over-
cooled and so not react before it has passed through the reactor, or if it does
react, become quenched after the hot-spot and so not react to completion.

The models described in Chapter 2 do demonstrate instabilities such as temperature runaway or lack of initiation perfectly adequately. On the verge of stability
the more detailed two-dimensional catalytic tubular models predict differences from
the one-phase-model, but within normal operating points the two models are reasonably
similar (Rase, 1977).

The theoretical study of the CSTR shows it to possess multiple steady states
where the heat balance is satisfied. The normal CSTR model solves non-linear
equations to find a root, that is, a steady state. Which steady state and whether
it is a stable or unstable state cannot be known from a single run with the model.
Repeated solutions from other initial values will identify whether more than one
state exists, and a perturbation study at each point will define whether it is a
stable or an unstable state. For a complete picture, a programme could be written
to construct Fig. 10.5 since all the data are available in the reactor model.

Where particularly difficult control problems are suspected, the model proposed
in Chapter 2 will not suffice and special differential control models may have to
be written which take into account transients, thermal capacities, and dead times
(Franks, 1972). It is, however, not usually necessary to go to this extreme in
determining the control scheme for a reactor.

10.5 EXAMPLES OF REACTOR CONTROL SYSTEMS

The Figs. 10.10 to 10.13 show a number of feasible reactor control schemes.
They are included here to show what comprises a control scheme for a reactor, and
to show the way in which these are drawn.

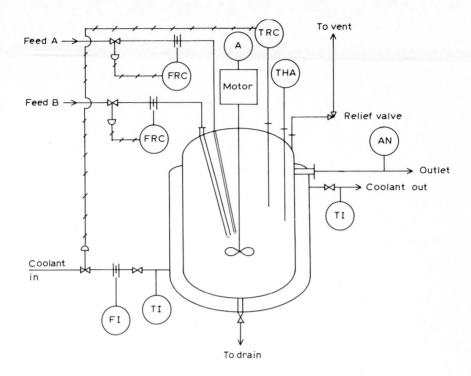

Fig. 10.10. Control system for a CSTR.

10.6   PROCESS STREAM ANALYSIS

Because the composition of the reactor outlet stream is the most useful control information for reactor operation, it is worth devoting a section to the automatic methods of analysis that are available, so that the reactor designer can decide whether his reactor outlet is analyzable automatically or not.

In principle, any analysis that is carried out in the laboratory can be automated, but in practice one learns to specify only developed, marketed instruments for process plants. To develop one's own instrument for a new process usually ends up in failure, leaving a process designed round a control scheme that does not exist.

The instrument manufacturer has had to spend hundreds of thousands of dollars developing an automatic analysis instrument that is simple and robust enough for process work. He does this only when there is an adequate market potential, e.g. atmospheric analyses, effluent water analysis, petroleum fraction quality determination, etc.

The following pages list instruments that have been developed for process use that may be of use in reactor outlet analysis. If you have a system outside these categories then the news is bad - there is probably no developed process analysis method available.

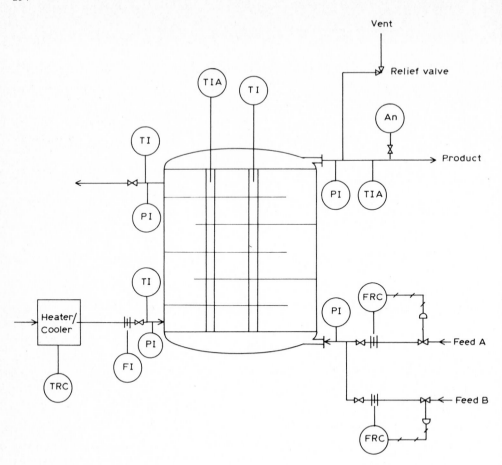

Fig. 10.11. Control system for a tubular reactor.

10.6.a   Physical methods

Table 10.1 lists the physical property measurements that are possible with automatic process instrumentation (Clevett, 1973). These can be useful if one or more physical properties can characterize the reactor outlet composition adequately.

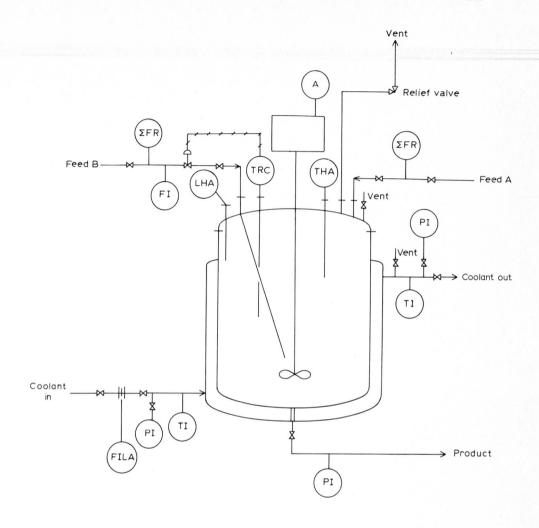

Fig. 10.12. Control system for a semi-batch reactor.

10.6.b   Chromatographic analysis

Gas phase chromatography (GC) is the most versatile tool for use in process-stream analysis. It is generally possible to analyze streams for all volatile components, and produce a quantitative analysis giving the stream composition.

The principle of chromatography is to input a pulse of a mixture to be analyzed into a gas stream (usually helium) which then passes through a long heated column. The column is packed with solid particles which have been treated with particular organic components. This column packing is selected so as to present different absorption affinities to the different components in the mixture to be analyzed. Hence, those weakly absorbed undergo quicker absorption/desorption cycles between

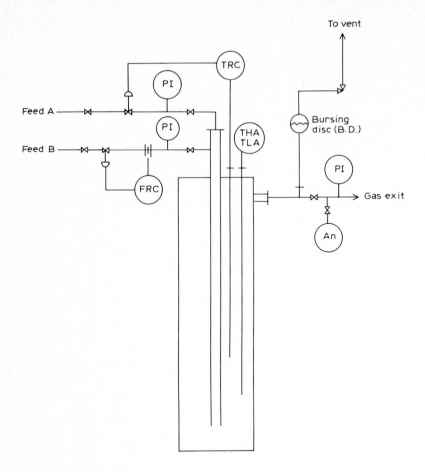

Fig. 10.13. Control system for an autothermal gas reactor.

TABLE 10.1

Physical properties which can be measured in process streams

Property	Notes
Liquid density	
Gas density	
Liquid refractive index	
pH	
Redox potential	
Electrical conductivity	
Gas thermal conductivity	(katherometer for H_2 analysis)
Liquid viscosity	
Pour point	
Flash point	
Vapour pressure	
Ultra-violet absorption	(can be considered for chromophoric groups)
Visible absorption	fluorescence and colour detection
Infra-red absorption	e.g. CO, CO_2, SO_2, C_2H_2, CH_4

the carrier gas and solid than do the components which are strongly absorbed. Hence, at the end of the column the components appear at varying times depending on their absorptivity, and so by detecting the components in the outlet gas a series of peaks appear, each peak representing one component. The area (or height) of the peak defines the total quantity present in the injected sample, and the time to appear identifies the component (Fig.10.14).

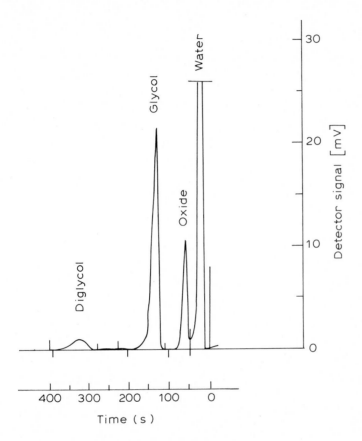

Fig. 10.14. Typical chromatogram.

By careful choice of packing, absorbent on the column, and temperature, adequate separations can be obtained for components in most mixtures, the expertise being in the selection of the packing and temperature for a particular mixture.

The detection is done by physical means and this again needs to be chosen after consideration of the components in the mixture. Thermal conductivity is the cheapest and most common method. Flame ionization detection is a second type which is more sensitive and can determine concentrations down to 1 ppm.

Gas chromatographs have been developed for use in process stream analysis, and they are now commonly used for distillation column product composition analysis. These process chromatographs have been developed from laboratory instruments, where the objective has been to make them more robust, flame-proof, and relatively maintenance-free compared to their laboratory counterparts.

Each chromatographic analysis is a batch process, whereas most other process analysis systems are continuous. This leads to comparatively complex sequencing controls and these are carried out automatically by pre-set cams or micro-computers associated with the instrument.

The instruments are expensive, but very versatile, and this has led to the use of one instrument for more than one stream by bringing all sample streams to the instrument and then having a sample selector valve preceding the sample injector valve (see section 10.6). To reduce the overall analysis time, temperature programming of the ovens and back-flushing of the less important, more strongly absorbed components is possible, and process chromatographs are available which carry out all these operations.

The signal is usually integrated electronically, and presented as total area/ hold-up time pairs of numbers, which, after calibration, can be interpreted as composition and component identification pairs.

Over the last few years LPC (liquid phase chromatography) has been developed for the analysis of non-vaporizable components. This method of analysis passes the liquid to be analyzed through a long, packed separation column requiring very high pressure, and a chromatographic separation occurs, leading to the different components reaching the end of the column at different times. The liquid leaving the column is analyzed with an IR or UV detector. If the component of interest cannot be satisfactorily analyzed by these methods, then the liquid leaving the column can be taken up on a thin wire and the wire passed through a combustion chamber. The heat generated by the combustion, or the IR, or UV signal is recorded as peaks giving a chromatogram similar to Fig. 10.14, which is interpreted in the same way.

At present there are no process LPC instruments on the market, but there should be such a market in the fine chemicals industry that it is to be expected that robust process instruments will be available in a few years. This will greatly increase the scope of on-line reactor performance monitoring in the organic chemicals manufacturing area.

10.6.c Automated laboratory analysis

Titration is a frequently met method of analysis in the laboratory, and process titrometers have been developed to enable such analytical techniques to be employed on-line. Such an instrument is a small plant in itself. The stirred titration flask is fed continuously with a flow of the stream to be analyzed, and a second stream of the titrating reagent is also added. A control system alters

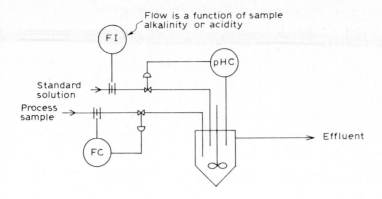

Fig. 10.15. A process titrometer.

this reagent flow so that the mixture in the flask reaches a required pH, (Fig. 10.15).

Needless to say, such equipment cannot be considered very robust, and considerable attention is required, for instance to keep the reagent reservoirs filled. On-line analysis by automatic titration is possible, but it is not welcomed and should only be used as a last resort.

Automatic analysis of oxygen concentration, both as gas and dissolved in liquid, has reached a very high degree of reliability because of its importance in burner control and in boiler feed water preparation.

Oxygen analyses for gases are usually based upon measuring the paramagnetic properties of the gas mixture, because oxygen has a much higher paramagnetic susceptibility than other gases. Instruments have also been developed based on measurement of heat evolved when the gas is burned catalytically in hydrogen.

Oxygen can be measured in solution by means of an "oxygen electrode", which consists of a galvanic cell which delivers a current which is proportional to the oxygen concentration in the liquid. These electrodes can be a little troublesome in plant conditions, particularly if they are used in complex reaction mixtures; they have been developed primarily for analysis of clean treated boiler feed water.

This brings to a close the survey of available on-line process stream analysis instruments that can find application in reactor control. The list is not long. If gas chromatography is not suitable for a particular application, then, unless an absorption method is suitable (and these are not very common), it is very likely that there is no developed method -a situation that is very commonly met.

300

10.6.d   Detection of reactor conversion by temperature measurement

Because of the dearth of automatic analytical methods the measurement of temperature takes on a particularly important role of being the indicator of reaction, and hence outlet concentration. In its simplest form the temperatures corresponding to optimum performance are defined by taking off-line samples, and then these temperatures are monitored as the normal control procedure. This method is made more complicated when the process continuously changes because of catalytic activity decay, for example.

With the increased application of computers associated with plants we can expect this method to increase in sophistication so that more temperature measurements are taken, full reactor heat balances are carried out, and conversions are displayed based on a thorough heat balance analysis.

10.6.e   Sampling of process streams

Process streams to be sampled normally have a continuous purge stream taken from them, and this flows through the instrument or the sample valve in the case of the gas chromatograph. The stream is normally so small that no effort is made to return it to the process stream, although in some cases it may be worth collecting the sample in a drum and returning it batchwise to the process.

The time delay between the sample leaving this main stream and reaching the instrument should not be more than 60 seconds. If the length of line to the instrume and the flow required by the instrument are not compatible with this 60 seconds maximum delay, then a faster stream can be taken through the sample feed pipe and a fraction of this taken to flow through the instrument, as shown in Fig. 10.16.

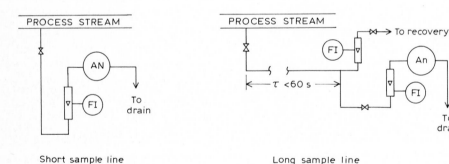

Short sample line   Long sample line

Fig. 10.16. Process sample removal system.

For successful sampling only a single phase should be present. When small quantities of liquid are entrained in a gas sample it is possible to install a filter or coalescer in the line to knock out the droplets, so ensuring that the analyzer receives a dry gas. Such a procedure must, however, be used with great caution, because the presence of the liquid may have so changed the gas composition that it no longer represents the gas composition leaving the reactor.

For example, a few drops of water condensing in a gas stream containing HCl will absorb such a substantial fraction of the HCl present that the final analysis will bear no resemblance to the HCl content of the gas before the water formed.

Chromatographs require intermittent samples and this is achieved by taking a small side stream from the process and passing it through a sampling valve (see Fig. 10.17).

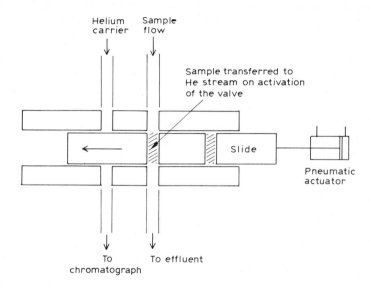

Fig. 10.17. A gas chromatograph sampling valve.

When the sample is to be taken the slide valve moves across (usually motivated by a compressed air stream, opened by an electromagnetic air valve by electrical impulse).

The small fluid sample in the slide, then in the column feed helium stream, is blown through into the chromatograph, where it vaporizes.

Such valves are reproducible for gas samples, and also for some liquid samples, but they are very temperamental and depend on the liquids used. If the liquids are difficult to vaporize, then peak tailing occurs in the chromatograph and other inaccuracies are introduced, possibly due to the slide valve being wet. For these reasons it is common to keep the sample valve hot, and this is done in process chromatographs by mounting the sample valve in the chromatograph oven. These problems are completely absent with the normal manual syringe injection technique.

10.7   HAZARDS ASSOCIATED WITH REACTORS

At the beginning of this chapter it was simply stated that the reactors must
not burst or leak. In principle these are the only two hazards associated with
reactor operation. Bursting results in personal injury through flying debris,
leaking results in poisoning or fire, depending on the reactor contents. Both involve
costly equipment replacement and production loss.

These catastrophies can occur either due to poor mechanical and process engineering, where the vessels have not been able to withstand the pressures and temperature
to which they were subjected during normal, or abnormal though not unusual, plant
operation, or due to the process being out of control, resulting in design temperatures and pressures being exceeded. The second of these two causes is the more
frequent, and it is here where the reactor engineer makes his contribution by knowing
what could possibly occur with the system and how it could be controlled if it did
occur.

Out-of-control reactions can either be spectacular, taking the form of an
explosion with very high pressures developing, or more modest, in which equipment
damage is caused by temperatures and pressures rising to exceed vessel design
ratings owing to too rapid reaction, or temperature runaway.

10.7.a   Explosions

Explosions are the result of fast chemical reactions which produce high pressures
due to very rapid rises in temperature, or increases in volume, on reaction.
Explosions can simply be fast reactions, where end temperatures and pressures are
calculable, assuming complete adiabatic reaction to combustion products, or they
can take the form of a detonation. Detonations are particularly fast reactions
where the pressure increase is unable to dissipate fast enough and a pressure wave,
many times the calculated end pressure, is propagated through the vessel with
disastrous consequences.

Although one could design vessels to withstand explosions outside the detonation
region, the general design philosophy is to avoid all types of explosion rather
than to design for them. Once the possibility of an explosion occurring is accepted
there is no guarantee that it will not be a detonation when it does occur, and this
cannot be designed against.

An explosive mixture is any mixture capable of a very rapid reaction. Most
common, of course, are the oxidations of hydrocarbons and other inflammable gases
and vapours with oxygen. Oxygen does not have to be involved, and any reactive gas
mixture must be considered capable of producing an explosion.

Also important in reactor engineering is the ability of organic liquids and
solids to explode. Reactors are, by definition, dealing with reactive compounds,
and explosions of organic mixtures, either due to the unwanted reaction becoming
out of control, or due to the appearance of obscure side reactions, are a constant

hazard in the organic chemical industry.

Explosions only occur over restricted composition ranges. Outside these ranges the composition of one or both of the reactants is too low to allow a self-sustaining violent reaction to occur. In gaseous processes these concentration limits are measurable and are of basic importance in the control of the reactor to ensure that concentration drifts into this region never occur. In liquid processes this information is not available and other techniques have to be employed to prevent explosions.

Gas mixtures in the explosive region do not explode without a source of ignition. This need only be a single spark, or a hot surface. The spark could come from an electrical switch, electrostatic discharge due to a moving fluid, or a spark from a hammer or other tool. Explosions can be prevented and flames quenched by the supply of a large surface area/unit volume, in the form of fine packing or wire mesh. Since explosions are fast free-radical reactions, the presence of surface area allows the free radicals to combine together and so the reaction is no longer self-propagating.

10.7.b Out-of-control reactions

One of the major objectives of reactor engineering is the removal of heat of reaction, since if this is not performed adequately, the reaction will increase in temperature, which increases the reaction rate, liberating more heat, and leading to temperature runaways and the achieving of near adiabatic temperature rises. In practice, the matter is much worse, because at higher temperatures new pyrolysis reactions often occur, which frequently involve even more heat evolution and a higher adiabatic temperature rise. These temperature rises usually go well beyond the design temperature of the reactor and reactor weakening and failure result.

The reactor must be designed to operate stably without temperature runaway for the normal design case, but there are abnormal, but not unusual, situations that can cause an adequately designed reactor to become out of control.

In tubular reactors this can occur if the feed concentration, temperature, or coolant temperature differ from design.

In batch and semi-batch reactors dangerous situations occur if there is a build-up in concentration of feed before the reaction begins to consume reactant as designed. This can occur if:
o the initial temperature is too low
o mixing is inadequate (i.e. stirrer failure)
o the reaction has not initiated properly, because of the presence of trace impurities.

10.8   SAFETY MEASURES TO BE TAKEN AT THE DEVELOPMENT STAGE

When a process or reaction is under development, the potential hazards associated with it should be thoroughly investigated. Maximum allowable working concentration for all components should be obtained to fully appreciate the health hazard due to exposure to the components involved, and flash points serve as an indication of the inflammability of the materials.

10.8.a   Gaseous systems

For a gaseous system the explosive limits should be known, not only of the binary mixture, but also of the multi-component reaction mixture. Since reactor development work is often dealing with novel mixtures it is usual that this information has to be obtained experimentally, and for this reason many research departments have their own equipment for measuring explosive limits.

Such equipment consists of a tube into which a gas of known composition is fed. The mixture is then ignited with a spark, and the pressure of a resulting explosion is registered by means of a pressure detector at the end of the tube. The experiment is repeated with different gas concentrations until a clear picture emerges over the boundary between exploding and non-exploding mixtures.

Some examples of upper and lower explosion limits of various gases and vapours in air are given in Table 10.2. Figure 10.18 shows a typical relationship between explosion limits and pressure in the presence of inert components.

TABLE 10.2

Some upper and lower inflammability limits for various gases in air

	Explosive limits (vol. %)	
	Lower	Upper
Ammonia	16.0	25.0
Butane	1.9	8.5
CO	12.5	74.2
Methyl ethyl ketone	1.8	10.0
H_2	4.0	74.2
CH_4	5.0	14.0
Octane	1.3	7.6

10.8.b   Liquid systems

For reactions in the liquid phase, it is necessary again to determine their explosive potential by direct experimentation. The explosion of a liquid mixture has a somewhat different characteristic to an explosion of a gas mixture. The reaction occurs slowly at first, generating heat which increases the temperature and hence the reaction rate. This picture can occur over a matter of hours before

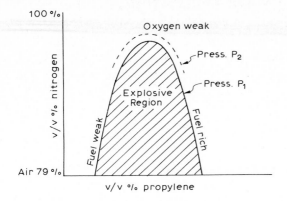

Fig. 10.18. Explosive limits for air/propylene.

the situation is reached where the temperature, and hence pressure, bursts the reactor with an explosive violence that causes bodily injury, poisoning, and fire.

It is the adiabatic nature of the large mass of liquid which induces the danger, and it is sometimes the slowness of the procedure which causes the specific reaction to occur that results in the final hazard.

Thus a laboratory investigation of the explosive potential of a reacting mixture must:

o be strictly adiabatic to simulate the conservation of heat that would occur on an industrial scale;

o be slow to allow hazardous substances to form by reaction as they might on the full scale;

o be small (e.g. 1 - 2 gram samples) to prevent damage in the case of an explosion occurring.

There are a number of standard test methods to determine the potential explosivity of liquids. Most of these determine the temperature at which the sample will decompose under adiabatic conditions. These include:

- Differential thermal analysis. This produces an indication of the temperature at which heat is liberated from the liquid sample.
- Geigy-Kühner decomposition temperature measurement. A 2 gram sample is placed in a block and the block is heated at a specific rate. The decomposition temperature is that where the sample temperature rises above that of a control sample in the same block.
- Reaction in a Dewar flask. The sample is slowly heated electrically, stagewise, in a Dewar flask. When the temperature rises without further heating, the decomposition temperature has been reached.
- Heat flow calorimeter. In this equipment, described in section 3.4.b, the coolant jacket temperature can be controlled to equal the reactor temperature, to

simulate adiabatic conditions, or to be a few degrees above it to heat at a constant rate. The plot of the heat developed from the reactor shows whether the reaction will be uncontrollable on the large scale. This equipment determines quantitative information and in addition it can be used to simulate any rate of temperature/time profile that the full-scale reaction may go through, so giving the same opportunity for hazardous intermediates to build up as is present on the full scale.

It must be emphasized that in an organic reaction containing a number of complex, reactive, organic compounds, it is far from clear what potential explosion hazards exist. The chemist is not able to categorically state that the process will have no explosion hazards. The novel nature of the mixture and the attainment of very high temperatures due to the adiabatic nature of the large reactors are often beyond the chemist's experience, and so it is essential to revert to the experimental methods described above to obtain information.

10.9   SAFETY MEASURES TO BE TAKEN AT THE DESIGN STAGE

10.9.a   Hazardous gas reactions

The golden rule is: never design a process that is to operate with gas mixtures inside the explosive region. It is true that explosions do occur without ignition, but simply to rely on the absence of a source of ignition is not an adequate safeguard.

The giant oil tankers were designed to work inside the explosive limit, since, on emptying, the fuel-rich vapour passes into the explosive region as air is drawn in to replace the oil pumped out. With monotonous regularity these monsters were reported to have mysteriously exploded and sunk at sea, and their owners expressed surprise and regret. The explanation is that the safety depends entirely on the absence of some source of ignition, but, from time to time, for reasons unexplained, a spark or hot surface is present which causes the explosion.

This rule can be broken, and processes designed inside an explosive limit, when the reaction occurs in a fluidized bed, because the large surface area prevents the propagation of an explosion. It has recently been reported that the phthalic anhydride fixed bed process is to be run inside the explosive limits. This is a departure from normal practice and it will be interesting to see whether the process can be run safely.

It is not sufficient simply to require that the operating conditions be outside the explosive region; care must be taken to see that the plant cannot be mal-operated and so drift into the explosive region. This has to be done by reliable instrumentation and an alarm system. The matter is made more difficult since many processes have their optimum operating conditions in, or very near, the explosive limits, and so the plant should be designed to operate as near to the inflammable region as the instrument accuracy will allow.

Flow meters have accuracies not normally better than ±5%, and they can drift when they become dirty and they do not have a very high reliability. When safety is involved, such instruments are often installed in triplicate, so that an instrument failure can be detected by disagreement between readings. Which instrument has failed is indicated by which reading is different from the other two.

In addition to flow rate measurement, gas concentration measurements can be made. Particularly useful is the automatic measurement of oxygen, as described in section 10.6.c.

A second instrument of use is the combustible gas meter. In these instruments a sample of gas is continuously passed over a heated catalyst and the potentially hazardous reaction proceeds, liberating heat. The heat liberated can be related to the nearness of the explosive limit for a particular gas mixture.

In practice, the heated catalyst is platinum wire, and the combustion on its surface alters its temperature, and hence resistance, compared with a standard wire. The imbalance of resistances is measured on a Wheatstone bridge and calibrated in percent approach to the explosive limit.

10.9.b   Hazardous liquid phase reactions

A large volume of reacting liquid becomes uncontrollable when the temperature rises, and full cooling will not bring the reactor temperature within control.

When the high temperature alarm rings, the foreman clears everybody out of the building and waits to see what happens. The designer can be of some assistance by defining some action, and providing some equipment that can be used in such a situation.

If all heat removal attempts have failed (the hazardous situation may have been the result of a cooling failure anyway) provision of further cooling is not very promising. Reaction rates are reduced by dilution, and this is usually used as the standard counter measure to an uncontrollable temperature rise. Dilution can be achieved by providing an oversized reactor, and provision for quick addition of a suitable diluent, or the reactor contents can be quickly discharged into a vessel containing the diluent - a "dump tank" (see Fig. 10.19).

The use of a dump tank, in practice, appears to be the preferred solution, particularly because it is effective even in the event of a stirrer failure, and it may have been the stirrer failure that initiated the hazardous situation originally.

The recognition of the imminence of a hazardous situation is given by the temperature rise, and so the sooner this can be recognized, the more chance there is of evasive action being taken.

A dangerous temperature rise exists when the rate of temperature rise, the second derivative, d^2T/dt^2, becomes positive (see Fig. 10.20). This has led to the development of an instrument, the Sikarex, (Hub, 1975), which monitors the reaction

308

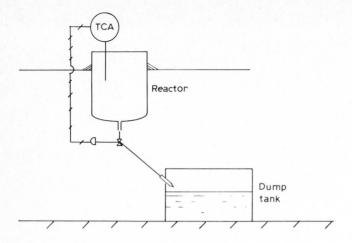

Fig. 10.19. The installation of a safety dump tank.

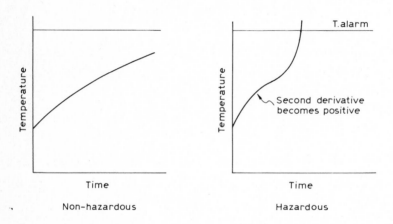

Fig. 10.20. Temperature profiles of a potentially hazardous batch reactor.

temperature, filtering the signal to remove noise, and so get a reliable measure
of the second derivative; this alarms when the second derivative is positive. This
gives 30-60 minutes extra warning over waiting for the reactor's high temperature
alarm to activate.

10.9.c Modelling sensitivity studies

It is not adequate to know that a reactor design is thermally stable at the
design conditions. It must be stable over the fluctuations of feed and coolant
temperature that are present in normal operation, and the effects of failures of
coolant flow, or stirrer, must be studied.

Such studies can be performed with the reactor model that has been developed

for the design of the reactor, by determining the results of the different feeds
and different cooling rates. Such a model will show whether the equilibrium state
is acceptable, or whether thermal runaway will occur.

This model will not show how fast these changes will occur, and what control
settings would be needed to provide a stable control system. Such questions could
be answered with a dynamic model composed of differential equations and partial
differential equations, including equipment volumes and heat capacities. It is not
normally necessary to go to such lengths to define a safe operating region for a
reactor, but if the matter was thought to be particularly critical for a certain
case, then the right approach would be to write a dynamic model of the system and
experiment further with that model.

Having tested the stability of the reactor for changes in feed and coolant
temperature, one is in the position to define the accuracy with which these should
be measured, and the settings that the high and low safety alarms should have. It
also enables instructions to be written for the operating manual on how to maintain
the reaction in stable operation and what actions to take in the event of a runaway.

10.9.d   The hazard and operability study

For many years design engineers have put their designs through "consistency
checks" before issuing them for fabrication.

In the consistency check, every item or remark written on the design document
is examined in turn to see that it does not conflict with information elsewhere
in the specification. During this thorough check, safety points such as:

o are all pipe flange ratings the same ?

o are all vessels fitted with safety valves ?

o are any pipe runs not protected for over pressure ?

are checked as the consistency check proceeds. With the move towards greater emphasis
on safety, these consistency checks are being replaced by the "hazard and
operability study".

This is a carefully designed formalized procedure that is carried out at five
levels of detail, depending on whether a process is in the research stage or has
reached the final design specification.

For a project in the research stage, the study requires only the hazards associated with the raw materials and products to be recognized. In the development work
the pilot plant must be shown to be safe, and in the detailed design stage a
thorough study taking a number of weeks is undergone to check what would happen
to the process for every conceivable malfunction.

Each controller is assumed to deliver more, or less, or fail closed or open,
and the consequences are followed through the rest of the plant.

As the study proceeds, potential hazards come to light, and design changes
normally have to be made to overcome these hazards. During the study the correct

action to take in the event of an emergency can be worked out, and this forms part of the operation instructions for the plant.

10.9.e   Standard design safety measures

Design temperatures and pressures. To specify the mechanical construction of all equipment, design temperatures and pressures must be defined. These are not the operating conditions, nor are they operating conditions plus a fixed margin; they are carefully considered conditions that might arise occasionally through abnormal operation.

The process must be studied, and various malfunctions envisaged, the outcome predicted (by model simulation, if necessary), and the resulting conditions used as the design temperature and pressure if they achieved maximum values.

It is convenient to take the design pressure as the maximum delivery pressure of the highest pressure feed, and the maximum design temperature as the maximum allowable temperature of an oil heating system.

If these limits pose difficult construction or material problems, then consideration can be given to the use of lower values plus the use of protective devices.

The designer must then specify a protective system that comes into action once the design temperature and pressures are overstepped.

Alarm and relief systems. The first measure is to install alarms that indicate that conditions have moved near to the design limit for the vessel. It is good practice to have the alarm sensor quite separate from the normal instrument sensor. Hence this covers the first question raised when an alarm sounds - is the instrument out of order, or is the alarm real ?

Following the alarm, automatic action should occur when design conditions are exceeded; for example, pressures should be automatically vented down, either by spring loaded relief valves, or bursting discs.

Where the contents are vented depends on the nature of the material involved. In some cases it is permissible to vent to the atmosphere at some high point in the plant. In oil refinery and petrochemical installations it is standard practice to vent to some sophisticated flare system which is composed of an extensive vents network to all parts of the works, plus a liquid knock-out and recovery section, and a continuously ignited flare stack, capable of handling the sudden surges from relief valve openings.

In cases of excess temperature, relieving pressure is not the correct action; other automatic actions must be devised with every case being considered on its own merits. Whether the coolant flow moves to maximum, the feeds are shut off, or other actions are devised, is dependent on each case in point.

Layout. Normal layout regulations should, of course, be adhered to because they are based on safety requirements: prevention of sources of ignition in inflammable vapour handling plant, correct segregation of types of equipment, and correct spacing of equipment. Most of these measures are based on prevention of an accident leading to a catastrophe, e.g. the small leak resulting in a fire, or the small fire resulting in an inferno.

A measure which is particularly relevent to reactors is the use of blast walls and explosion relief hatches in buildings containing reactors potentially capable of exploding. The blast proof wall separates the operators from the reactor, and all controls are carried out remotely from the other side of the wall to the reactor. The wall itself is expensive, and an additional expense is that every single valve and measuring device must be remotely controlled from behind the wall.

Explosion hatches are often found in research and development buildings, where each potentially dangerous reactor occupies a cubicle with the back or ceiling being an outward opening door or hatch. Any explosion opens this hatch, (rather like the principle of a bursting disc), the blast can be directed in a harmless direction, and the relief means that the cubicle itself is not destroyed.

On the industrial scale the use of these measures is questionable. Processes must be designed so that the reactor will not explode. If the designer is not sufficiently convinced that he has a safe process and he must install these extra measures, then he has not yet devised an operable process.

After all, the explosion of an industrial-scale reactor behind a blast wall is likely to result in fire, extensive poisoning of the neighbourhood, and lengthy production loss, none of which is acceptable to a serious-minded company.

10.10   SAFETY MEASURES TO BE TAKEN DURING OPERATION

Purging with inerts. Many reactor accidents occur that are completely beyond the control of the process developer and designer, but are due to operational failure. Any process handling inflammable vapours is potentially hazardous even if oxygen is not present in the process, because when the process is started-up, air is present in the equipment, and also when it is shut-down, inflammable vapours again have an opportunity to mix with incoming air. This means the purging of equipment must be a standard procedure to overcome these hazards. Nitrogen purging is one method, where N_2 is used to displace the air before start-up, and to displace vapours before opening up for maintenance. On equipment that is, or can be, hot, steam can be used for this purging operation.

Prevention of contamination. Complex organic reactions, in particular, are susceptible to trace contaminants which can catalyze violent reactions. Contamination between campaigns of different products, contamination from inadvertent addition of raw materials from the wrong drum, contamination from lubricants containing colloidal copper in organic processes, or oil lubricants in liquid oxygen processes, have all caused major accidents.

Standardization procedures. The thorough training of all operators, who should know the standard procedures to take for whatever situation arises on the plant, is a general principle of good plant operation and will not be discussed here. The development of such operation procedures should have occurred during the hazard and operability study described in section 10.9.d.

Perhaps one operational technique for use with hazardous reactions is worthy of note, if only for interest's sake. Traditionally, the operators in charge of TNT manufacture used to observe the reaction sitting on one-legged stools, because this at least ensured that they did not fall asleep.

FURTHER READING

Clevett, K.J., 1973. Handbook of Process Stream Analysis. Ellis Horwood. - A summary of available instruments up to 1973.
Franks, R.G.E., 1972. Modelling and Simulation in Chemical Engineering. Wiley. - Details on construction of dynamic models.

REFERENCES

Abdul-Kareem, H.K., Jain, A.K., Silverston, P.L. and Hudgins, R.R., 1980. Chem. Eng. Sci., 35:273.
Hub, L., 1975. Entwicklung von Prüfmethoden zur Erhöhung der Sicherheit bei der Durchführung chemischer Prozesse Diss. No. 5577, ETH Zürich.
Rase, H.F., 1977. Chemical Reactor Design for Process Plants. Wiley.

QUESTIONS TO CHAPTER 10

(1) What problems arise in specifying a method of controlling reactor feed flows ?
(2) What is the physical explanation for there being multiple steady-states in a CSTR reactor ?
(3) Why are tubular reactors very sensitive to the temperature of the coolant around the tubes ?
 How do "hot-spots" develop and how are they controlled ?
(4) How can the reactor model be useful in determining a control scheme for the reactor ?
(5) What is the most general form of process stream analysis instrument ? List the other general methods that are possible.
 How are process streams analyzed when the method of analysis does not appear on the list of developed methods ?
(6) What conditions are needed for gaseous explosions to occur ?
 What is the design philosophy used to prevent them ?
(7) What are the characteristics of organic liquid explosions ?
 What safety measures can be installed to prevent them ?
(8) What is a "Hazard and operability study" ?

Chapter 11

HEAT-TRANSFER FLUID SYSTEMS AND MATERIALS OF CONSTRUCTION

Two problems that beset all reactor designs are what is a suitable choice of heat-transfer fluid and what should be the materials of construction of the reactor and associated equipment.

The heat-transfer fluid problem arises when the reactor temperatures are too high for water/steam systems, that is when the steam pressure becomes unacceptably high.

Materials of construction raise problems when corrosive systems are involved: the choice of which non-corrosive alternative can often be difficult.

This chapter will simply list the various heat-transfer fluids that are suitable, with comments on their use, and discuss ancillary equipment required for an independent cooling circuit. In the section on materials of construction the various possibilities will be enumerated for coping with systems which are not compatible with mild steel.

11.1   HEAT-TRANSFER FLUIDS

Table 11.1 summarizes the heat-transfer fluids that are commonly used, giving their working temperature range.

One normally starts by choosing water, and moves from this up or down the list if the upper or lower temperature limits are not compatible with water. The further away from water one moves in the list, the more practical difficulties are encountered.

Water is the standard, most convenient medium; it is readily available, cheap, corrosion-free (when treated), has high specific and latent heats, and can conveniently be used for heat recovery by sending vaporized water into the works' steam main.

Heat-transfer oils have the inconvenience associated with not being able to use directly any recovered heat without further equipment, and they need to be replaced periodically because of deterioration due to oxidation and sludge formation.

Dowtherm A is, theoretically, an excellent medium, and it is one of the most commonly used fluids in its temperature range, but it is not generally welcomed by operating personnel because of its odour and the difficulty of keeping its system leak-tight. It has a decomposition problem should it become overheated and this demands careful design of its heater.

Inorganic salts (see Table 11.1) have the serious disadvantage that they solidify at about 150°C, and so special facilities must be allowed for in the design to

314

TABLE 11.1
Commonly used heat-transfer media and their applicable temperature range.

Media	Range	Use
Various fluorocarbons, methylene chloride	-40 - +20°C	low temperature duties
NaCl-brine	-20 - +20°C	low temperature duties, aqueous coolant
Water	0 - 200°C (10 bar) (exceptional designs to 50 bar, 280°C)	used as liquid phase or liquid/vapour system
Heat-transfer oils	30 - 320°C (certain grades to 400°C)	stable, low viscosity mineral oils for liquid circulation e.g. Mobiltherm 600
Dowtherm A	20 - 400°C	diphenyl oxide/diphenyl mixture used as liquid or liquid/vapour system
Inorganic salts	150 - 450°C (or to 600°C if slow decomposition is accepted)	mixture of sodium nitrite, nitrate, and potassium nitrate

drain equipment that will cool down under this temperature on shutdown, or to provide standby heating so that the salt will not solidify. Above 450°C the salt slowly decomposes evolving nitrogen, and the resulting mixture solidifies at progressively higher temperatures.

For systems above 500°C the available fluids are even more disagreeable than molten salts. Molten sodium/potassium alloys form low melting point mixtures, and these can be used from 25°C to around 800°C. The material must be kept oxygen-free, since oxygen causes corrosion and the formation of metal oxides which can be hazardous. The material is, of course, dangerous and obnoxious to work with whenever maintenance has to be carried out.

In addition to the systems mentioned there are numerous other fluids which duplicate the temperature ranges already covered. Since these fluids are less common than those mentioned it can be assumed that they do not possess any particular strong advantage over those detailed here for general use. They are discussed in more detail in Perry (1973).

Table 11.2 shows the physical properties of the fluids discussed in this section. These are of course necessary for the design of the heat-exchange equipment which they serve. More precise physical properties, particularly their relationship to temperature, can be obtained from the suppliers of the fluids.

TABLE 11.2

Physical properties of heat-transfer fluids.

	Water	Dowtherm A[1]	Fused salt[2] NaNO$_3$ NaNO$_2$ KNO$_3$	Oil: Mobiltherm 600[3]	Oil: Mobiltherm light[3]	NaK
Mol. wt.	18	165	92	---	---	---
Density (100 °C) (kg m^{-3})	958	997	1980	900	930	840
Melting point °C	0	12	150	-5	-5	20
C_p liquid (kJ kg^{-1} K^{-1})	4.18	2.20	1.56	2.43	2.21	1.04
Heat of vaporization (kJ kg^{-1})	2253	290	---	---	---	---
Cubic expansion coefficient (K^{-1})	0.0043	0.00077	0.00036	0.00063	0.00063	---
Viscosity of liquid (kg m^{-1} s^{-1})	0.284x10^{-3} (100°C)	0.30x10^{-3} (300°C)	1.7x10^{-3} (450°C)	0.595 (260°C)	0.873x10^{-3} (150°C)	0.24x10^{-3} (300°C)
Thermal conductivity of liquid (kW m^{-1} K^{-1})	6.80x10^{-4}	1.31x10^{-4}	6.06x10^{-4}	1.16x10^{-4}	1.13x10^{-4}	0.027

Suppliers: (1) Dow Chemical
(2) Du Pont de Nemours & Co.
(3) Mobil Oil Corp.

11.2   THE DESIGN OF HEAT-TRANSFER FLUID SYSTEMS

11.2.a   The heater

Even when the duty of the heat exchange medium is to be the cooling of the reactor, there remains the need to provide some form of heating equipment so that the apparatus can be started up. The duty of the heater to be installed can be worked out from the thermal capacities of the total system that will be heated to operating temperature, i.e. the total heat-transfer medium system and its equipment and the total reactor and its contents. Then, after discussion with operating personnel on how long they consider it reasonable to wait for temperature to be attained, (4-6 hours will seem reasonable if start-ups are rare), a heating duty for the heater can be defined after allowing for heat losses. When the system is

316

endothermic, and the heater must continuously supply heat to the process, it may
be that this continuous heating demand is greater, or less, than the heating rate
to achieve a reasonable start-up time. Both the start-up and continuous heat load
must be calculated and the greater of these taken for rating the heater.

Heating above available site steam temperatures can be achieved by electrical
immersion heaters, or gas- or oil-fired furnaces (see Fig. 11.1).

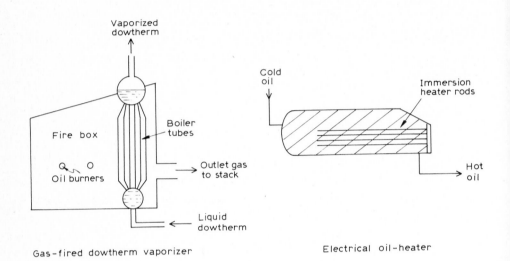

Fig. 11.1. Forms of heater unit.

Electrical heating is by far the most convenient, but for continuous use it
will be the most expensive and should be considered only for start-up heating. It
may also be the most expensive investment-wise when one takes into account the extra
offsite cost for the additional electrical load. If the heater substantially
increases offsite power investment, this should be seen as the cost of convenience,
and electrical heating used only if it is considered worth paying this extra cost.
If it can be shown that start-up will never occur at peak loads, then there may be
no additional offsite electrical cost. For continuous heating duties, oil- or gas-
fired furnaces should be employed.

Consideration can always be given to direct heating of stirred tanks with gas
or oil. If a kettle requires only heating, with no cooling processes being involved
and the equipment is not in a flame-proof area, it is possible to obtain a kettle
fitted with its own oil or gas burner. This may look like a witches cauldron but
it is cheap and adequate under these special circumstances.

The heat-exchange surface area in the heater is normally sized by the heat flux
that is passed across it. Heat fluxes of the order of 100-140 kW m^{-2} are acceptable

rates which do not cause tube overheating and decomposition of heat-transfer fluid. A similar heat flux should also be used when heating with electrical immersion heaters.

11.2.b The cooler

All heat-exchange medium systems require a cooler in the circuit as well as a heater. If the process is exothermic, the continuous heat load on the cooler will be given by the heat evolved by the process. If the process is endothermic, the cooler is used only to cool the system down for maintenance or catalyst changing. The cooler size to do this is determined by the allowable time to cool the equipment down for maintenance to be carried out, considering the thermal capacity of the whole system, analogous to the start-up heat load calculation.

Cooling of heat-transfer media is usually done by air coolers if heat recovery is not being contemplated, since if the process is not already cooled by water, temperature levels are too high for normal process cooling water (tube scaling would occur). If heat recovery is intended then the hot fluid should be cooled with boiler feed water which can be vaporized and be sent to the works' steam main. To prevent tube scaling, boiler feed water or condensate must be used, and this inevitably complicates any heat recovery scheme.

11.2.c The expansion tank

Since the fluid will initially be at 20°C and then be heated to 200-400°C, there is inevitably a considerable volume change due to thermal expansion. This variation is usually handled by installing an expansion tank in the circuit. This tank is installed at a high point in the process (see Fig. 11.2) and this serves the following purposes:
o it is the expansion vessel that absorbs the changes in volume;
o it fixes the pressure at a point in the system which can be chosen so that the whole system is above atmospheric pressure, hence there is no possibility of ingress of air which could cause oxidation and corrosion;
o since inerts collect at high points in circulating systems, any inerts in the system find their way to the expansion tank, from which they can be conveniently vented. Various heat-transfer fluids evolve gaseous by-products with time, and these should be continuously removed from the system; the expansion tank provides a means of doing this.

11.2.b The bulk storage tank

In the event of maintenance or emergency it is necessary to be able to drain the heat-transfer fluid out of the equipment. Capacity for this is provided by a storage tank which has sufficient volume to accept the total heat-transfer fluid charge. Extra volume to this should be supplied to provide reserve capacity for

318

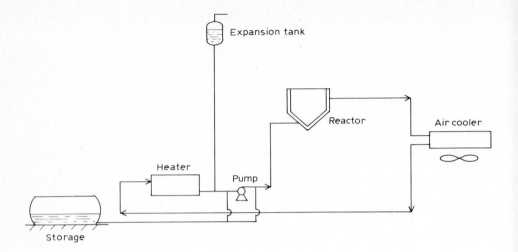

Fig. 11.2. Flow diagram for a simple heat-transfer fluid circuit.

make-up fluid to make good losses.

11.2.e   The total system

Figure 11.2 gives the line diagram for a single phase heating medium system
serving one reactor at one temperature. It is often necessary to have a number
of users on the same system, but each requiring its own temperature control. This,
for instance, is necessary in a multiproduct batch plant which may have a number
of reactors manufacturing different products and each requiring a different
temperature. This can be done by having a number of circuits, each of different
temperature. The heater is part of the highest temperature circuit, and the lowest
temperature circuit takes the high temperature fluid as and when necessary to
maintain its set temperature. This displaces the low temperature fluid into the
high temperature circuit where it is compensated for by the temperature control of
the heater (Fig. 11.3). In this way it is possible to have a number of different
temperature levels continuously working off the same heater.

With the Dowtherm A fluid, as with water, there is the possibility of vaporizing
and heating by condensing vapour, or cooling by boiling liquid. In these cases no
expansion vessel is required because the system is not completely full with liquid.
The circuit is shown in Fig. 11.4. With vaporizing methods, systems with multiple
temperature levels can be devised using vaporization/condensation at the highest
temperature level, and at the lower levels using the fluid in the liquid phase only,
with a series of circulating systems similar to those in Fig. 11.3 (see Fig. 11.5).

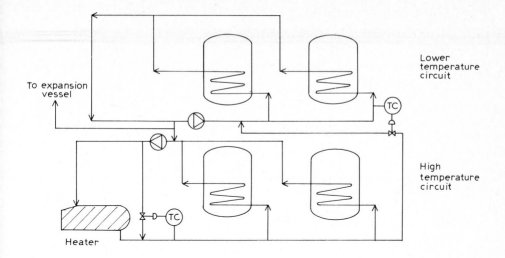

Fig. 11.3. Multiple temperature levels on the same system.

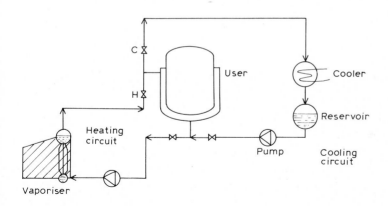

Fig. 11.4. Dowtherm vapour/liquid heating/cooling system.

Special attention should be paid to prevention of equipment damage due to the development of hydrostatic pressures which occur when liquids completely fill equipment, are sealed in, and the temperature then increases. All pipe-runs and equipment must be protected by relief valves and there must be no way of isolating equipment or pipe-runs from relief valves by the installation of too many valves in the system.

A number of contracting companies specialize in the design and erection of heat-transfer fluid systems, and when a project reaches a stage where it is fairly certain to proceed to full scale, consideration should be given to consulting such

a specialized firm. Their experience gained from their previous designs is
generally worth paying for, and if they are consulted before the process development
is completed, they may be able to make a positive contribution at this stage. For
instance, they could be invited to provide the heat-transfer fluid circuit for the
pilot plant.

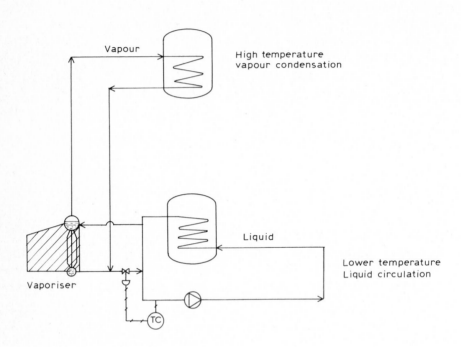

Fig. 11.5. A Dowtherm multiple temperature with vapour and liquid heating system.

11.3   MATERIALS OF CONSTRUCTION

11.3.a   Corrosive liquid systems

   Very many liquid reactions take place in acid media. This means that the reactor
should either be made of corrosion-resistant material, or lined in a corrosion-
resistant material.

   The choice of material depends on the degree of corrosiveness of the reactor
contents. Carbon steel is corroded by water in the presence of oxygen, but this
corrosion is not fast, and water can be handled quite satisfactorily by carbon
steel by allowing for a given degree of corrosion to occur by specifying a
corrosion allowance. Steel is also corroded by most concentrations of sulphuric
acid, but in this case the corrosion is so fast that simply including a corrosion
allowance in the design is no solution to the problem. For mildly corrosive
conditions, coating the reactor may be an adequate measure. For the highly corrosive

condition a new material of construction must be sought. This means that in choosing materials of construction for reactors one must firstly determine the corrosion rate for the medium: often it is so strongly corrosive that one should talk of the medium dissolving rather than corroding; sometimes the medium is mildly corrosive in that traces of wet acid are present, but only in very small quantities, e.g. a chlorinated solvent system may have traces of water present which cause hydrolysis leading to traces of moist HCl in the system.

When conditions are mildly corrosive some form of surface treatment of the mild steel (e.g. resin coating or enamelling), or the use of stainless steel or nickel alloys are possible solutions. In fact, in this case, the problem may be more one of preventing product contamination than protection of the reactor. When the conditions are highly corrosive other solutions have to be found which are more expensive, and all have disadvantages.

When temperatures are not high, plastic-lined equipment can be employed if a suitable plastic which withstands the organics in the system can be found. PTFE, polypropylene, and PVC are frequently used plastic lining materials.

At high temperatures plastic linings cannot be used and other lining materials must be sought. Glass-lined vessels are very commonly used for reactors, although they are fragile and accidental damage can be catastrophic for a production schedule. But the fact that they are so frequently used indicates that this problem cannot be too serious. Glass linings (and enamel) are susceptible to thermal shock and problems of differential thermal expansion. Hence, temperature differences across the reactor wall should be limited to 60°C.

For equipment that cannot be glass-lined (where certain geometrical forms and certain radii of curvature must be adhered to), or for equipment with no reserve should the lining be damaged, a more desirable but more expensive solution is to clad the equipment with a corrosion resistant metal such as tantalum. Titanium is also possible as cladding but it is not as generally resistant to acid media. Titanium is cheaper than tantalum and it may be economical to make certain parts of the equipment out of pure titanium without the need for cladding the metal on a less resistant base.

Table 11.3 lists the various acid resistant materials of construction and Table 11.4 lists the possible lining materials that can be used to line carbon steel reactors.

11.3.b Corrosive gaseous systems

These systems are very much rarer than corrosive liquid systems, and in fact much corrosion in gaseous systems is liquid phase corrosion caused by dew formation in the gas. This can occur at higher temperatures than the calculated dew point of the gas mixture when the surface contains traces of hygroscopic material. For example if a surface contains ferric chloride from previous corrosion, this will become

322

TABLE 11.3

Various acid resistant materials of construction

Reactor materials	Notes
Stainless steels	limited corrosion resistance
Hastalloys (Ni-alloys)	limited corrosion resistance
Glass, silica	fragile, size limitation
Carbon, resin-impregnated carbon (Karbate), graphite	fragile, suitable for absorbers and towers
Tantalum	very expensive
Titanium	corrosion by chlorides, expensive
Glass filled epoxi-resins	temperature limitation
Plastics, rubbers etc.	low mechanical strength, temperature limitations (better used as a lining material)

TABLE 11.4

Various acid resistant lining and coating materials

(a) Lining	Notes
Glass	fragile, susceptible to thermal shock (1)
Tile	loosened by temperature fluctuations (1)
PTFE	temperatures < 260°C (1)
Polypropylene, PVC, chlorinated PVC, various rubbers	low temperature range 100-150°C, various degrees of swelling with organic solvents (1)
Tantalum cladding	expensive

(b) Coating (2)	Notes
Enamel	fragile
Resin-based e.g. epoxi-resins Plastic-based e.g. PTFE, rubber	various degrees of chemical resistance (Perry, 1973)

Note (1) A general disadvantage of the non-metallic linings is their insulating property which seriously handicaps heat removal through the vessel walls.

Note (2) No coating can be completely perfect, and so it cannot protect equipment from highly corrosive media. It can however improve product quality for mildly corrosive media.

moist at a higher temperature than the expected dew point, and this moist ferric chloride will then be acidic and a centre for further corrosion. The correct dew point calculation should involve the vapour pressure of water over saturated ferric chloride solution.

True gaseous corrosion occurs, for example, with chlorine, which reacts, even when dry, with carbon steel at elevated temperatures. Often the result of gas reactions is a surface layer which remains on the metal and renders it passive to further corrosion. When this is the case, care must be taken to see that the conditions are not also erosive, because this protected area can then be worn off, which will result in extensive corrosion.

Corrosion with gases is so component-specific that no generalizations can be made. Materials of construction tables must be consulted whenever a new system is under`investigation. Internal tile- or brick-lining a vessel is sometimes a useful technique when corrosion occurs at operating temperature but not at lower temperatures. This inner lining, besides producing a fairly impervious barrier, can be designed to reduce the wall temperature to below that at which corrosion occurs.

It is also possible to coat the reactor surface or line it with an impervious plastic before erecting the brick lining to ensure that the corrosive reactor contents do not contact the reactor through imperfections in the brickwork, as long as temperatures are not too high.

11.3.c   Temperature limitations of normal materials of construction

Carbon steels are the normally used construction materials unless changes have to be made for reasons of corrosion. It is not recommended that carbon steel is used for temperatures below -50°C, but special grades of low temperature steel can be specified which can be used to -100°C. Below these temperatures nickel steels or stainless steel must be specified.

As temperatures increase, the strength of materials of construction decrease. Carbon steel, however, still has an adequate strength for use as a construction material up to temperatures well over 500°C. However, at this temperature carbon steel corrodes badly in oxidizing conditions because at this temperature scaling occurs and the protective scale detaches from the metal surface. Hence, for temperatures above 500°C various austentic stainless steels are used because they are more resistant to corrosion (up to 1100°C).

The higher the design temperature, the weaker are the materials and so the greater is the required wall thickness. When temperatures become out of control and increase considerably above design temperature, equipment damage occurs because the equipment is no longer strong enough to withstand its designed pressure and stresses, and so creep, vessel deformation, and failure occur. These occur, not because the temperature has risen beyond that suitable for the material of construction, but because the design has assumed that the construction materials were stronger than was the case because of the higher temperature involved.

Brick linings extend the temperature range of steels. Refractory brick is specifically designed to insulate reactor walls from very high temperatures within the reactor and this enables reaction temperatures of up to 1600°C to be used with carbon steel vessels. Brick and tile lining can also be used in large tank reactors and empty volume gas reactors to prevent corrosion. The temperature of brick-lined equipment should be kept as constant as possible because of the damage caused by temperature cycling and the differential thermal expansion that results.

Temperature limitations for other types of lined equipment are generally fairly restrictive. Glass-lined tanks are suitable up to a temperature of 250°C. Higher temperatures are limited by the increased possibility of fracture of the lining due to thermal shock. Special glass/ceramic linings are, however, obtainable with improved thermal shock characteristics enabling temperatures up to 650°C to be used.

Linings of plastic materials or coating with organic films have temperature limitations corresponding to the maximum working temperatures for the organic material forming the coating. The best is PTFE with an upper temperature of 260°C.

11.3.d   Specifying the corrosion allowance

Most metallic materials of construction corrode to some extent in service and this takes place as an even loss of material over the whole surface, resulting in a thinning of the walls with time.

The process design should specify the design temperature and pressure and also a corrosion allowance. The mechanical engineer can then determine the vessel thickness necessary to have adequate mechanical strength at the design temperature and pressure, and then add on the corrosion allowance to provide the vessel thickness for its manufacture. Hence the vessel is still mechanically adequate at the end of its life since the corrosion allowance is specified as the thickness that will have corroded away over the expected life of the vessel.

Consider that a carbon steel vessel for carbon tetrachloride is to be specified, and the corrosion rate for this medium is 0.25 mm year^{-1}. Assuming a 10 year life for the reactor, a 2.5 mm corrosion allowance would be specified, given that the outside surface was to be protected by painting.

Values for corrosion rates are available in the corrosion engineering literature or are measured in the laboratory when a novel process for which no literature data are available is under development.

Corrosion must be even. If pitting or localized corrosion occurs the material of construction is not suitable and another must be sought.

The selection of economical materials of construction can be made by carrying out a NPW calculation for the various types of material under consideration. A cheaper material of construction with a high corrosion rate could be replaced, say, every four years, and this compared with the extra initial investment for the better material of construction and no replacement. If the economic calculation indicates that the better material is justified, it can be specified in the design. If it cannot be economically justified then additional advantages of the better material (less product contamination, greater reliability, less maintenance), should be subjectively weighed against the negative NPW before a final decision is made.

FURTHER READING

Perry, R.H. and Chilton, C.H., 1973. Chemical Engineer's Handbook (5th Edition). McGraw-Hill. - Details on both H.T. fluids and materials of construction.
Mantell, C.C., 1958. Engineering Materials Handbook. McGraw-Hill. - Constructional materials treated in great detail.

QUESTIONS TO CHAPTER 11

(1) What heat-transfer media would be suitable if a reaction in a tubular reactor had to be controlled at:
    (a)   90°C
    (b) 400°C
    (c) -10°C ?
(2) Consider a stirred tank reactor operating with an acid medium. How does the materials of construction problem affect the heat removal problem ? What are possible solutions ?
(3) How are the design temperature, design pressure, and corrosion allowance used in determining the thickness of material to be used in the vessel walls ?
(4) What are the normal temperature limits of carbon steel as a construction material, and what materials of construction are used outside these limits ?
(5) Under what circumstances is it advantageous to line vessels with ceramic tiles or bricks ?

Chapter 12

ADVANCED REACTOR DESIGNS

Chapters 7, 8, and 9 have discussed the design of fairly standard forms of chemical reactor, which are well understood and capable of being scaled up with relative confidence.

However, once any process becomes well-established, developments are made to the reactors which result in a larger, cheaper, more suitable design for the specific reaction in question, but the reactor design has then become so specific that it can hardly ever be transferred to another reaction without a great deal of pilot work and a number of scale-up stages.

These reactors are therefore less interesting as methods of carrying out reactions in new situations, but they are nevertheless very interesting and instructive in that they do indicate the ingenuity that goes into reactor design when enough effort and incentive is provided to improve the standard types.

This chapter contains short descriptions of some of these special reactors to illustrate this point. It does not contain details of the design of these reactors; those interested in such details should refer to more extensive texts.

12.1   STEAM CRACKERS

When hydrocarbon fractions such as naphtha or gas oil are heated to a high temperature, the long molecules crack into C_1, C_2, and C_3 molecules. It is possible to control this pyrolysis so as to maximize the yield of particular products. Ethylene and propylene are the normally required products and optimal conditions for these are low partial pressures, temperatures between 750°C and 950°C, and low contact times, e.g. 0.05 to 1 second.

The low partial pressures are achieved by adding steam to the feed, hence the name "steam cracking". The reaction is endothermic, so heat is required both to raise the temperature of the feed to reaction temperature and then to keep it at this temperature.

Ethylene is required in vast quantities, since it is the major raw material for the whole organic chemical industry, and so very large and specific designs have evolved for this reactor.

To attain the required temperature a direct oil-fired furnace is used, and this means that the reactor is in effect the tubes of the furnace inside the furnace fire box (Fig. 12.1).

The furnace in itself presents a difficult design problem to produce a distinct reactor temperature without overheating. This is done by carefully selecting fluid velocities to prevent high tube wall temperatures which would result in the

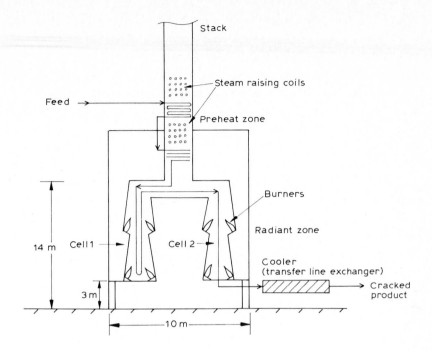

Feed →

Typically: 4 burners per cell
         2 cells per stream
         6 cells per furnace (3 parallel streams, one common stack)

Fig. 12.1. A steam cracker furnace.

formation of carbon in the tubes and lead to tube failure. The reactions involved
are extremely complex and the resulting product distribution is very dependent on
the particular reactor conditions chosen. Add to this the need to have a flexibility
in feed stock composition and it is possible to see that the design of such reactors
is a detailed, specialized procedure.

12.1.a   Furnace design
    The furnace design must include all principles of good furnace design to achieve
good heat economy, and in addition ensure that the reactants achieve a distinct
temperature/time profile and a narrow residence time distribution.
    Furnace correlation data are available to enable tube diameters and heat fluxes,
of both conductive and radiative heat to the tubes, to be determined for specific
furnace layouts. Pressure drop calculations allow for various velocities and
parallel pass combinations to be considered to obtain a design with a skin wall
temperature that will not initiate coking reactions on the wall.

With a number of parallel tube passes, there is often difficulty in achieving equal flow, and hence temperatures, in each pass, particularly when the furnace has been operating some time. This is usually solved by independent temperature control by feed flow to each parallel pass separately.

The design of these monster oil-fired furnaces, which can be over 30 m high, demands a great deal of specialized knowledge which has been developed over the years by a limited number of engineering contractors and petrochemical companies. The design of such equipment is generally left to such specialists who use their own in-house experience and correlations in arriving at a design.

12.1.b Chemical kinetics

When hydrocarbons are heated many of the species present dissociate to form free radicals. These free radicals then propagate chain reactions with other species present, by means of hydrogen abstraction reactions, radical additions, radical decompositions, radical isomerisations, chain termination and molecular reactions. The large number of components in the feed petroleum fraction, together with the combinatoric characteristics of free radical reactions, result in full descriptions of the reactions occurring in a steam cracker requiring some 2000 reactions.

The product distribution for the output of a steam cracker is very dependent on its temperature and reaction time. High temperatures and short contact times favour high selectivities to ethylene, whereas low temperatures for longer times favour C_3 and C_4 olefine production. Hence the optimum operating conditions depend upon the sales requirement and the feed stock used (be it light   naphtha or gas oil, and from which crude oil).

Because of the need to revise operating conditions, depending on the feed material and required product slate, almost all operators and designers have a kinetic model with which to carry out their design and operation planning.

Different companies have different types of model. The simplest are purely empirical regression type models based on past operating experience. Other models are semi-empirical ignoring the free radical mechanism, but based on simple first-order decompositions of the major hydrocarbons present in the feed (Schwarz, 1970).

A model consisting of the 2000 free radical reactions has also been developed and shown to give accurate predictions of steam cracker performance over a wide range of conditions and feed stocks (Denti, 1979). This model has been reported as producing savings of the order of 1-4% in ethylene production cost owing to better furnace operations, feed stock selection, and production optimization.

Besides being of considerable economic importance, this model is very interesting from the modelling point of view. The work represents a successful application of theoretical, rather than semi-empirical modelling, which is unusual. Also interesting is the way in which the enormous number of possible reactions were grouped together to provide a more limited number of reaction types for which data had to be found.

The data for most of these reactions were obtained from literature reported
values, predicted values, and analogies between similar reactions. Only a limited
amount of reaction data was obtained by fitting to experimental information. In
addition, they used a novel way of integrating the set of kinetic equations
produced, without the need to eliminate the radicals by Bodenstein stationary
principle. The work is the exception that breaks the rule that successful models
are very simple models. It must be noted, however, that 10 man-years' effort was
invested in its development, and that much simpler steam cracker models have been
found "adequate" for design and operating purposes (Rase, 1977).

12.2   FLAME REACTORS

Certain reactions occur very quickly and liberate much heat, but because of
the stability of the reaction products there is no need to control the temperature
in any way. Such reactions produce very high temperatures resulting in flames
which are no longer kinetically controlled but entirely mass-transfer controlled.
Hence the technology of such reactors reduces to burner and nozzle design to
achieve short, stable flames. Flame instability is caused when the velocity of
the incoming gas is greater than the rate of heat transfer from the flame into
this gas. Too low a nozzle velocity for the gas mixture must also be avoided
because this causes the flame to travel back into the inlet nozzle resulting in
excessive nozzle temperatures and equipment damage.

To achieve stable flames, various internals can be used which induce eddies
and provide a heat input to incoming gas through back mixing of already reacted
hot gas. The design of flame reactors and the need for back mixing generally has
to be determined by pilot trials.

The simplest form of flame reactor is the HCl burner where HCl is produced by
burning hydrogen in chlorine. The burner consists of a tall cylinder, which has
to be brick-lined to reduce wall temperatures to below those where HCl corrodes
steel (Fig. 12.2). The H_2 and Cl_2 are fed by means of a jet into the bottom of the
cylinder. Control can be achieved by H_2 analysis by katherometer in the exit gas
when the burner operates at 1 or 2% excess hydrogen. Unreacted chlorine is
undetectable in the reactor exit.

Care must be taken at start-up because of the possibility of explosive mixtures
of H_2/air or H_2/Cl_2 if the correct start-up procedure is not employed before the
flame is ignited. Fig. 12.3 indicates the purging facilities required.

In the same category would come the H_2/air burned for a nitrogen plant (Fig. 12.4).
In this burner, H_2 is burned stoichiometrically with air so that after condensation
of the water produced, nitrogen, essentially free of oxygen, is produced and can
be used for purging purposes during plant operation and start-up.

A number of hydrocarbon flame processes have been developed where, by flame
quenching, low reaction times of the order of 1/1000 s produce high concentrations

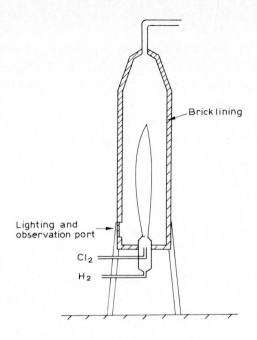

Fig. 12.2. An HCl burner flame reactor.

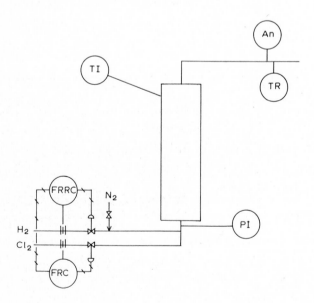

Fig. 12.3. Start-up and control of an HCl burner.

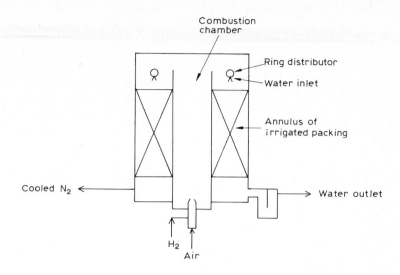

Fig. 12.4. N_2 producer by H_2/air combustion.

of valuable products. A number of acetylene processes have been developed where methane or naphtha is burned in a specially developed burner where the flame is quenched with a water stream after a very short reaction time. The reaction chamber is a small cylinder about 500 mm in diameter and 500 mm tall, which then supports an extensive acetylene plant which separates the quenched reaction products and isolates acetylene and the other useful olefines as separate product streams (Fig. 12.5). Such reactors can only be developed by extensive laboratory work backed up by pilot trials of a large number of designs to achieve good mixing and optimum yield of acetylene.

Flame type reactors have also been developed for the production of synthesis gas from hydrocarbons and O_2. At high temperatures the CO_2 and H_2O formed in the combustion have their equilibrium shifted to CO and H_2. Hence the heat from the hydrocarbon burning produces the high temperatures required for good yields of synthesis gas. Rapid quenching is then necessary to prevent the equilibrium reverting to CO_2 as the temperature falls. Reaction temperatures of 1500°C are used, with rapid cooling to 500°C being achieved within one second by means of water injection.

12.3   FLUID BED COMBUSTORS

A special case of fluid bed reactor application is that to the combustion of solid fuels such as pulverized coal. The reactor consists of a bed of sand, fluidized with air, into which the pulverized coal is fed (Fig. 12.6). The coal particles burn and the heat produced is removed by boiler tubes which are installed

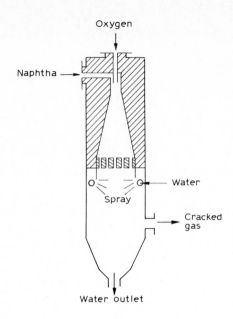

Fig. 12.5. A quenched flame acetylene producer.

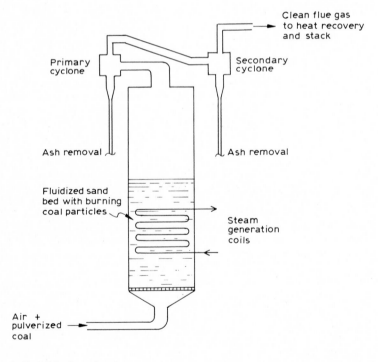

Fig. 12.6. A fluid-bed combustor.

in the bed. Conditions are generally chosen so that the combustion is not kinetic, but mass-transfer controlled, and the bed is essentially fluidized sand and ash with some circulating burning coal particles. The coal concentration is very low in the bed and so overall bed temperatures are much lower than with conventional coal firing.

Since such a coal combustor can operate at low temperatures this reduces the nitric oxide formation, with consequent environmental advantages. More important is the ability to reduce SO_2 emission by 70-95% by adding limestone or dolomite to the bed, whereby the SO_2 is held as sulphates in the bed and does not appear in the flue gas from the plant.

Fluidized combustion of coal promises a cleaner, more economical, and more efficient process than existing coal combustion processes (Locke, 1975).

Some theoretical aspects of solid combustion in fluidized beds is given by Levenspiel (1972) and Kunii (1969). It is clear, however, that the development of this type of equipment must lie heavily on laboratory and pilot plant experimentation.

12.4   COAL GASIFICATION

With the natural oil and gas reserves becoming scarce, there is a revived interest in producing replacement fuels from coal, of which there is an abundance.

By heating coal in the presence of oxygen and steam to temperatures of the order of 1000°C, the coal can be completely gasified, except for the ash content, producing CO, CH_4, H_2 as useful fuels, and CO_2, H_2O, and SO_2 as diluents. If replacement liquid fuels are required this gas can be converted into methanol or liquid hydrocarbons by later processes.

The complete gasification of coal is an old process with plants operating as far back as 1936; the present development effort is devoted to improving the efficiency of the process, as measured by the heat of combustion of the gas produced ratioed to the heat of combustion of the original coal.

A coal gasifier is a form of solid/gas reactor in which H_2 and other gases evolve from the coal on heating and carbon reacts with the O_2 and H_2 to form CO, CO_2 and CH_4. The vessel requires to be in heat balance; the required heat can be produced by partial burning of the coal or by supplying heat from an external source.

The design of modern coal gasifiers attempts to meet these requirements and the resulting designs take the form of a moving bed working at around 30 bar pressure with O_2/H_2O being blown up the bed and ash being removed from the bottom (Fig. 12.7) (Rudolf, 1976). The design of such a reactor, 4 m in diameter, water-jacketed and capable of withstanding operating pressures of 30 bar, being fed with solid particles and solid ash being removed by means of a locked hopper system, operating at around 1000°C with rotating internals, is undoubtedly a Very Advanced Design, which can only be arrived at as a result of years of pilot trials and modest increases in scale.

334

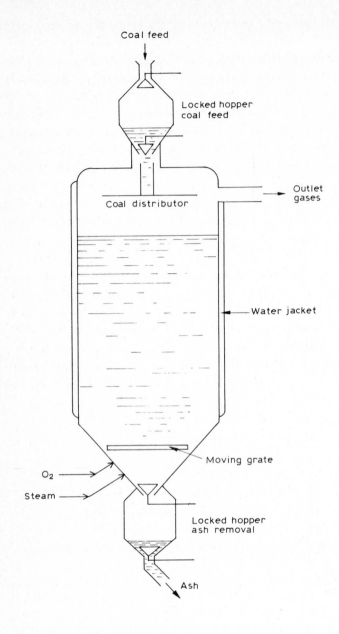

Fig. 12.7. Lurgi coal gasifier.

12.5   MULTIFUNCTIONAL REACTORS

Chemical engineering made great progress in the 1920s by using the principle of unit operations whereby the various processes were broken down into unit operations and design methods for each unit operation were then defined. Hence the chemical engineer, trained in a limited number of unit operations, could tackle any process.

Over the last 50 years intense research has been carried out in all of these unit operations, reactors being one of them. It is very likely that the major advances in each unit operation have now been made. It may well be that now sophisticated engineering should be devoted to moving in the opposite direction, to developing specific tailor-made equipment as some of the reactors already described in this chapter. Further advantages could be made by combining processes, so that one piece of equipment can carry out a greater part of a process than just one unit operation.

Reaction-and-separation is a very suitable candidate for such schemes. Not to be content with designing only a reactor, but designing it in such a way that product leaves it and reactants are returned to it for further reaction would be a mark of advanced engineering, as well as contributing markedly to reducing the capital cost of the plant.

There are isolated examples of this in existence. The chemist's reflux column condenser on his batch reactor is simultaneously separating products, returning reactant and removing heat of reaction (Fig. 12.8). Such a stirred tank plus reflux column and condenser is part of a number of industrial processes, although the columns are not there out of any advanced chemical engineering decision, but just because they were scaled up that way from the laboratory.

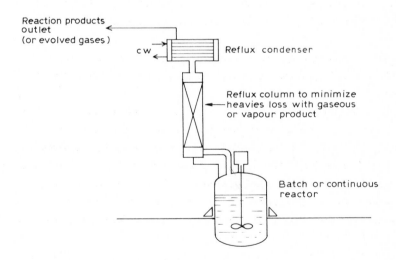

Fig. 12.8. Reactor with reflux column and condenser.

A second example is quoted by Horak (1978) of the esterification of terephthalic acid by methanol in a plate column. The molten terephthalic acid passes down the column, the methanol vapour passes up. The vapour strips out the water at the bottom of the column so the reaction equilibrium is displaced to increase the quantity of ester produced.

A third example is a process for the chlorination of benzene in a refluxing distillation column so that at the bottom of the column benzene is stripped out of the mono- and dichlorobenzenes and returned to the centre of the column where it is further chlorinated. In this way the degree of chlorination for a given mono/dichlorobenzene ratio can be increased beyond the plug-flow limits, and the one piece of equipment is taking the place of reactor, distillation column, pumps, tanks, and so on.

A fourth example is the loop pressure cycle fermenter developed by ICI for production of single cell protein from methanol (Matthews, 1978). The fermenter consists of an enormous loop standing some 30 m high (Fig. 12.9). The culture requires to be fed with oxygen to enable growth to proceed, and to have CO_2 removed from the system. The hydrostatic pressure of liquid in the system produces a high pressure at the bottom of the loop, and at this point air is pumped into the system, and the high pressure enables comparatively high O_2 concentrations in the liquid to be achieved. At the top of the loop, pressures

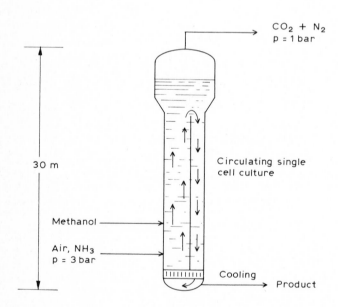

Fig. 12.9. Pressure cycling fermenter.

are atmospheric, and the gas disengages at this point, taking with it the CO_2 formed by the metabolism. Hence, when the culture circulates it undergoes pressure cycles which enable it to grow in a much more efficient way than by blowing extra air through a stirred tank.

The design of such multifunctional equipment, still maintaining plant flexibility, is very demanding and presumably that is why so few examples of operating plant are reported. There is still clearly much work to be done on improving our ability to design low cost flexible reactors, but whether it can be achieved by breaking the processes down still further, as is the present vogue, by studying tortuosity factors, bubble wakes, droplet size, and film thicknesses, or whether it will be best achieved by combining process steps together, is a matter worthy of consideration.

FURTHER READING

Rase, H.F., 1977. Chemical Reactor Design for Process Plants. Wiley. - Much detail on steam crackers and other specialized forms of reactor.
Rudolf, P.E.H., 1976. 4th ISCRE, p. 537, DECHEMA, Frankfurt, West Germany. - Practical survey of available coal gasification processes.

REFERENCES

Denti, M., Ranzi, E. and Goossens, A.G., 1979. CACE '79, E.F.C.E., Montreux, April 1979.
Horak, J. and Pasek, J., 1978. Design of Industrial Chemical Reactors from Laboratory Data. Heyden.
Kunii, D. and Levenspiel, O., 1969. Fluidization Engineering. Wiley.
Levenspiel, O., 1972. Chemical Reactor Engineering (2nd Edition). Wiley.
Locke, H.B. and Lunn, H.G., 1975. The Chemical Engineer, p. 667.
Matthews, J.F., 1978. 1st Mediterranean Congress on Chemical Engineering, E.F.C.E., Barcelona, p. 1382.
Schwarz, P., Di Massa, I. and Ragain, V., 1970. E.F.C.E. Symposium on Routine Computer Programs, Florence.

Appendix 1

SOME USEFUL CONVERSIONS TO SI UNITS

	SI unit	Divide by this factor to convert to SI unit
Length	1 m	= 3.281 ft
		= 39.372 inches
Area	1 m^2	= 10.765 ft^2
		= 1550 inches^2
Volume	1 m^3	= 35.31 ft^3
		= 220.0 imperial gallons
		= 264.2 US gallons
		= 6.290 barrels
Mass	1 kg	= 2.205 lbs
		= 1×10^{-3} tonnes
		= 0.9842×10^{-3} long tons
Temperature	K	$T°C = (T + 273.15)$ K
		$T°F = \dfrac{(T - 32.0)}{1.8} + 273.15$ K
Pressure	1 bar	= 0.1×10^6 kg m^{-1} s^{-2} (Pascal, N m^{-2})
		= 0.9870 atm
		= 14.51 psi
		= 750.1 mmHg
		= 401.5 inches water
Diffusivity	1 m^2 s^{-1}	= 1×10^4 cm^2 s^{-1}
Surface tension	kg s^{-2} (N m^{-1})	= 1000 dynes cm^{-1}
Density	1 kg m^{-3}	= 0.001 g cm^{-3}
		= 0.06243 lb ft^{-3}
		= 0.01002 imperial gallons
		= 0.008345 US gallons
Energy	1 kJ (N m)	= 1 kW s = 1 kg m^2 s^{-2}
		= 0.2778×10^{-3} kW h
		= 0.239 kcal
		= 0.9478 BTU
		= 737.6 ft lbs
Power	1 kW	= 1 kJ s^{-1} = 1 N m s^{-1} = 1 kg m^2 s^{-3}
		= 0.001 amp x volt
		= 1.340 HP
		= 860 kcal h^{-1}
		= 3412 BTU h^{-1}

	SI unit	Divide by this factor to convert to SI unit
Force	1 N	= 1 kg m s^{-2} = 0.1 x 10^6 dynes = 0.1020 kilopond = 0.2248 lb force
Volumetric flowrate	1 m^3 s^{-1}	= 3600 m^3 h^{-1} = 2119 ft^3 min^{-1}(cfm) = 13200 imperial gallons min^{-1} (gpm) = 543.6 x 10^3 barrels day^{-1}
Henry's gas solubility constant	1 bar m^3 kmol^{-1}	= 0.987 atm (mol l^{-1})$^{-1}$ = 750.1 mmHg (mol l^{-1})$^{-1}$
Heat-transfer coefficient	1 kW m^{-2} K^{-1}	= 860 kcal m^{-2} h^{-1} °C^{-1} = 176 BTU h^{-1} ft^{-2} °F^{-1} = 79.9 kcal h^{-1}ft^{-2}°C^{-1} = 0.02391 cal cm^{-2} s^{-1} °C^{-1}
Thermal conductivity	1 kW m^{-2} K^{-1}	= 2.391 cal cm^{-1} s^{-1} °C^{-1} = 557.7 BTU h^{-1} ft^{-2} (°F ft^{-1})
Heat capacity	1 kJ kg^{-1} K^{-1}	= 0.239 cal g^{-1} °C^{-1} = 0.239 BTU lb^{-1} °F^{-1}
Gas mass-transfer coefficient	1 kmol m^{-2} bar^{-1} s^{-1}	= 3647 kmol m^{-2} atm^{-1} h^{-1} = 0.1013 mol cm^{-2} atm^{-1} s^{-1} = 747.0 lb mol ft^{-2} atm^{-1} h^{-1}
Liquid-phase mass-transfer coefficient	1 m s^{-1}	
Viscosity (dynamic)	1 kg m^{-1} s^{-1} (N s m^{-2})	= 10 poise = 1000 centipoise = 10 g cm^{-1} s^{-1} = 0.672 lb ft^{-1} s^{-1}
Velocity	1 m s^{-1}	= 3.281 ft s^{-1}

Appendix 2

SOME PHYSICAL PROPERTIES OF COMMON SUBSTANCES

The tables in this Appendix are intended to give values of physical properties in SI units for those properties involved in the various correlations in the text.

This should enable typical values to be worked out from the correlations and exercises to be undertaken with the minimum of exasperation over data and units.

It is not intended to be an authoritative source of physical properties and no guarantee of its accuracy is given. The data were obtained from the EURECHA teaching Physical Property Data Bank CHEMCO.

TABLE A2.1

Basic data

Common liquids

Component	Mol. wt.	b.p. (K)	$(H_f)_g^\circ$ (kJ kmol^{-1})	Lat. ht. at b.p. (kJ kmol^{-1})	$(H_f)_L^\circ$ (kJ kmol^{-1})
Acetic acid	60.1	391.1	-438463.	24326.	-468182.
Acetone	58.1	329.7	-216845.	30257.	-248905.
Acrolein	56.1	325.8	- 74487.	28346.	-104576.
Benzene	78.1	353.3	82986.	30788.	49380.
Carbon tetrachloride	153.8	349.9	-100488.	29983.	-133147.
Chloroform	119.4	334.9	-100400.	29498.	-131729.
Cyclohexane	84.2	353.9	-123223.	29979.	-155151.
Dichloromethane	84.9	312.9	- 95464.	28003.	-124191.
Ethanol	46.1	351.4	-234589.	38751.	-276180.
Hexane	86.2	341.9	-167312.	29037.	-198608.
Methanol	32.0	337.8	-201311.	35255.	-239375.
Pyridine	79.1	388.4	140264	35137.	101280.
Trichloroethylene	131.4	360.4	- 5862.	31403.	- 40220.
Chlorobenzene	112.6	405.4	51769.	35660.	10650.
Dichlorobenzene	147.0	447.1	22387.	38900.	- 24814.
Water	18.0	373.2	-241992.	40685.	-285812.

Common gases

Component	Mol. wt.	b.p. (K)	$(H_f)_g^\circ$ (kJ kmol^{-1})	Lat. ht. at b.p. (kJ kmol^{-1})	$(H_f)_L^\circ$ (kJ kmol^{-1})
Ammonia	17.0	239.8	- 45890.	23334.	- 66647.
Benzene	78.1	353.3	82986.	30788.	49380.
Butane	58.1	272.7	-124823.	22410.	-146084.
Carbon dioxide	44.0	185.1	-393796.	25262.	-414362.
Chloromethane	50.5	248.9	- 86378.	21551.	-105886.
Ethane	30.1	184.5	- 84728.	14714.	- 95391.
Ethylene	28.1	169.5	52321.	13552.	45347.
Ethylene oxide	44.1	283.7	- 52672.	25545.	- 77647.
Hydrogen	2.0	20.4	0.	904.	-
Hydrogen chloride	36.5	188.1	- 92378.	16120.	-104956.
Hydrogen sulphide	34.1	211.4	- 20160.	18687.	- 35011.
Nitrogen	28.0	77.4	0.	5577.	10325.
Oxygen	32.0	90.2	0.	6779.	- 388.
Propene	42.1	225.5	20428.	18444.	5056.
Sulphur dioxide	64.1	263.1	-297055.	24934.	-320315.

342

TABLE A2.2

Some liquid properties at 293 K

Component	Heat capacity (kJ kmol^{-1} K^{-1})	Viscosity (N s m^{-2})	Molar volume (m^3 kmol^{-1})	Surface tension (N m^{-1})	Thermal conductivity (kW m^{-1} K^{-1})
Acetic acid	.121 E+03	.122 E-02	.572 E-01	.421 E-01	-
Acetone	.131 E+03	.317 E-03	.728 E-01	.251 E-01	-
Acrolein	.127 E+03	.344 E-03	.669 E-01	-	-
Benzene	.135 E+03	.639 E-03	.891 E-01	.280 E-01	-
Carbon tetrachloride	.131 E+03	.968 E-03	.950 E-01	.258 E-01	.104 E-03
Chloroform	.113 E+03	.556 E-03	.836 E-01	.278 E-01	.118 E-03
Cyclohexane	.155 E+03	.962 E-03	.109 E+00	.235 E-01	-
Dichloromethane	.999 E+02	.426 E-03	.681 E-01	.212 E-01	.152 E-03
Ethanol	.113 E+03	.115 E-02	.552 E-01	.415 E-01	.172 E-03
Hexane	.196 E+03	.306 E-03	.131 E+00	.183 E-01	-
Methanol	.119 E+03	.581 E-03	.393 E-01	.440 E-01	.212 E-03
Pyridine	.135 E+03	.976 E-03	.831 E-01	.318 E-01	-
Trichloroethylene	.125 E+03	.570 E-03	.875 E-01	.120 E+00	.118 E-03
Chlorobenzene	.141 E+03	.809 E-03	.104 E+00	.332 E-01	.149 E-03
Dichlorobenzene	.141 E+03	-	.189 E+00	-	-
Water	.746 E+02	.835 E-03	.180 E-01	.710 E-01	.594 E-03

TABLE A2.3

Some vapour properties at 473 K

Component	Heat capacity (kJ kmol^{-1} K^{-1})	Viscosity (N s m^{-2})	Thermal conductivity (kW m^{-1} K^{-1})
Acrolein	.892 E+02	.139 E-04	-
Ammonia	.410 E+02	.164 E-04	-
Benzene	.131 E+03	.122 E-04	-
Butane	.143 E+03	.120 E-04	.368 E-04
Carbon dioxide	.440 E+02	.234 E-04	-
Carbon tetrachloride	.964 E+02	.165 E-04	.119 E-04
Chlorine	.359 E+02	.214 E-04	-
Chlorodifluoromethane	.715 E+02	.204 E-04	-
Chloroform	.788 E+02	.171 E-04	-
Chloromethane	.537 E+02	.167 E-04	.248 E-04
Dichloromethane	.658 E+02	.173 E-04	.156 E-04
Ethane	.754 E+02	.146 E-04	.473 E-04
Ethylene	.616 E+02	.160 E-04	.467 E-04
Ethylene oxide	.727 E+02	.165 E-04	-
Hydrogen	.292 E+02	.104 E-04	-
Hydrogen chloride	.292 E+02	.271 E-04	-
Hydrogen sulphide	.366 E+02	.188 E-04	-
Methanol	.653 E+02	.119 E-04	.313 E-04
Nitrogen	.293 E+02	.246 E-04	-
Oxygen	.307 E+02	.288 E-04	-
Propene	.919 E+02	.137 E-04	.389 E-04
Pyridine	.124 E+03	.121 E-04	-
Sulphur dioxide	.460 E+02	.208 E-04	-
Water	.350 E+02	.161 E-04	.315 E-04

Appendix 3

EXERCISES TO RUN IN PARALLEL TO A PRACTICAL REACTOR DESIGN COURSE

The following nine exercises were devised for 2 hour class sessions to run in parallel with the lectures.

The nine exercises are given in the following section, and the last part of this Appendix is devoted to the material that goes with the exercises, in particular the instructions and listings of the computer programs necessary.

Most of the exercises were carried out on an interactive computer. This demands a lot of preparatory work, but the increased interest level and experience gained by teaching with simulation models more than justifies the effort in their preparation.

A3.1   THE EXERCISES

Exercise 1

1. It is postulated that the following mechanism describes the chlorination of methylene chloride:

$$Cl_2 \xrightarrow{k_1} Cl\cdot + Cl\cdot$$

$$CH_2Cl_2 + Cl\cdot \xrightarrow{k_2} CHCl_2\cdot + HCl$$

$$CHCl_2\cdot + Cl_2 \longrightarrow CHCl_3 + Cl\cdot$$

$$CHCl_3 + Cl\cdot \longrightarrow CCl_3\cdot + HCl$$

$$CCl_3\cdot + Cl_2 \longrightarrow CCl_4 + Cl\cdot$$

$$CHCl_2\cdot + CHCl_2\cdot \rightarrow C_2H_2Cl_4$$

Derive the kinetic equations for an isothermal PFR. How do these equations simplify if it is assumed that the radical concentrations are very low ?

2. Propylene is partially oxidized to acrolein by oxygen over a solid catalyst. The kinetics are described by equation 1.39 of the text. What is the apparent fractional order when $K_{A_{O_2}}$ and $K_{A_{pr}}$ have values:

(a) 0.001 bar^{-1}
(b) 10 bar^{-1}
(c) 1 bar^{-1}

What would the rate equation be if acrolein competed with propylene for sites,

344

i.e. the equilibrium equation for propylene becomes:

$$(1 - \Theta_P - \Theta_{Ac}) \, K'_P \, P_P = K''_P \, \Theta_P$$

Exercise 2

Methyl chloride is produced by the reaction of methanol and HCl in a tubular reactor:

$$CH_3OH + HCl \xrightarrow[400°C]{cat.} CH_3Cl + H_2O$$

The heat of reaction is H_f kJ mol^{-1}, activation energy E kJ mol^{-1} and the reaction is simple second-order. The reaction rate is:

$$A_1 \, e^{-E_1/RT} \, C_1 \, C_2 \quad m^3 \, kmol^{-1} \, s^{-1}$$

Develop the equations which model the reactor assuming a constant OHTC between gas and tube wall. Neglect volume changes and pressure drop.

Draw a flowchart of the calculation for a computer program of the model.

For those possessing both enthusiasm and ability, program the model, using an integration subroutine available on your computer.

Exercise 3

You are given two sets of results from an experimental reactor which produces product C with side products B and D, and one set of repeat experiments and results from a multi-response non-linear regression program run to determine whether the mechanism is:

A → B → C → D
or
A → B
A → C
B → D
C → D

Interpret the results:
 - what is the correct mechanism ?
 - what are the parameter values ? how accurate are they ?
 - are both sets of experiments equally good for interpretation ? if not, why not ?

Sets of experimental results.

	End conc. of B c_B	End conc. of C c_C	Initial conc. of A c_{A_0}	Reaction time
Set 1				
1/1	.1870	0.00	4.00	5.0
1/2	.3059	.0524	4.00	10.0
1/3	.4922	.0576	4.00	20.0
1/4	.8460	.1936	4.00	40.0
1/5	1.104	.5619	4.00	80.0
1/6	.9898	1.095	4.00	160.0
1/7	.4144	.9210	4.00	320.0
1/8	.0454	.1528	4.00	640.0
Set 4				
4/1	.0850	.0021	1.00	5.0
4/2	.0983	.0020	1.00	10.0
4/3	.0842	.0382	1.00	15.0
4/4	.1256	.0708	1.00	20.0
4/5	.1787	.0529	1.00	30.0
4/6	.2465	.0136	1.00	40.0
4/7	.2665	.0797	1.00	50.0
4/8	.2913	.0845	1.00	60.0
Repeated experiments				
Set 8				
8/1	.2421	.3018	1.00	200.0
8/2	.2279	.2594	1.00	200.0
8/3	.2029	.2635	1.00	200.0
8/4	.1779	.2176	1.00	200.0
8/5	.2105	.3071	1.00	200.0
8/6	.1963	.2646	1.00	200.0
8/7	.2018	.2704	1.00	200.0
8/8	.2126	.2280	1.00	200.0

Regression output for experimental set 1.

Fitting Model 1

XTX Matrix

```
1   .342 +05
2   .169 +04   .115 +05
3   .454 +04   .149 +04   .158 +05
```

Determinant   5.8535 +12

XTX-inverse matrix

```
        1          2          3
1   .308 -04
2   .570 -05   .887 -04
3   .938 -05   .100 -04   .670 -04
```

Correlation matrix

```
        1        2        3
1   1.0000
2    .1091   1.0000
3    .2065    .1297   1.0000
```

Ind. Var(I)	Name	Coef. B(I)	S.E. Coef.	T-value (calc.)
1	K1	7.896 -03	1.20 -04	65.8
2	K2	1.276 -02	2.04 -04	62.7
3	K3	9.014 -03	1.77 -04	51.0

T-value (calc.) = B(I)/S.E. coef.

No. of observations	16
No. of coefficients	3
Residual degrees of freedom	13
Residual root mean square	2.16 -02
Residual mean square	4.67 -04
Residual sum of squares	6.07 -03

────────── Ordered by computer input ──────────

Ident.	Obs. Y	Fitted Y	Residual (Obs. Y-fitted Y)
1/1 C_B	1.870 -01	1.499 -01	3.702 -02
1/2 C_B	3.059 -01	2.848 -01	2.102 -02
1/3 C_B	4.922 -01	5.139 -01	-2.178 -02
1/4 C_B	8.460 -01	8.370 -01	8.912 -03
1/5 C_B	1.104 +00	1.112 +00	-8.667 -03
1/6 C_B	9.898 -01	9.924 -01	-2.660 -03
1/7 C_B	4.144 -01	4.093 -01	5.074 -03
1/8 C_B	4.540 -02	3.960 -02	5.798 -03
1/1 C_C	0.000	4.795 -03	-4.795 -03
1/2 C_C	5.240 -02	1.825 -02	3.414 -02
1/3 C_C	5.760 -02	6.617 -02	-8.570 -03
1/4 C_C	1.936 -01	2.173 -01	-2.371 -02
1/5 C_C	5.619 -01	5.867 -01	-2.481 -02
1/6 C_C	1.095 +00	1.074 +00	2.070 -02
1/7 C_C	9.210 -01	9.180 -01	2.962 -03
1/8 C_C	1.528 -01	1.790 -01	2.626 -02

Regression output for experimental set 1.

Fitting Model 2

XTX Matrix

```
1    .115 +06
2   -.159 +05    .180 +05
3   -.121 +05    .000        .104 +05
```

Determinant   1.6244 +13

XTX-inverse matrix

```
        1          2          3
1    .115 -04
2    .102 -04    .645 -04
3    .134 -04    .118 -04    .112 -03
```

Correlation matrix

```
        1          2          3
1   1.0000
2    .3741    1.0000
3    .3730    .1395    1.0000
```

Ind. Var(I)	Name	Coef. B(I)	S.E. Coef.	T-value (calc.)
1	K1	4.639 -03	7.60 -04	6.1
2	K2	8.585 -03	1.80 -03	4.8
3	K3	1.042 -02	2.37 -03	4.4

T-value (calc.) = B(I)/S.E. coef.

No. of observations	16
No. of coefficients	3
Residual degrees of freedom	13
Residual root mean square	2.23 -01
Residual mean square	5.00 -02
Residual sum of squares	6.50 -01

───────Ordered by computer input───────

Ident.	Obs. Y	Fitted Y	Residual (Obs. Y-fitted Y)
1/1	1.870 -01	8.936 -02	9.763 -02
1/2	3.059 -01	1.721 -01	1.337 -01
1/3	4.922 -01	3.194 -01	1.727 -01
1/4 C_B	8.460 -01	5.504 -01	2.955 -01
1/5	1.104 +00	8.198 -01	2.842 -01
1/6	9.898 -01	9.215 -01	6.824 -02
1/7	4.144 -01	6.123 -01	-1.979 -01
1/8	4.540 -02	1.604 -01	-1.150 -01
1/1	0.000	8.977 -02	-8.977 -02
1/2	5.240 -02	1.737 -01	-1.213 -01
1/3	5.760 -02	3.252 -01	-2.676 -01
1/4 C_C	1.936 -01	5.704 -01	-3.768 -01
1/5	5.619 -01	8.784 -01	-3.165 -01
1/6	1.095 +00	1.048 +00	4.717 -02
1/7	9.210 -01	7.641 -01	1.568 -01
1/8	1.528 -01	2.221 -01	-6.932 -02

Regression output for experimental set 4.

Fitting Model 1

XTX Matrix

```
1    .245 +04
2   -.423 +03    .207 +03
3   -.231 +02   -.157 +02    .389 +01
```

Determinant   2.7605 +05

XTX-inverse matrix
```
        1           2           3
1    .208 -02
2    .742 -02    .334 -01
3    .414 -01    .175 +00    .118 +01
```

Correlation matrix
```
       1         2         3
1   1.0000
2    .8902    1.0000
3    .8344     .8806    1.0000
```

Ind. Var(I)	Name	Coef. B(I)	S.E. Coef.	T-value (calc.)
1	K1	8.920 -03	1.24 -03	7.2
2	K2	1.224 -02	4.95 -03	2.5
3	K3	2.073 -02	2.94 -02	.7

T-value (calc.) = B(I)/S.E. coef.

No. of observations	16
No. of coefficients	3
Residual degrees of freedom	13
Residual root mean square	2.70 -02
Residual mean square	7.32 -04
Residual sum of squares	9.52 -03

————————Ordered by computer input————————

Ident.	Obs. Y	Fitted Y	Residual (Obs.Y-fitted Y)
4/1	8.500 -02	4.230 -02	4.269 -02
4/2	9.830 -02	8.024 -02	1.805 -02
4/3	8.420 -02	1.141 -01	-2.997 -02
4/4 C_B	1.256 -01	1.443 -01	-1.879 -02
4/5	1.787 -01	1.948 -01	-1.618 -02
4/6	2.465 -01	2.338 -01	1.266 -02
4/7	2.665 -01	2.630 -01	3.455 -03
4/8	2.913 -01	2.840 -01	7.201 -03
4/1	2.100 -03	1.273 -03	8.265 -04
4/2	2.000 -03	4.751 -03	-2.751 -03
4/3	3.820 -02	9.973 -03	2.822 -02
4/4 C_C	7.080 -02	1.654 -02	5.425 -02
4/5	5.290 -02	3.242 -02	2.047 -02
4/6	1.360 -02	5.023 -02	-3.663 -02
4/7	7.970 -02	6.844 -02	1.125 -02
4/8	8.450 -02	8.599 -02	-1.492 -03

Regression output for experimental set 4.

Fitting Model 2

XTX Matrix

```
1   -.369 +04
2   -.163 +02    .127 +01
3   -.686 +03    .000         .158 +03
```

Determinant   1.0221 +05

XTX-inverse matrix
```
       1           2           3
1    .196 -02
2    .252 -01    .111 +01
3    .850 -02    .109 +00    .431 -01
```

Correlation matrix
```
      1         2         3
1   1.0000
2    .5396   1.0000
3    .9239    .4986    1.0000
```

Ind. Var(I)	Name	Coef. B(I)	S.E. Coef.	T-value (calc.)
1	K1	7.534 -03	1.27 -03	5.9
2	K2	1.069 -01	3.02 -02	3.5
3	K3	6.327 -03	5.94 -03	1.1

T-value (calc.) = B(I)/S.E. coef.

No. of observations	16
No. of coefficients	3
Residual degrees of freedom	13
Residual root mean square	2.86 -02
Residual mean square	8.18 -04
Residual sum of squares	1.06 -02

────────Ordered by computer input────────

Ident.	Obs. Y	Fitted Y	Residual (Obs. Y-fitted Y)
4/1 ⎫	8.500 -02	3.638 -02	4.861 -02
4/2 ⎪	9.830 -02	7.029 -02	2.800 -02
4/3 ⎪	8.420 -02	1.018 -01	-1.765 -02
4/4 C_B 1.256 -01		1.311 -01	-5.581 -03
4/5 ⎪	1.787 -01	1.836 -01	-4.901 -03
4/6 ⎪	2.465 -01	2.284 -01	1.807 -02
4/7 ⎪	2.665 -01	2.664 -01	7.876 -05
4/8 ⎭	2.913 -01	2.983 -01	-7.017 -03
4/1 ⎫	2.100 -03	2.858 -02	-2.648 -02
4/2 ⎪	2.000 -03	4.427 -02	-4.227 -02
4/3 ⎪	3.820 -02	5.244 -02	-1.424 -02
4/4 C_C 7.080 -02		5.624 -02	1.455 -02
4/5 ⎪	5.290 -02	5.737 -02	-4.475 -03
4/6 ⎪	1.360 -02	5.499 -02	-4.139 -02
4/7 ⎪	7.970 -02	5.162 -02	2.807 -02
4/8 ⎭	8.450 -02	4.808 -02	3.641 -02

Exercise 4

You are faced with a complex reaction producing product C from reactant A, with a number of side reactions being possible. The object is to optimize the yield of C, and important variables might be:

1. initial concentration of A (range 0 to 4 mol 1^{-1});
2. catalyst concentration (range 0.0 to 0.01 mol 1^{-1});
3. space time (range 0-800 min.).

Plan a set of two-level experiments to investigate the importance of these effects, carry out these experiments, and analyze the results to determine the importance of the individual parameter values using a linear model.

The existing laboratory work has been carried out at

$A_0 = 2.0$ mol 1^{-1}

$t = 50$ s

$C_{cat} = 0.002$ mol 1^{-1}

Program EXPT4 is a computer simulation of the reaction mechanism which includes randomly generated experimental error. It can therefore be used to generate results for this exercise.

Input data required are:

group no., expt. no., C_{A_0}, C_{cat}, t (2I5 3F10.0).

The group and experiment numbers seed the random number generator. No experiments should have identical group and experimental numbers.

Exercise 5

We now require to find the optimum conversion to C for the reaction introduced in Exercise 4. We have no model and no mechanism, but must satisfy the demands of a research manager who wants the information tomorrow.

Try Box-Wilson.

Exercise 6

You are supplied with a simulation model of a methyl chloride reactor, detailed below.

It is necessary to determine optimum reactor conditions so that the process can be defined and the material balance completed.

The product level required is of the order of 10000 tons yr^{-1}.

Determine the optimum feed MeOH/HCl ratio, reactor temperature (limited from 270 to 523 K), pressure (0.5 to 5 bar), space time, and tube length by the trial and error method.

The objective function is cost per ton methyl chloride, defined as:

(Annual raw material costs + $\dfrac{\text{reactor capital cost}}{4}$) / (Annual production of)
methyl chloride

Cost data:

MeOH $100 per ton

HCl $ 25 per ton

Reactor costs = $(100a + 15.0n), where a is the tube wall area (m^2) and n is the number of tubes.

Description of the model for an isothermal methyl chloride reactor

A PF model has been developed for the reaction

$$CH_3OH + HCl \xrightarrow[\substack{150°C \\ 1 \text{ bar}}]{\text{cat}} CH_3Cl + H_2O$$

with competing side reactions

$$CH_3OH + CH_3OH \longrightarrow CH_3-O-CH_3 + H_2O$$

and a slower reaction

$$CH_3-O-CH_3 + 2HCl \rightarrow 2CH_3Cl + 2H_2O$$

It is assumed that the reaction proceeds isothermally. The model of the reactor has been developed as a subroutine for use with MINUN, a small flowsheet program.

User instructions

Unit number	1
Number of components	5
Component order	(1) Methanol
	(2) HCl
	(3) CH_3Cl
	(4) CH_3-O-CH_3
	(5) H_2O
No. of inlet streams	1 (mixed reactor feed)
Number of parameters	
o Integer	0
o Real	4

	Laboratory conditions
1. Reactor space time	5 s
2. Temperature of reactor	423 K
3. Inlet pressure of reactor	1 bar
4. Reactor tube length (2.5 cm diameter tubes are assumed)	0.33 m

352

Results summary sheet for Exercise 6.

Experiment number	Independent variables (input)					from computer output				scaled for 10,000 ton yr^{-1}			
	Feed of methanol	Feed of HCl	Space time	Temperature	Pressure	Length	Number of tubes	MeCl$_3$ output	Reactor yield	Reactor cost	Methanol cost	HCl cost	Cost per ton

User instructions for MINUN

Description. MINUM is a small flowsheet program which communicates between the user and the unit equipment model (reactor). It requests component details, input flow details and model parameter details that define the operation of the equipment model. It passes this data over to the equipment model in the standard format and receives the output stream information from the model after the simulation.

A model can be run in cascade by requesting cascade operation. In this case the output stream from the last model is taken as the first input stream to the next unit in the cascade.

Data requirement.
1. Number of components in the system.
2. Component names, in the order required by the subroutine.
3. Number of inlet streams to the unit.
4. Stream values of each stream; kmol h^{-1} each component, temperature (K), pressure (bar), phase (0.0 liquid, 1.0 gas).
5. Number of integer parameters required by model.
6. The value of these parameters (omit if there are no integer parameters).
7. Number of real parameters required by the model.
8 The value of these parameters (omit if there are no real parameters).
9. Unit number (defines which model will be used).
The simulation is carried out and the results printed, then the following questions are asked:

"Do you want to continue (Yes/No)?"

If the answer is "Yes", then:

"Do you want to run unit as a cascade?"

"Yes" puts this output stream as input stream to the next unit and returns to point 3 above. If the answer is not "Yes", then the program returns to point 1 above.

Exercise 7

The liquid phase chlorination of benzene is to be carried out by a cascade of CSTR reactors. The reaction is consecutive:

$$C_6H_6 \xrightarrow{\;+Cl_2\;} C_6H_5Cl \xrightarrow{\;+Cl_2\;} C_6H_4Cl_2 \xrightarrow{\;+Cl_2\;} C_6H_3Cl_3$$

Use the supplied model to locate a suitable cascade arrangement to maximize the production of monochlorobenzene.

Define how many reactors are required in the cascade, determine their volumes, the Cl_2 flow to each reactor, and estimate their heat-exchange areas.

The reaction is very exothermic. How could this affect your decisions ($\Delta H = -80,000$ kJ kmol^{-1}, $U \simeq 1$ kW m^{-2} K^{-1}, assume ΔT between jacket and reactor is 10°C).

Monochlorobenzene output required is 5000 tons yr^{-1}.

Suggested approach

(1) Make a rough mass balance by hand.

(2) Use large residence times (high chlorine conversions) and investigate selectivity as a function of number of reactors and Cl_2 distribution.

(3) Having fixed the number of reactors, investigate the effect of shorter residence times to size each reactor.

Write a one page summary giving your conclusions.

Description of the CSTR benzene chlorination reactor model

This CSTR model has been written to represent the liquid phase reactions:

$$benzene \xrightarrow[]{+Cl_2} monochlorobenzene \xrightarrow[]{+Cl_2} dichlorobenzene \xrightarrow[]{+Cl_2} trichlorobenzene$$
$$+ HCl \qquad + HCl \qquad + HCl$$

Cl_2 is added as a gas and it is assumed that the process is kinetic-, not mass-transfer-controlled.

The reaction is catalyzed by equimolar proportions of $FeCl_3 + H_2O$, both at 0.001 mf concentration.

The model has been written as a subroutine of MINUN.

User instructions

Unit number	2
Number of components	8
Component order	(1) C_6H_6
	(2) C_6H_5Cl
	(3) $C_6H_4Cl_2$
	(4) Cl_2
	(5) HCl
	(6) $FeCl_3$
	(7) H_2O
	(8) $C_6H_3Cl_3$
No. of inlet streams	2
	(1) the liquid stream, excluding $Cl_2 + HCl$
	(2) the gas stream (Cl_2)
No. of output streams	2
	(1) the liquid stream (used in cascade)
	(2) the gas stream ($HCl + Cl_2$)

Number of parameters
o Integer   0
o Real   4

	Typical values
	Typical values
1. Reactor temperature	318 K
2. Operating pressure	1.0 bar
3. Residence time	0.2 h
4. Cl_2 solubility coefficient	0.87 bar $kmol^{-1}$

The model can be run as a cascade with MINUN.

Exercise 8

You are supplied with a model of a methyl chloride gas phase reactor. Use this model to investigate the effect of bath temperature, tube length, and catalyst time on reactor performance.

Design a stable reactor for 10,000 tons yr^{-1} methyl chloride.

The suggested approach is to carry out one run at the standard conditions, and then one successful run at higher reactor temperature, one with smaller tube length, and one with increased contact time.

Consider these results carefully and then try a number of runs moving the three variables simultaneously to obtain a higher conversion.

Write a one page summary giving your conclusions.

Description of model for a methyl chloride reactor model with temperature profile

A PF model has been developed for the reaction

$$CH_3OH + HCl \xrightarrow{cat} CH_3Cl + H_2O \qquad E_1 = 84,000 \text{ kJ } kmol^{-1}$$

with competing side reaction

$$CH_3OH + CH_3OH \longrightarrow CH_3-O-CH_3 + H_2O \qquad E_2 = 125,000 \text{ kJ } kmol^{-1}$$

The model of the reactor has been developed for use with MINUN.

User instructions
Unit number   3
Number of components   5

Component order   (1) Methanol
                          (2) HCl
                          (3) CH_3Cl
                          (4) CH_3-O-CH_3
                          (5) H_2O

No. of inlet streams 1 (mixed reactor feed)

Number of parameters

o Integer   0

o Real   5

	Typical values
1. Space time	5 s
2. Coolant temperature	560 K
3. Reactor inlet pressure	1.0 bar
4. Tube length	3 m
5. Tube diameter	0.025 m

Exercise 9

A reaction between acrolein and ammonia (feed ratio 2:1) to form pyridine is to be carried out at 1.5 atm pressure and 400°C in a fluidized bed because 5% of the acrolein forms carbon on the catalyst, which can be burned off in a fluidized bed regenerator.

The reaction yield based on acrolein to pyridine is 70% of the theoretical with a 5 second space time and fresh catalyst, dropping 2% for every 1% w/w carbon content of the catalyst.

The reactor kinetics at 500°C have been found to be:

$$\frac{dp_{O_2}}{dt} = 88.2 \ C_C \ p_{O_2} \ e^{-5000/T} \quad \text{bar s}^{-1}$$

i.e. first-order with respect to carbon level (C_C) (expressed as kmol m^{-3}). A temperature limit of 500°C is set by catalyst sintering.

Size the reactor and regenerator, determine a suitable % carbon on the catalyst in the reactor and determine the catalyst circulation rate and air flow to the regenerator. The output should be 5000 tons yr^{-1} pyridine. Will your system be in heat balance ?

Data c_p air = 29.4 kJ kmol^{-1} K^{-1}
         c_p catalyst = 15.2 + 0.100 T kJ kmol^{-1} K^{-1}
         catalyst is SiO_2 (M wt. 60)
         ΔH C-burning = -393,000 kJ kmol^{-1}
         cat. bulk density = 1100 kg m^{-3}

A3.2   NOTES ON THE SUPPORT MATERIAL FOR THE EXERCISES

Exercise 1:   requires no support material.

Exercise 2:   can be run without support material, but to supply user instructions
for an integration routine available to the students would give them
experience in using library subroutines and also enable the more keen
ones to complete the model and compile the program.

Exercise 3:   Tables enclosed with Exercise 3 contain two sets of experimental data
and a set of repeated experiments from the simulated batch reactor
(including simulated experimental error). Further tables give the outputs of a regression from the two experimental sets, using a modified
form of NLWOOD program (Daniels and Wood, 1971. Fitting Equations to
Data. Wiley).

This exercise can also be used to introduce the students to using
non-linear regression packages by giving them the instructions for an
in-house package and asking them to prepare the subroutine and data
to carry out the regression. Ideally, the instructor should hand out
outputs from the in-house program so that the students are not confused
by seeing details of programs they will never use.

Exercise 4:   This exercise requires the running of reactor model EXPT4, preferably,
though not essentially, on-line. Instructions on how to start the
program must be given to the students; the prompts from the program
describe how the input data must be presented. The reactor model is
given in Listing A3.1. As it is intended to be a "Black Box" exercise,
there is no need to understand what the model is doing, though this can
be deciphered by an enthusiast from the coding.

Exercise 5:   This is the repeated use of the model given in Exercise 4.

Exercise 6:   An interactive, or very quick response, batch computer system is essential for this exercise. The student must have instructions on how to start
the program, and a description of the reactor model. The data input is
defined by the prompts given by the program.

The main program that handles the reactor models for Exercises 6, 7,
and 8 is called MINUN (MINI-UNICORN). This is a very small "flowsheet
program" which enables standard flowsheet subroutine models to be run
as single or cascade units. This has the double advantage that to the
students the input format for the three exercises is identical (which
reduces the errors made), and secondly, the models can be used directly
in a large flowsheet program as part of a complete process if the
flowsheet program can handle the EURECHA standard interface, as can
for instance the full UNICORN flowsheet program (D.W.T. Rippin,
Technische-Chemisches Labor, E.T.H., Zürich).

358

Hence, for this exercise, the student requires:
- instructions for using MINUN
- instructions for using the isothermal methyl chloride reactor
  subroutine
- a blank form in which to summarize his results
  (the exercise can be very confusing to the student and such a form
  helps him to organize his thoughts).

To prepare the exercise, MINUN, UNIT1, FT1, OUTP1, RGKS must be loaded
on the computer. These are given as Listings A3.2, A3.3, and A3.4.

Exercise 7: This, like Exercise 6, is a reactor simulation exercise using MINUN
as the communication between the student and the reactor model. The
student therefore again needs the instructions on starting the program,
a description of the benzene chlorination reactor model, and MINUN
instructions. The program MINUN and subroutines UNIT2, FCT, and SECT
must be loaded. These are given as Listings A3.2 and A3.5.

Exercise 8: This is very similar to Exercise 6, except that the reactor model
includes a heat balance and this increases the difficulty in looking for
good operating conditions. It again works through MINUN and instructions
for the methyl chloride reaction with temperature profile must be
supplied. The program MINUN (Listing A3.2), UNIT3, FT3, and OUTP3
(Listing A3.6) and RGKS (Listing A3.3) must be loaded.

Exercise 9: Needs no support material.

359

Listing A3.1   EXPT4   (Exercises 3 and 4).

```
        PROGRAM EXPT4 (INPUT,OUTPUT,TAPE1=INPUT,TAPE2=OUTPUT)
        DIMENSION C(4)
C
C   SET REACTION CONSTANTS
        RK1=.008
        RK2=.013
        RK3=.009
C
5       WRITE(2,1000)
1000    FORMAT(/88 H GIVE YOUR GROUP NO,CONC OF A, CONC OF CAT, REACTION
      *TIME,(2I5,3F10.0)
        READ(1,101)  IGP,IEXP,A,CAT,T
101     FORMAT(  2I5,3F10.0)
        WRITE(2,62 )  IGP,IEXP,A,CAT,T
62      FORMAT(11H GROUP NO,         ,I3, 10H EXPT NO       ,I3/11H A,CAT,T,
      *             ,2F10.4,F10.0)
        IF(CAT .GT. 0.01) GO TO 110
        IF(T    .GT.1000.) GO TO 110
        IF(A    .GT.4.00 ) GO TO 110
        IF(IGP .LE. 0 .OR. IEXP .LE. 0) GO TO 110
C
C   FORM SEED FOR RANF AND INITIALIZE
C    RANSET IS SYSTEM SUBROUTINE  TO INITIALISE RANDOM NO GENERATOR.
C    RANF DELIVERS RANDOM NO.
        IX=2*IGP*100+IEXP+1
        X=RANSET(FLOAT(IX))
C  SET STD DEV
        SD=0.01
C
13      CALL RACT2(A,T ,RK1,RK2,RK3,C)
C
12      DO 60 N=1,4
        X=-6.0
        DO 1 I=1,12
1       X=X+RANF(0.)
        V=X*SD + C(N)
        IF(V)52,53,53
52       V=0.
53      CONTINUE
60      C(N)=V
        CAT=CAT/0.01
        X=CAT-CAT*CAT
        C(3)=C(3)*X*4.0
        WRITE(2,63)  C(3)
63      FORMAT(/25H RESULTING CONC OF C =    ,F10.4
65      WRITE(2,120)
120     FORMAT(40H DO YOU WISH TO CONTINUE (YES,NO)
        READ(1,130)YES
130     FORMAT(1A1)
        IF(YES .NE. 1HY) STOP
        GO TO 5
110     WRITE(2,1010)
1010    FORMAT (30H INPUT DATA OUT OF RANGE
        GO TO 65
        END
        SUBROUTINE RACT2(A,T,X,Y,Z,C)
```

```
C MODEL FOR 3 CONSECUTIVE REACTIONS,
C    EXPT4 READS ONLY 1 RESPONSE
      DIMENSION C(4)
      P=EXP(-X*T)
      Q=EXP(-Y*T)
      R=EXP(-Z*T)
      E=X-Y
      F=X-Z
      G=Y-Z
C
      C(1)=A*P
      C(2)=A*X*(Q-P)/E
      C(3)=A*X*Y*(P/E/F-Q/E/G+R/F/G)
      C(4)=A-A*(Y*Z*P/E/F-X*Z*Q/E/G+X*Y*R/F/G)
      RETURN
      END
```

Listing A3.2   MINUN   (Exercises 6, 7, and 8)


```
      PROGRAM MINUN(INPUT,OUTPUT,TAPE1=INPUT,TAPE2=OUTPUT
     *,DBIN,TAPE14=DBIN)
C  SMALL FLOWSHEET PROGRAM FOR ATTACHING INDIVIDUAL MODELS OR CASCADES
C    DESIGNED WITH THE EURECHA STANDARD INTERFACE
C
C   COMPATABLE WITH MODELS REQUIRINC CHEMCO DATA BANK
C    COMPATABLE WITH 40 EURECHA FLOWSHEET SUBROUTINE MODELS
      COMMON/BANK1/ KOUT,KIN,NASTRC,NTOT,NCOAC,KSTAN,NAME(3,10)
      DIMENSION STRIN1(25),STRIN2(25),STRIN3(25),STOUT1(25),STOUT2(25),
     1STOUT3(25),EN (30), IEN(10  ), NORD(10)
      DIMENSION S(5)
      DIMENSION STRIN(25,3)
      DIMENSION NAMEC(3,10)                  ,STOUT(25,3)
      REAL NAMEC
      EQUIVALENCE (STRIN,STRIN1),(STRIN(1,2),STRIN2),(STRIN(1,3),STRIN3)
      EQUIVALENCE (STOUT,STOUT1),(STOUT(1,2),STOUT2),(STOUT(1,3),STOUT3)
      COMMON/FLAGS/MODE,JFORM,LABEL,IR,IW
      DATA NAMEC/30*4H
      IR=1
      IW=2
      KOUT=IW
      KIN=IR
      NASTRO=14
5     DO 1 I=1,25
      STRIN1(I)=0.0
      STRIN2(I)=0.0
      STRIN3(I)=0.0
      STOUT1(I)=0.0
      STOUT2(I)=0.0
      STOUT3(I)=0.0
1     CONTINUE
      NTOT=0
      NCOAC=0
      DO 2 I=1,30
2     EN(I)=0.0
      DO 3 I=1,10
      IEN(I)=0
3     CONTINUE
      JFORM=3
      JJ=1
      IJS=1
      WRITE(IW,101)
      READ(IR,99          ) NCOMP
      NCP=NCOMP+1
      NT=NCOMP+2
      LS=NCOMP+5
      WRITE (IW,125)
      DO 15 I=1,NCOMP
15    READ (IR,104) (NAMEC(J,I),J=1,3),NORD(I)
      IF(NORD(1) .NE. 0) WRITE(IW,140) (NORD(I),I=1,NCOMP)
      NAMEC(1,NCP)=4HTEMP
      NAMEC(1,NT)=4HPRES
      WRITE(IW,103    )
16    CONTINUE
      READ(IR,99          ) NIN
      IF (NIN.LT.1 .OR. NIN.GT.3 ) WRITE (IW,132) NIN
      IF (NIN.LT.1 .OR. NIN.GT.3) GO TO 16
```

```
18      CONTINUE
        IF (IJS.GT.NIN) GO TO 30
        DO 20 IJ=IJS,NIN
        WRITE(IW,106) IJ
        READ(IR,100          ) (STRIN(I,IJ),I=1,NT),STRIN(NT+2,IJ)
20      CONTINUE
21      WRITE(IW,113)
        READ(IR,99           ) LIEN
        IF(LIEN .EQ. 0) GO TO 25
        IF (LIEN.LT.0 .OR. LIEN.GT.10) WRITE (IW,133) LIEN
        IF (LIEN.LT.0 .OR. LIEN.GT.10) GO TO 21
        WRITE(IW,111)
        READ(IR,99           ) (IEN(I),I=1,LIEN)
25      CONTINUE
        IF (LIEN.EQ.0) LIEN=1
26      CONTINUE
        WRITE(IW,110)
        READ(IR,99           ) L1
        IF(L1 .EQ. 0) GO TO 30
        IF (L1.LT.0 .OR. L1.GT.30) WRITE (IW,134) L1
        IF (L1.LT.0 .OR. L1.GT.30) GO TO 26
        WRITE(IW,111)
        READ(IR,100          ) (EN(I),I=1,L1)
30      CONTINUE
        IF (L1.EQ.0) L1=1
            WRITE (IW,126) (I,I=1,3)
            DO 32 I=1,NCOMP
   32       WRITE (IW,127) (NAMEC(IJ,I),IJ=1,3),(STRIN(I,IJ),IJ=1,NIN)
        DO 34 IJ=1,NIN
        S(IJ)=.0
        DO 34 I=1,NCOMP
34      S(IJ)=S(IJ)+STRIN(I,IJ)
        DO 35 IJ=1,NIN
35      STRIN(NCOMP+3,IJ)=S(IJ)
            WRITE (IW,131) (S(I),I=1,NIN)
            DO 36 I=NCP,NT
   36       WRITE (IW,127) (NAMEC(IJ,I),IJ=1,3),(STRIN(I,IJ),IJ=1,NIN)
            WRITE (IW,128) (EN(I),I=1,L1)
            WRITE (IW,129) (IEN(I),I=1,LIEN)
        WRITE (IW,122)
        MODE=-1
        LABEL=1
        READ (IR,99          ) IU
40      MODE=MODE+2
        IF(IU .LT. 1 .OR . IU .GT.20) GO TO 38
        IF (IU.EQ. 1)CALL UNIT01
       1(STRIN1,STRIN2,STRIN3,STOUT1,STOUT2,STOUT3,NCOMP,LS,
       2EN ,L1,IEN,LIEN,JJ,NORD)
        IF (IU.EQ. 2)CALL UNIT02
       1(STRIN1,STRIN2,STRIN3,STOUT1,STOUT2,STOUT3,NCOMP,LS,
       2EN ,L1,IEN,LIEN,JJ,NORD)
        IF (IU.EQ. 3)CALL UNIT03
       1(STRIN1,STRIN2,STRIN3,STOUT1,STOUT2,STOUT3,NCOMP,LS,
       2EN ,L1,IEN,LIEN,JJ,NORD)
        IF (IU.EQ. 4)CALL UNIT04
       1(STRIN1,STRIN2,STRIN3,STOUT1,STOUT2,STOUT3,NCOMP,LS,
       2EN ,L1,IEN,LIEN,JJ,NORD)
        IF (IU.EQ. 5)CALL UNIT05
       1(STRIN1,STRIN2,STRIN3,STOUT1,STOUT2,STOUT3,NCOMP,LS,
       2EN ,L1,IEN,LIEN,JJ,NORD)
```

```
      IF (IU.EQ. 6)CALL UNITC6
     1(STRIN1,STRIN2,STRIN3,STOUT1,STOUT2,STOUT3,NCOMP,LS,
     2EN  ,L1,IEN,LIEN,JJ,NORD)
      IF (IU.EQ. 7)CALL UNITC7
     1(STRIN1,STRIN2,STRIN3,STOUT1,STOUT2,STOUT3,NCOMP,LS,
     2EN  ,L1,IEN,LIEN,JJ,NORD)
      IF (IU.EQ. 8)CALL UNITC8
     1(STRIN1,STRIN2,STRIN3,STOUT1,STOUT2,STOUT3,NCOMP,LS,
     2EN  ,L1,IEN,LIEN,JJ,NORD)
      IF (IU.EQ. 9)CALL UNITC9
     1(STRIN1,STRIN2,STRIN3,STOUT1,STOUT2,STOUT3,NCOMP,LS,
     2EN  ,L1,IEN,LIEN,JJ,NORD)
      IF (IU.EQ.10)CALL UNIT10
     1(STRIN1,STRIN2,STRIN3,STOUT1,STOUT2,STOUT3,NCOMP,LS,
     2EN  ,L1,IEN,LIEN,JJ,NORD)
      IF (IU.EQ.11)CALL UNIT11
     1(STRIN1,STRIN2,STRIN3,STOUT1,STOUT2,STOUT3,NCOMP,LS,
     2EN  ,L1,IEN,LIEN,JJ,NORD)
      IF (IU.EQ.12)CALL UNIT12
     1(STRIN1,STRIN2,STRIN3,STOUT1,STOUT2,STOUT3,NCOMP,LS,
     2EN  ,L1,IEN,LIEN,JJ,NORD)
      IF (IU.EQ.13)CALL UNIT13
     1(STRIN1,STRIN2,STRIN3,STOUT1,STOUT2,STOUT3,NCOMP,LS,
     2EN  ,L1,IEN,LIEN,JJ,NORD)
      IF (IU.EQ.14)CALL UNIT14
     1(STRIN1,STRIN2,STRIN3,STOUT1,STOUT2,STOUT3,NCOMP,LS,
     2EN  ,L1,IEN,LIEN,JJ,NORD)
      IF (IU.EQ.15)CALL UNIT15
     1(STRIN1,STRIN2,STRIN3,STOUT1,STOUT2,STOUT3,NCOMP,LS,
     2EN  ,L1,IEN,LIEN,JJ,NORD)
      IF (IU.EQ.16)CALL UNIT16
     1(STRIN1,STRIN2,STRIN3,STOUT1,STOUT2,STOUT3,NCOMP,LS,
     2EN  ,L1,IEN,LIEN,JJ,NORD)
      IF (IU.EQ.17)CALL UNIT17
     1(STRIN1,STRIN2,STRIN3,STOUT1,STOUT2,STOUT3,NCOMP,LS,
     2EN  ,L1,IEN,LIEN,JJ,NORD)
      IF (IU.EQ.18)CALL UNIT18
     1(STRIN1,STRIN2,STRIN3,STOUT1,STOUT2,STOUT3,NCOMP,LS,
     2EN  ,L1,IEN,LIEN,JJ,NORD)
      IF (IU.EQ.19)CALL UNIT19
     1(STRIN1,STRIN2,STRIN3,STOUT1,STOUT2,STOUT3,NCOMP,LS,
     2EN  ,L1,IEN,LIEN,JJ,NORD)
      IF (IU.EQ.20)CALL UNIT20
     1(STRIN1,STRIN2,STRIN3,STOUT1,STOUT2,STOUT3,NCOMP,LS,
     2EN  ,L1,IEN,LIEN,JJ,NORD)
      IF(MODE .EQ. 1) GO TO 40
      IF(LABEL .NE. 1) WRITE(IW,155)
      WRITE (IW,150)IU
      DO 45 I=1,NCOMP
   45 WRITE (IW,127) (NAMEC(IJ,I),IJ=1,3),(STOUT(I,IJ),IJ=1,3)
      DO 47 IJ=1,3
      S(IJ)=.0
      DO 47 I=1,NCOMP
   47 S(IJ)=S(IJ)+STOUT(I,IJ)
      WRITE (IW,131) (S(I),I=1,3)
      DO 48 I=NCP,NT
   48 WRITE (IW,127) (NAMEC(IJ,I),IJ=1,3),(STRIN(I,IJ),IJ=1,NIN)
```

```
49      WRITE(IW,135)
        READ(IR,102) IEND
        IF (IEND.NE.1HY) STOP
        WRITE( IW,123)
        READ(IR,102) ICAS
        IF(ICAS .NE. 1HY) GO TO 5
        DO 60 I=1,25
        STRIN1(I)=STOUT1(I)
        STRIN2(I)=0.0
        STRIN3(I)=0.0
        STOUT1(I)=0.0
        STOUT2(I)=0.0
        STOUT3(I)=0.0
60      CONTINUE
        MODE=3
        JFORM=3
        LABEL=1
        JJ=1
        IJS=2
        WRITE (IW,124)
        READ (IR,99          ) NIN
        GO TO 18
38      CONTINUE
        WRITE (IW,121) IU
        GO TO 49
99      FORMAT( 16I5)
100     FORMAT( 8F10.0)
101     FORMAT (20H HOW MANY COMPONENTS)
102     FORMAT (A1)
103     FORMAT (23H HOW MANY INPUT STREAMS)
104     FORMAT(3A4,8X,I3)
106     FORMAT (7H STREAM,I3,29H :KMOL EACH COMP, TEMP, PRES,
       ^    ,28H  PHASE (0.0=LIQUID 1.0=GAS))
111     FORMAT (22H GIVE THESE PARAMETERS)
110     FORMAT (36H HOW MANY REAL MODEL PARAMETERS (EN))
113     FORMAT (34H HOW MANY INTEGER PARAMETERS (IEN))
121     FORMAT (5H UNIT,I3,17H IS NOT AVAILABLE)
122     FORMAT (17H GIVE UNIT NUMBER)
123     FORMAT (36H IF YOU WANT TO RUN UNITS IN CASCADE,/,
       ^    ,53H  (INPUT STREAM1 WILL BE REPLACED BY OUTPUT STREAM 1),/,
       ^    ,27H GIVE YES OTHERWISE GIVE NO)
124     FORMAT (49H HOW MANY INPUT STREAMS (CASCADE STREAM INCLUDED))
125     FORMAT (38H GIVE COMPONENT NAMES ONE IN EACH LINE
       E ,30H AND CHEMCO NO IN FIELD 21-23      ,/38H ENTER ZERO IF CHEMCO
       *NOT TO BE USED             )
126     FORMAT (21H INPUT TABLE            ,/,16H COMPONENT NAMES,
       ^    3(5X,6HSTREAM,I1))
127     FORMAT (1X,3A4,(1X,5G12.4))
128     FORMAT (22H REAL MODEL PARAMETERS,(2X,5G12.4))
129     FORMAT (25H INTEGER MODEL PARAMETERS,(2X,5I10))
131     FORMAT (1X,5HTOTAL, 6X,(1X,5G12.4))
132     FORMAT (1X,I4,36H IS NOT PROPER FOR NUMBER OF STREAMS)
133     FORMAT (1X,I4,47H IS NOT PROPER FOR NUMBER OF INTEGER PARAMETERS)
134     FORMAT (1X,I4,44H IS NOT PROPER FOR NUMBER OF REAL PARAMETERS)
135     FORMAT (35H DO YOU WANT TO CONTINUE (YES/NO)
140     FORMAT(23H CHEMCO BANK NOS ARE         ,10I5)
150     FORMAT (1H0,28H OUTPUT STREAMS FROM UNIT   ,I5 ,///,
       ^  1X,15HCOMPONENT NAMES,3(5X,6HSTREAM,I1))
155     FORMAT(/45H ********ERROR DETECTED IN SUBROUTINE MODEL
        END
```

```
SUBROUTINE CHGPAR(A,B,C)
RETURN
END
```

Listing A3.3   Integration Subroutine RKGS   (Exercises 6 and 8)

```
C           SUBROUTINE RKGS
C     RUNGE-KUTTA INTEGRATION SUBROUTINE
C           TO SOLVE A SYSTEM OF FIRST ORDER ORDINARY DIFFERENTIAL
C           EQUATIONS WITH GIVEN INITIAL VALUES.
      SUBROUTINE RKGS(PRMT,Y,DERY,NDIM,IHLF,FCT,OUTP,AUX)
C
C
      DIMENSION Y(1),DERY(1),AUX(8,1),A(4),B(4),C(4),PRMT(1)
      DO 1 I=1,NDIM
    1 AUX(8,I)=.06666667*DERY(I)
      X=PRMT(1)
      XEND=PRMT(2)
      H=PRMT(3)
      PRMT(5)=0.
      CALL FCT(X,Y,DERY)
C
C     ERROR TEST
      IF(H*(XEND-X))38,37,2
C
C     PREPARATIONS FOR RUNGE-KUTTA METHOD
    2 A(1)=.5
      A(2)=.2928932
      A(3)=1.707107
      A(4)=.1666667
      B(1)=2.
      B(2)=1.
      B(3)=1.
      B(4)=2.
      C(1)=.5
      C(2)=.2928932
      C(3)=1.707107
      C(4)=.5
C
C     PREPARATIONS OF FIRST RUNGE-KUTTA STEP
      DO 3 I=1,NDIM
      AUX(1,I)=Y(I)
      AUX(2,I)=DERY(I)
      AUX(3,I)=0.
    3 AUX(6,I)=0.
      IREC=0
      H=H+H
      IHLF=-1
      ISTEP=0
      IEND=0
C
C
C     START OF A RUNGE-KUTTA STEP
    4 IF((X+H-XEND)*H)7,6,5
    5 H=XEND-X
    6 IEND=1
C
C     RECORDING OF INITIAL VALUES OF THIS STEP
    7 CALL OUTP(X,Y,DERY,IREC,NDIM,PRMT)
      IF(PRMT(5))40,8,40
    8 ITEST=0
    9 ISTEP=ISTEP+1
```

```
C
C
C      START OF INNERMOST RUNGE-KUTTA LOOP
       J=1
   10  AJ=A(J)
       BJ=B(J)
       CJ=C(J)
       DO 11 I=1,NDIM
       R1=H*DERY(I)
       R2=AJ*(R1-BJ*AUX(6,I))
       Y(I)=Y(I)+R2
       R2=R2+R2+R2
   11  AUX(6,I)=AUX(6,I)+R2-CJ*R1
       IF(J-4)12,15,15
   12  J=J+1
       IF(J-3)13,14,13
   13  X=X+.5*H
   14  CALL FCT(X,Y,DERY)
       GOTO 10
C      END OF INNERMOST RUNGE-KUTTA LOOP
C
C
C      TEST OF ACCURACY
   15  IF(ITEST)16,16,20
C
C      IN CASE ITEST=0 THERE IS NO POSSIBILITY FOR TESTING OF ACCURACY
   16  DO 17 I=1,NDIM
   17  AUX(4,I)=Y(I)
       ITEST=1
       ISTEP=ISTEP+ISTEP-2
   18  IHLF=IHLF+1
       X=X-H
       H=.5*H
       DO 19 I=1,NDIM
       Y(I)=AUX(1,I)
       DERY(I)=AUX(2,I)
   19  AUX(6,I)=AUX(3,I)
       GOTO 9
C
C      IN CASE ITEST=1 TESTING OF ACCURACY IS POSSIBLE
   20  IMOD=ISTEP/2
       IF(ISTEP-IMOD-IMOD)21,23,21
   21  CALL FCT(X,Y,DERY)
       DO 22 I=1,NDIM
       AUX(5,I)=Y(I)
   22  AUX(7,I)=DERY(I)
       GOTO 9
C
C      COMPUTATION OF TEST VALUE DELT
   23  DELT=0.
       DO 24 I=1,NDIM
   24  DELT=DELT+AUX(8,I)*ABS(AUX(4,I)-Y(I))
       IF(DELT-PRMT(4))28,28,25
C
C      ERROR IS TOO GREAT
   25  IF(IHLF-10)26,36,36
   26  DO 27 I=1,NDIM
```

```
   27 AUX(4,I)=AUX(5,I)
      ISTEP=ISTEP+ISTEP-4
      X=X-H
      IEND=0
      GOTO 18
C
C     RESULT VALUES ARE GOOD
   28 CALL FCT(X,Y,DERY)
      DO 29 I=1,NDIM
      AUX(1,I)=Y(I)
      AUX(2,I)=DERY(I)
      AUX(3,I)=AUX(6,I)
      Y(I)=AUX(5,I)
   29 DERY(I)=AUX(7,I)
      CALL OUTP(X-H,Y,DERY,IHLF,NDIM,PRMT)
      IF(PRMT(5))40,30,40
   30 DO 31 I=1,NDIM
      Y(I)=AUX(1,I)
   31 DERY(I)=AUX(2,I)
      IREC=IHLF
      IF(IEND)32,32,39
C
C     INCREMENT GETS DOUBLED
   32 IHLF=IHLF-1
      ISTEP=ISTEP/2
      H=H+H
      IF(IHLF)4,33,33
   33 IMOD=ISTEP/2
      IF(ISTEP-IMOD-IMOD)4,34,4
   34 IF(DELT-.02*PRMT(4))35,35,4
   35 IHLF=IHLF-1
      ISTEP=ISTEP/2
      H=H+H
      GOTO 4
C
C
C     RETURNS TO CALLING PROGRAM
   36 IHLF=11
      CALL FCT(X,Y,DERY)
      GOTO 39
   37 IHLF=12
      GOTO 39
   38 IHLF=13
   39 CALL OUTP(X,Y,DERY,IHLF,NDIM,PRMT)
   40 RETURN
      END
```

- Runge-Kutta-Gill Integration
Subroutine of IBM's Scientific
Subroutine Package Library.

Listing A3.4   UNIT01 Methyl chloride reactor without temperature profile
              (UNIT01, FN1, OUTP1)   (Exercise 6)

```
C
C
C          CH3CL   FORMATION REACTOR WITHOUT TEMPERATURE PROFILE
C
C
       SUBROUTINE UNIT01
      1(STRIN1,STRIN2,STRIN3,STOUT1,STOUT2,STOUT3,NCOMP,LS,
      2EN1 ,L1,IEN,LIEN,JJ,NORD)
       COMMON/FLAGS/MODE,JFORM,LABEL,IR,IW
       DIMENSION STRIN1(LS),STRIN2(LS),STRIN3(LS),STOUT1(LS),STOUT2(LS),
      1STOUT3(LS),EN1(L1), IEN(LIEN), NORD(10)
       DIMENSION  YI(5),DY(5),Y(5),E(5),PRMT(5),AUX(8,5)
       COMMON/MECL/C(3)
       EXTERNAL FN1   ,OUTP1
       IF(MODE-2)3,4,4
   3   IF(NCOMP-5)9,2,9
   9     WRITE(IW,102)NCOMP
 102   FORMAT(48HREACTOR MODEL ONLY FOR 5 COMPONENTS YOU SPECIFY ,I2)
       LABEL=2
       GOTO 2
C      PARAMETERS CLASS ONE ARE,  CONTACT TIME(SECS)
C      TEMP(K), AND PRESSURE(BARS )     ,TUBE LENGTH(M)
C      FIRST WE HAVE TO CONVERT THE FEED STREAM IN
C       K MOLES/SEC.
   4   T=EN1(2)
       P=EN1(3)
C   FREQUENCY FACTORS
       B(1)=17.8 E+11
       B(2)=47.4 E+15
       B(3)=11.5 E+11
       DO 7 I=1,NCOMP
   7   YI(I)=(STRIN1(I)+STRIN2(I)+STRIN3(I))/3600.
C   VELOCITY CONSTANTS M3/KMOL
       C(1)=B(1)*EXP(-20000./(1.98*T))
       C(2)=B(2)*EXP(-30000./(1.98*T))
       C(3)=B(3)*EXP(-20000./(1.98*T))
       WRITE (IW,150) (C(I),I=1,3)
       WRITE (IW,151) (YI(I),I=1,NCOMP)
C  TOTAL VOLUME FLOW RATE (M3/S)
       F=(YI(1)+ YI(2)+ YI(3)+ YI(4)+ YI(5))*0.08314*T/P
       N=4
       PRMT(1)=0.
       PRMT(2)= EN1(1)
       PRMT(3)=0.5
       PRMT(4)=0.00001*YI(1)
       PRMT(5)=0.0
       DO 30 J=1,N
       DY(J)=1.0
  30   Y(J)=YI(J)
C  CONVERT MOLES/SEC TO CONC UNITS  -COMBINE WITH VELOCITY CONSTANT
C  TO REDUCE COMPUTATION WITHIN INTEGRATION
       DO 40 J=1,3
  40   C(J)=C(J)/F
       CALL RKGS (PRMT,Y,DY,N,IHLF,FN1,OUTP1,AUX)
C  CALCULATE ALL EXIT FLOWS
       IF(IHLF-10)  45,45,80
```

```
45      Y(5)=YI(5)+(Y(3)-YI(3))+(Y(4)-YI(4))
        NTUBE = IFIX( EN1(1)*F/(0.0005*EN1(4) ) + 0.9)
        PD=0.1*(EN1(4))**3/EN1(1)
        DO 10 I=1,NCOMP
10      STOUT1(I)=Y(I)*3600.
        STOUT1(NCOMP+1)=T
        STOUT1(NCOMP+2)=P -PD
        STOUT1(NCOMP+4)=1.0
        STOUT1(NCOMP+5)=0.0
        IF(MODE.EQ.2)GOTO 2
        WRITE(IW,60)(Y(I),I=1,5)
60      FORMAT(30H  K MOLES/SEC LEAVING REACTOR /
       $47H         MEOH        HCL        MECL        DME        H2O/
       $ 5G10.4)
C   CALCULATE MECL/MEOH YIELD AND DME/MEOH YIELD.
        EFFMC=100.0*(Y(3)-YI(3))/(YI(1))
        EFFDME = 100.0*(Y(4)-YI(4))/YI(1)
        WRITE(IW,70)  EFFMC,EFFDME
70      FORMAT(  36H MECL YIELD BASED ON METHANOL FED=          ,F10.3,/
       *36H  DME YIELD BASED ON METHANOL FED=          ,F10.3   )
        WRITE(IW,152) NTUBE,EN1(4), PD
152     FORMAT(/16H REACTOR DESIGN   /I10, 6H TUBES  ,F10.1, 14H METERS L
       1NG     ,4HPD=    ,G10.2,4HBAR.   //)
        GO TO 2
80      WRITE(IW,90)
90      FORMAT( 20H INTEGRATION FAILURE
        LABEL=2
2       RETURN
150     FORMAT (37H REACTION RATE CONSTANTS(KMOL/S/M3)     ,(2X,6G12.4))
151     FORMAT (16H COMPONENT FEEDS,(2X,6G12.4))
        END



        SUBROUTINE  FN1(X,Y,DY)
        DIMENSION  Y(5),DY(5)
        COMMON/MECL/C(3)
        DY(1)= -C(1)*Y(1)*Y(2) - 2.0*C(2)*Y(1)*Y(1)
        DY(2)=-C(1)*Y(1)*Y(2)   -  2.0*C(3)*Y(2)*Y(4)
        DY(3)= C(1)*Y(1)*Y(2)   +  2.0*C(3)*Y(2)*Y(4)
        DY(4)=  C(2)*Y(1)*Y(1)   -  C(3)*Y(2)*Y(4)
        RETURN
        END



        SUBROUTINE  OUTP1  (X,Y,DY,IHLF,N,PRMT  )
        DIMENSION  DY(5),Y(5),PRMT(5)
C
        RETURN
        END
```

Listing A3.5   UNIT02   Benzene chlorination model   (UNIT02, FCT, SECT)
              (Exercise 7)

```
C
C     BENZENE CHLORINATION - CSTR
C
C
      SUBROUTINE UNIT02
     1(STRIN1,STRIN2,STRIN3,STOUT1,STOUT2,STOUT3,NCOMP,LS,
     2EN1 ,L1,IEN,LIEN,JC,NORD)
      COMMON/FLAGS/MODE,JFORM,LABEL,IR,IW
      DIMENSION STRIN1(LS),STRIN2(LS),STRIN3(LS),STOUT1(LS),STOUT2(LS),
     1STOUT3(LS),EN1 (L1),IEN(LIEN),NORD(1)
      COMMON/CBR/ TOTGAS, CBO, CMO, CDO, CTO, RK1, RK2, RK3, CB, CM, CD,
     * CT, CCO, CG, CGO, BH, VTOT, P
      EXTERNAL FCT
      JJ=JC
      IF(MODE-4)  1,11,11
1     GO TO (2,3,3), MODE
2     MRET=1
      IF(EN1(1).LT.100..OR.EN1(1).GT. 600.) GO TO 111
      IF(EN1(2).LT..01.OR.EN1(2).GT.1000.) GO TO 111
      GO TO 112
111   WRITE(IW ,113)
      LABEL=2
113   FORMAT( 25H MODEL PARAMETER ERROR
112   CONTINUE
      GO TO 11
3     CONTINUE
C     DENSITIES IN KMOL/M3
      RHO1= 10.26
      RHO2= 10.26
      RHO3= 10.26
C     CALC TOTAL INLET VOLUME FLOWS
      V1=STRIN1(1)/RHO1
      V2=STRIN1(2)/RHO2
      V3=STRIN1(3)/RHO3
      V4=STRIN1(8)/RHO3
      VTOT=V1+V2+V3+V4
C     COMPONENT INLET FLOWS KMOL/H
      CBO=STRIN1(1)
      CMO=STRIN1(2)
      CDO=STRIN1(3)
      CTO=STRIN1(8)
      CCO=STRIN1(4)
      FECL3=STRIN1(6)
      H2O =STRIN1(7)
      TOTGAS=STRIN2(4) + STRIN2(5)
      CGO=STRIN2(4)
C     TRANSFER MODEL PARAMETERS
      T=EN1(1)
      P=EN1(2)
      TH=EN1(3)
      BH=EN1(4)
      XK=(4.66*EXP(-1.57*H2O/FECL3)+0.03)*FECL3
      RK1=2.7756E11*EXP(-5364./T    )*XK
      RK2=5.256E6*EXP(-2526./T    )*XK
      RK3=0.1*RK2
```

```
C    REACTOR MODEL PROPER AS FUNCTICN FOR SECANT EQUATION SOLVER
C
C    COMBINE RATES WITH REACT. VCL AND FLOWS TO REDUCE ITERATIVE COMP.
      RK1= RK1*TH/VTOT
      RK2= RK2*TH/VTOT
      RK3= RK3*TH/VTOT
C  SOLVE TO FIND Z   KMOL /H CL2 IN EXIT LIQUID
C    PREPARE FOR CALLING SECANT EQUATION SOLVER
      IEND=30
      XL=0.0
      XH=VTOT*CGO*P/(BH*TOTGAS)
      XST=XH/10.
      XND=1.1*XST
      EPS=0.0001
      CALL SECT(Z,XST,XND,XL,XH,FCT,EPS,IEND,RES,IER)
      IF(IER.NE.0) LABEL=2
      STOUT1(1)=CB
      STOUT1(2)=CM
      STOUT1(3)=CD
      STOUT1(4)=Z
      STOUT1(5)=0.0
      STOUT1(6)=STRIN1(6)
      STOUT1(7)=STRIN1(7)
      STOUT1(8)=CT
      DO 20 I=1,8
20    STOUT2(I)=0.
      STOUT2(4)=CG
      STOUT2(5)= STRIN1(5) + STRIN2(5) + (CM-CMO) + 2.*(CD-CDO) + 3.*(C
     * -CTO)
      STOUT1(9)=T
      STOUT2(9)=T
      STOUT1(10)=P
      STOUT2(10)=P
      IF(MODE-3)   10,12,12
10    MRET=2
11    RETURN
12    CONTINUE
      CONB=100.*(STRIN1(1) - STOUT1(1))/STRIN1(1)
      CONC=100.*(STRIN2(4) - STOUT2(4))/STRIN2(4)
      CONM=100.*(STOUT1(2) - STRIN1(2))/STRIN1(1)
      COND=100.*(STOUT1(3) - STRIN1(3))/STRIN1(1)
      RVOL=TH*VTOT
      WRITE(IW,22) CONB,CONC,CONM,COND,RVOL
22    FORMAT(36H CONVERSION OF BENZENE AND CHLORINE ,2F10.4/
     *42H CONVERSION TO MONO- AND CI- CHLORBENZENE ,2F10.4/
     * 21H REACTOR VOLUME (M3) ,F10.5)
      MRET=3
      RETURN
      END


      SUBROUTINE FCT(Z,RES)
      COMMON/CBR/ TOTGAS, CBO, CMO, CDO, CTO, RK1, RK2, RK3, CB, CM, CD
     * CT, CCO, CG, CGO, BH, VTOT, P
C
C  EXPRESS ALL SPECIES IN TERMS OF CL2(LIQ),(Z)   USING C S T R  MASS
C  BALANCE EQUATIONS
C  CL2(GAS) BY MASS BALANCE
C  COMPARE Z ASSUMING Z IN EQUILIBRIUMWITH CL2(GAS)(IE KINETIC LIMITED
```

```
      CB=CBO/(1. + RK1*Z)
      CM= (RK1*Z*CB + CMO)/(1. + RK2*Z)
      CD=(RK2*Z*CM + CDO)/(1. + RK3*Z)
      CT=RK3*Z*CD + CTO
      CG=CGO-(CM-CMO)-2.*(CD-CDO) - 3.*(CT-CTO) + CCO -Z
      RES=Z-VTOT*P*CG/(BH*TOTGAS)
      RETURN
      END


      SUBROUTINE SECT(Z,XST,XND,XL,XH,FCT,EPS,IEND,RES,IER)
      DIMENSION R(2),X(2)
C   SECANT METHOD FOR ROOT OF ONE EQUATION,BETWEEN BOUNDS XL AND XH.
C      TWO INITIAL GUESSES ,XST AND XND REQUIRED
C         IER RETURN CODE ERROR,    =0 NO ERROR
      H=XST-XND
       K=0
5      Z=XND
       CALL FCT(Z,RES)
       R(2)=RES
       X(2)=Z
      X(1)=XST
      IT=0
10     IT=IT+1
      IF(IEND-IT)20,20,40
20     IER=1
      GO TO 100
40     Z=X(1)
       CALL FCT(Z,RES)
       R(1)=RES
       Z=X(1)-R(1)*(X(1)-X(2))/(R(1)-R(2))
       IF(Z-XL)50,50,60
50     IF(K-1)70,80,80
80     IER=2
      GO TO 100
70     XST=XH
      XND=XH+H
      K=1
      GO TO 5
60     IF(Z-XH)85,90,90
90     IF(K-1)95,80,80
95     XST=XL
      XND=XST+H
      K=1
       GO TO 5
85     E=ABS((Z-X(1))/Z)
       IF(E-EPS)97,97,96
97     IER=0
       Z=X(1)
       RES=R(1)
      GO TO 110
96     X(2)=X(1)
       X(1)=Z
       R(2)=R(1)
      GO TO 10
100    IW=2
      WRITE(IW,120)
110    RETURN
120    FORMAT( 33H   SECT NOT CONVERGED
      END
```

374

Listing A3.6   UNITO3   Methyl chloride reactor with temperature profile
             (UNITO3, FT3, OUTP3)   (Exercise 8)

```
C
C
C          CH3CL FORMATION REACTOR WITH TEMPERATURE PROFILE
C
C
       SUBROUTINE UNITO3
      1(STRIN1,STRIN2,STRIN3,STOUT1,STOUT2,STOUT3,NCOMP,LS,
      2EN1 ,L1,IEN,LIEN,JJ,NORD)
       INTEGER ANTUB
       COMMON/FLAGS/MODE,JFORM,LABEL,IR,IW
       DIMENSION STRIN1(LS),STRIN2(LS),STRIN3(LS),STOUT1(LS),STOUT2(LS),
      1STOUT3(LS),EN1(L1), IEN(LIEN), NORD(10)
       DIMENSION  YI(5),DY(5),Y(5)      ,PRMT(5),AUX(8,5)
       COMMON/MECH/ P, A(3), HT(3), CP, FM,UA, AREA, TC
       EXTERNAL FN3  ,OUTP3
       IF(MODE-2)3,4,4
   3   IF(NCOMP-5)9,2,9
   9    WRITE(IW,102)NCOMP
       LABEL=2
       GOTO 2
C      MAIN CALCULATION STARTS HERE
C      PARAMETERS CLASS ONE ARE,  CONTACT TIME(SECS)
C      TEMP(K), AND PRESSURE(BARS )    ,TUBE LENGTH(M)
C      TUBE DIAMETER (M)
   4   TC=EN1(2)
       P=EN1(3)
C        HEAT TRANSFER COEFFICIENT IS RELATED TO GAS VELOCITY
       U=0.116*(EN1(4)/EN1(1))**0.7
       HT(1)= 92000.
       HT(2)= 92000.
       CP=75.
C      FIRST WE HAVE TO CONVERT THE FEED STREAM IN
C      KMOLES/SEC.
       FM=0.0
       DO 7 I=1,NCOMP
       FM=FM+STRIN1(I)/3600.
   7   YI(I)=STRIN1(I)/3600.
C      CROSS SECTION AREA   M**2
       AREA= 3.1416*EN1(5)*EN1(5)/4.
       UA=U*3.1416*EN1(5)
C  TOTAL VOLUME FLOW RATE  M**3/S
       F=FM*0.0814*STRIN1(6)/P
       TVOL=AREA*EN1(4)
C      NUMBER OF TUBES TO SATISFY CONTACT TIME
       ANTUB=F*EN1(1)/TVOL
C      FM IN KMOL/(S*TUBE)
       FM=FM/FLOAT(ANTUB)
       N=5
       DO 30 J=1,5
       DY(J)=1.0
C      ALL FLOWS IN KMOL/(S*TUBE)
       YI(J)=YI(J)/FLOAT(ANTUB)
```

```
30      Y(J)=YI(J)
        Y(5)=STRIN1(6)
        DY(5)=1.E-5
        PRMT(1)=0.
        PRMT(2)= EN1(4)
        PRMT(3)=0.1
        PRMT(4)=0.0001*YI(1)
        PRMT(5)=0.0
C       PREEXPONENTIAL FACTORS
        A(1)=6.23E7
        A(2)=2.56E12
        CALL RKGS (PRMT,Y,DY,N,IHLF,FN3,OUTP3,AUX)
C   CALCULATE ALL EXIT FLOWS
        IF(IHLF-5)   45,45,80
45      IF (PRMT(5).GT.0.) GO TO 80
C       Y(5) CALCULATED BY MASS BALANCE
        Y(5)=YI(5)+(Y(3)-YI(3))+(Y(4)-YI(4))
        DO 10 I=1,NCOMP
 10     STOUT1(I)=Y(I)*3600.*FLOAT(ANTUB)
        STOUT1(NCOMP+1)=Y(5)
        STOUT1(NCOMP+2)=P
        STOUT1(NCOMP+4)=1.0
        STOUT1(NCOMP+5)=0.0
        IF(MODE.EQ.2)GOTO 2
C   CALCULATE MECL/MEOH YIELD AND DME/MEOH YIELD.
        EFFMC=100.0*(Y(3)-YI(3))/(YI(1))
        EFFDME = 100.0*(Y(4)-YI(4))/YI(1)
        WRITE(IW,70)   EFFMC,EFFDME
        TMC=STOUT1(3)*404.
        WRITE (IW,121) ANTUB,EN1(5),EN1(4),EN1(1),EN1(2),TMC
        GO TO 2
80      WRITE(IW,90)
        LABEL=2
 2      RETURN
C
70      FORMAT(   36H MECL YIELD BASED ON METHANOL FED=          ,F10.3,/
       *36H   DME YIELD BASED ON METHANOL FED=              ,F10.3    )
90      FORMAT (20H MCL PYROLYSES                           )
102     FORMAT(48HREACTOR MODEL ONLY FOR 5 COMPONENTS YOU SPECIFY ,I2)
121     FORMAT (22H NUMBER OF TUBES      =,I6,/
       ^   ,22H TUBE DIAMETER   (M) =,G10.3,/
       ^   ,22H TUBE LENGHT     (M) =,G10.3,/
       ^    ,22H CONTACT TIME    (S) =,G10.3,/
       ^     ,22H REACTION TEMP.  (K) =,G10.3,/
       ^      ,22H MC PRODUCTION (T/A) =,G10.3)
        END


        SUBROUTINE   FN3(X,Y,DY)
        DIMENSION  Y(5),DY(5),C(3)
        COMMON/MECH/ P, A(3), HT(3), CP, FM,UA, AREA, TC
        T=Y(5)
```

```
C      VOLUME FLOW/S
       F=0.0814*FM*T/P
       C(1)=A(1)*EXP(-20000./(1.98*T))
       C(2)=A(2)*EXP(-30000./(1.98*T))
       C(1)=C(1)*AREA/(F*F)
       C(2)=C(2)*AREA/(F*F)
       DY(1)= - C(1)*Y(1)*Y(2) - 2.0*C(2)*Y(1)*Y(1)
       DY(2)=-C(1)*Y(1)*Y(2)
       DY(3)= C(1)*Y(1)*Y(2)
       DY(4)=  C(2)*Y(1)*Y(1)
       DY(5)= C(1)*Y(1)*Y(2)*HT(1) + C(2)*Y(1)*Y(1)*HT(2)
     ^ - UA*(Y(5)-TC)
       DY(5)=DY(5)/(CP*FM)
       RETURN
       END


       SUBROUTINE OUTP3 (X,Y,DY,IHLF,N,PRMT
       DIMENSION  DY(5),Y(5),PRMT(5)
       COMMON/FLAGS/MODE,JFORM,LABEL,IR,IW
       DATA IZEK/0/
       DATA IP/0/
       IF (IZEK.EQ.0) WRITE (IW,160)
       IZEK=1
       IP=IP+1
       IF (IP.LT.5) RETURN
       WRITE(IW,150)X, Y(1), Y(2), Y(3), Y(4), Y(5)
     ^ ,DY(1),DY(2),DY(3),DY(4),DY(5)
       IF (Y(5).GT.750.0) PRMT(5)=1.
       IP=0
       RETURN
150    FORMAT(12G10.4)
160    FORMAT (3X,1HX, 9X,4HY(1),6X,4HY(2),6X,4HY(3),
     ^ 6X,4HY(4),6X,4HY(5),5X,5HDY(1),5X,5HDY(2),5X,
     ^5HDY(3),5X,5HDY(4),5X,5HDY(5)
     ^ /,3X,1HM,4X,4(4X,6HKMOL/S),5X,1HK,
     ^ 5X,4(2X,8HKMOL/S*M),3X,3HK/M)
       END
```

INDEX